Sustainable Agroecosystems - Principles and Practices

Edited by Vijay Singh Meena,
Ram Swaroop Bana, Ram Kishor Fagodiya
and Mohammad Hasanain

Published in London, United Kingdom

Sustainable Agroecosystems - Principles and Practices
http://dx.doi.org/10.5772/10.5772/intechopen.1004984
Edited by Vijay Singh Meena, Ram Swaroop Bana, Ram Kishor Fagodiya and Mohammad Hasanain

Contributors
Abhik Patra, Adina Berbecea, Alexandrius de Moraes Barbosa, Ali Raza, Ameer Favas V, Archana Verma, Areeba Rehman, Arien Jean M. Lopez, Ashish Rai, Banoth Bixapathi Nayak, Bareera Munir, Bienvenu A. Dagoudo, Charles Ssekyewa, Chebrolu Sravani, Chhavi Sirohi, Dibyajyoti Panda, Dinesh Jinger, Dwaram Mrudula, Felix K. Abagale, Godofredo P. Lubat Jr, Humaira Rizwana, Imran Khan, Iqra Fatima, Isidora Radulov, Jaseem KP, John Paul R. Pacyao, Karin Ferretto Santos Collier, Kiran Kumar Mohapatra, Leonardo Santos Collier, Marlyn B. Llameg, Michael Aide, Mohamed Joseph Sesay, Mohammad Afrose, Mude Ramesh Naik, Muhammad Bilal Chattha, Muhammad Talha Aslam, Muhammad Umair Hassan, Muhammad Umer Chattha, Muqarrab Ali, Najat A. Bokhari, Nandha Kumar N, Pankaj Panwar, Plabani Roy, Rajesh Kaushal, Sandeep Kumar, Sanjay Singh Rathore, Sayon Mukherjee, Shaibu Abdul-Ganiyu, Shama Sharif, Shana Sherin, Silvère D. Tovignan, Sonia Waqar, Sripriya Das, Tanu Sharma, Tehmina Anjum, Thulisegari Prasanna, Vandana Kumari, Vijay Singh Meena, Vinod Kumar Singh, Waheed Akram, Zill-e-Huma Aftab

First published in London, United Kingdom, 2025 by IntechOpen
IntechOpen is the global imprint of INTECHOPEN LIMITED, registered in England and Wales, registration number: 11086078, 167-169 Great Portland Street, London, W1W 5PF, United Kingdom

For EU product safety concerns: IN TECH d.o.o., Prolaz Marije Krucifikse Kozulić 3, 51000 Rijeka, Croatia, info@intechopen.com or visit our website at intechopen.com.

British Library Cataloguing-in-Publication Data
A catalogue record for this book is available from the British Library

Sustainable Agroecosystems - Principles and Practices
Edited by Vijay Singh Meena, Ram Swaroop Bana, Ram Kishor Fagodiya and Mohammad Hasanain
p. cm.

This title is part of the Agricultural Sciences Book Series, Volume 19
Topic: Agronomy and Horticulture
Series Editor: W. James Grichar
Topic Editor: Ibrahim Kahramanoglu
Associate Topic Editors: Murat Helvaci and Olga Panfilova

Print ISBN 978-0-85014-718-6
Online ISBN 978-0-85014-717-9
eBook (PDF) ISBN 978-0-85014-719-3
ISSN 3029-052X

IntechOpen Book Series

Agricultural Sciences

Volume 19

Aims and Scope of the Series

The importance of agriculture cannot be overstated. It helps sustain life, as it gives us the food we need to survive and provides opportunities for economic well-being. Agriculture helps people prosper around the world and combines the creativity, imagination, and skill involved in planting crops and raising animals with modern production methods and new technologies. This series includes two main topics: Agronomy and Horticulture, and Animal Farming. This series will help readers better understand the intricacies of production agriculture and provide the new knowledge that is required to be successful. The success of a farmer in modern agriculture requires knowledge of events happening locally as well as globally that impact input decisions and ultimately determine net profit.

Meet the Series Editor

W. James Grichar has been employed with Texas A&M AgriLife Research for over 45 years with an emphasis on research in agronomy, plant pathology, and weed science. He obtained his BS from Texas A&M in 1972 and his Masters of Plant Protection in 1975. He has published 195 journal articles, over 330 research reports and briefs, 11 book chapters, and over 300 abstracts of profession meetings. He also directs research in many crops including corn, grain sorghum, peanuts, and sesame. He has held various positions in different professional societies including the American Peanut Research and Education Society, Southern Weed Science Society, and Texas Plant Protection Conference in addition to being Associate Editor for Peanut Science and Weed Technology. Significant accomplishments have included spearheading efforts to determine the optimum planting time for soybean production along the upper Texas Gulf Coast. These efforts have shown growers that soybean yields can be improved by 10 to 20% by following a late March to early April plant date. He also has been instrumental in developing a herbicide program for peanut production in the south Texas growing region. Through the development and use of herbicides that are effective against major weed problems in the south Texas region, peanut yields have increased by 25 to 30%.

Meet the Volume Editors

Dr. Vijay Singh Meena is a Senior Scientist at ICAR-Indian Agricultural Research Institute, specializing in climate-resilient agriculture, soil health, and sustainable farming. His research focuses on improving crop productivity, water management, and precision farming. Recognized as a Top Agri-Pioneer by the World Food Foundation, he has received prestigious awards, including the INSA and NAAS Young Scientist Awards. He has edited 12 books with Springer and Elsevier and is an editor for leading journals. Previously, he worked at the International Maize and Wheat Improvement Center-Borlaug Institute for South Asia (CIMMYT-BISA), contributing to climate-smart agriculture. An active member of the FSN Forum, he collaborates on global initiatives to advance sustainable farming and agri-food and nutritional security.

Dr. Ram Swaroop Bana is a Senior Scientist at the Division of Agronomy, Indian Council of Agricultural Research – Indian Agricultural Research Institute (ICAR–IARI) in New Delhi, India. He specializes in precision conservation agriculture, dryland farming, and crop system modeling. A recipient of prestigious awards, including the Endeavour Fellowship (Australia) and ISA-PS Deshmukh Young Scientist Award, he has published 116 research papers, 16 edited books, and numerous book chapters and extension materials. He has delivered over 165 TV/Radio talks and 70 national and international training lectures. He has been actively involved in training and conferences and has guided Ph.D. and M.Sc. students. His contributions to sustainable agriculture and research have significantly advanced agronomic practices in India and beyond.

Dr. Ram Kishor Fagodiya has over 10 years of research and teaching experience in natural resources management in agriculture. His work focuses on climate change modeling, greenhouse gas emissions, carbon sequestration, and the food-water-energy-carbon nexus. He has contributed to 13 research projects on conservation agriculture, degraded land management, and sustainable irrigation. With over 50 research publications (h-index: 22, i10-index: 28, 1,800+ citations), he plays a key role in national-level database development on global temperature change potential and nutrient budgeting. An active member of scientific societies and editorial boards, he also engages in farmer outreach, promoting climate-resilient and resource-efficient agricultural practices.

Dr. Mohammad Hasanain is a Scientist at ICAR-IARI, R.S. Bihar in India, specializing in cropping system optimization, precision nutrient management, conservation agriculture, and greenhouse gas quantification. He has published over 20 research articles, six reviews, 50 popular articles, and one edited book. He has secured top national ranks in agricultural entrance exams and received multiple awards, including the Best Ph.D. Thesis Award (ISA 2023) and Dr. K.L. Sharma Young Scientist Award (ISDA 2023). He handles multiple funded research projects and is a reviewer for reputed journals. His contributions focus on advancing sustainable and climate-resilient agricultural practices.

Contents

Preface **XIII**

Section 1
Sustainable Soil and Nutrient Management **1**

Chapter 1 **3**
Nurturing Soil Health through Conservation Agriculture Practices
by Dwaram Mrudula, Banoth Bixapathi Nayak, Chebrolu Sravani, Thulisegari Prasanna, Mude Ramesh Naik and Mohammad Afrose

Chapter 2 **27**
Nutrient Management for Sustainable Soil Fertility
by Isidora Radulov and Adina Berbecea

Chapter 3 **57**
Organic Fertilizers and Biopesticides: Eco-Innovations, Circular Economy, and Agribusiness Developed by Organic Farmers in Uganda
by Bienvenu A. Dagoudo, Charles Ssekyewa and Silvère D. Tovignan

Chapter 4 **71**
Evaluating Land Management Options to Optimize Soil Organic Matter Retention in Oxisols: Case Studies Involving Brazil and Hawaii
by Michael Aide

Chapter 5 **91**
Cover Crops for More Sustainable Soil Fertility Management: Examples from Brazilian Cerrado
by Leonardo Santos Collier and Karin Ferretto Santos Collier

Section 2
Precision Agriculture and Agroecological Practices **117**

Chapter 6 **119**
Advancements in Precision Nitrogen Management for Sustainable Agriculture
by Sandeep Kumar, Tanu Sharma, Sanjay Singh Rathore and Vinod Kumar Singh

Chapter 7 141
Perspective Chapter: Redesigning Agroecological Practices for Enhanced Resource Use Efficiency in Agroecosystems
by Muhammad Talha Aslam, Imran Khan, Muhammad Umer Chattha, Ali Raza, Muhammad Bilal Chattha, Muqarrab Ali, Muhammad Umair Hassan and Iqra Fatima

Chapter 8 165
Perspective Chapter: Agroforestry Strategies for Integrated Soil and Water Conservation
by Dinesh Jinger, Nandha Kumar N, Chhavi Sirohi, Archana Verma, Pankaj Panwar and Rajesh Kaushal

Section 3
Water and Irrigation Management 191

Chapter 9 193
Agricultural Drainage Systems in Irrigated Ecologies
by Mohamed Joseph Sesay, Felix K. Abagale and Shaibu Abdul-Ganiyu

Chapter 10 221
Agrometeorological Aspects of Sugarcane
by Alexandrius de Moraes Barbosa

Section 4
Biological and Ecosystem Interactions 237

Chapter 11 239
Biological Pest Control and Sustainable Agroecosystems
by Zill-e-Huma Aftab, Tehmina Anjum, Waheed Akram,Najat A. Bokhari, Areeba Rehman, Shama Sharif, Humaira Rizwana and Bareera Munir

Chapter 12 273
Bio-Organic Management of Root-Knot Nematodes in Sustainable Agroecosystems
by Jaseem KP, Ameer Favas V, Shana Sherin and Sonia Waqar

Chapter 13 297
Soil Microbiome and Vegetable Crop Interaction in Climate Change Scenario
by Sripriya Das, Vandana Kumari, Kiran Kumar Mohapatra, Sayon Mukherjee, Plabani Roy, Dibyajyoti Panda, Abhik Patra, Ashish Rai and Vijay Singh Meena

Chapter 14 325
Utilizing Alternative Carbon Sources for Biofloc System for Growth and Survival of Pacific Whiteleg Shrimp (*Litopenaeus vannamei*)
by Arien Jean M. Lopez, Marlyn B. Llameg, John Paul R. Pacyao and Godofredo P. Lubat Jr

Preface

The sustainable management of agroecosystems is critical for ensuring global agricultural development, agri-food security, and environmental resilience in the face of climate change, resource depletion, and population growth. This edited volume, *Sustainable Agroecosystems – Principles and Practices*, brings together insights from leading experts, offering innovative strategies that support the long-term sustainability of agri-food systems.

The book begins by emphasizing soil health as a foundation for sustainable agriculture. Conservation practices, nutrient management, and organic inputs enhance soil fertility while integrating circular economy principles for eco-friendly farming. Advancements in precision agriculture and agroecological practices further optimize resource efficiency, promoting sustainability through data-driven and holistic approaches.

Water and irrigation management are central themes, addressing global water challenges through innovative techniques that improve efficiency and resilience to climate variability. Additionally, the book explores biological and ecosystem interactions, highlighting nature-based solutions for pest control, soil microbiome management, and sustainable aquaculture.

By integrating scientific research, technological advancements, and practical applications, this volume provides a comprehensive framework for transforming agricultural systems. It is a valuable resource for researchers, policymakers, and practitioners working toward resilient, productive, and environmentally sustainable agriculture.

We extend our sincere gratitude to the contributing authors, reviewers, and editorial team for their efforts in making this volume possible. We hope this book is an essential reference for advancing sustainable agroecosystems worldwide.

Vijay Singh Meena and Mohammad Hasanain
ICAR-Indian Agricultural Research Institute,
Regional Station,
Pusa, Samastipur (Bihar), India

Ram Swaroop Bana
Department of Agronomy,
ICAR – Indian Agricultural Research Institute,
New Delhi, India

Ram Kishor Fagodiya
ICAR-Central Soil Salinity Research Institute,
Karnal, Haryana, India

Section 1

Sustainable Soil and Nutrient Management

Chapter 1

Nurturing Soil Health through Conservation Agriculture Practices

Dwaram Mrudula, Banoth Bixapathi Nayak, Chebrolu Sravani, Thulisegari Prasanna, Mude Ramesh Naik and Mohammad Afrose

Abstract

Conservation agriculture (CA) is an innovative approach to managing agricultural ecosystems aimed at achieving sustainable productivity, enhanced profitability, and food security, while conserving resources and protecting the environment. This chapter explores the core principles of CA, focusing on minimum soil disturbance, permanent land cover, and crop rotation, which together significantly improve soil health by enhancing its physical, chemical, and biological properties. Through global and Indian research, CA practices have been shown to improve soil structure, boost natural resource availability, and enhance nutrient cycling and biodiversity. In addition to its environmental benefits, CA plays a key role in mitigating climate change by sequestering carbon and reducing greenhouse gas emissions. The chapter discusses the broader economic, environmental, and social benefits of CA, including its contribution to sustainability and resilience. Practical applications, such as site-specific strategies, appropriate equipment, and farmer training programs, demonstrate CA's potential. However, challenges remain, particularly lower yields for staple crops and higher labor demands. Looking ahead, CA presents significant opportunities for innovation and development. The chapter concludes by underscoring CA's crucial role in promoting soil health, enhancing climate resilience, and establishing sustainable food systems, while emphasizing the need for continuous organizational support and strategic alignment for successful implementation.

Keywords: soil health, conservation agriculture, crop rotation, nutrient cycling, soil organic matter

1. Introduction

Soil health affects agricultural production, environmental quality, and ecological resilience; it is essential for sustainable agriculture. Soil quality is essential for plant growth, water infiltration, carbon sequestration, and nutrient cycling, providing a basis for profitable agricultural systems but conventional farming methods often break down the condition of the soil due to over-use of chemicals, monocropping, and over-cultivation [1]. Conservation agriculture (CA) has become a potent collection of practices that help to mitigate the negative consequences over time caused by conventional agriculture. CA seeks to establish robust and sustainable agricultural systems by improving soil structure, increasing soil resources, and promoting biological activity [2]. Three basic principles that represent conservation agriculture are crop rotation, permanent land cover, and minimal or no disturbance [3]. These techniques provide the soil with physical, chemical, and all-natural attributes for increasing soil health and productivity. Moreover, CA is important in global issues such as food security, water scarcity, and climate change [4]. In addition, California significantly reduces the impacts of climate change by improving soil carbon storage and reducing greenhouse gas emissions [5]. CA is an important countermeasure to global warming because it maintains land cover and reduces soil disturbance [6]. In addition, it makes agricultural systems resistant to flash floods and droughts by improving water retention and penetration [7]. Due to the increase in the world's population, the need for food production rises, putting more pressure on the agriculture system to produce more food by providing fewer inputs.

Conventional farming methods produce more yield within a short period but cause soil and land degradation as well as loss of biodiversity which leads to long-term deterioration. Due to this unsustainable perspective emphasizes the need for different models which in turn produce sustainable yields. CA provides a practical answer to this problem through an integrated approach that further improves soil health, water retention, and productivity without damaging the environment. However, the use of CA is not a one-size-fits-all approach but must be adapted to the unique circumstances and constraints of each community. For example, in Brazil the practice of no-till farming was implemented successfully and it results in improving soil health and productivity [8]. In addition to this, in Asia and Africa, farmers are facing issues like a lack of knowledge about CA training and equipment usage [9]. Getting better results from CA methods requires some considerations like soil type, climate, and socio-economic conditions of farmers.

Although conservation agriculture originated from traditional agricultural practices, it became increasingly popular in the mid-twentieth century in response to increasing soil erosion and degradation [10] and the American Dust Bowl which occurred in the 1930s highlighted the importance of early land conservation. The use of CA has increased worldwide over the years, with notable increases in Australia, North and South America, and many European regions [11]. Brazil and Argentina, these two countries have adopted CA, and their resources are used for stability, and soil health, and there are significant benefits in terms of economic benefits. Given the importance of reducing soil erosion, improving nutrients, and improving water use efficiency, India and other Asian countries are gradually adopting CA [12]. Globally it is estimated that about 180 million hectares are covered by CA, and increased profitability and potential for sustainable agriculture Rationale. CA is becoming increasingly popular in India, especially in Indo-Gangetic plains, where environmental health and food security depend on sustainable energy. The arid regions of Rajasthan and the fertile Ganga Plains of Bharat are the only two agroecological regions of India where the CA method has shown the possibility of land conservation and residues changing in use when CA technologies with Happy Seeder and zero-till seed drills [13].

This chapter looks at the origins of conservation agriculture (CA) and outlines its expansion and widespread adoption. It looks at how CA affects soil properties are effected by CA methods how it is put up to mitigate climate change, and how it can be applied to real-world situations. A broader perspective on the impact of CA on soil health is sought, using case studies from multiple sources to illustrate its environmental, economic, and social impacts. The chapter also discusses real-world challenges when farmers use CA and looks at possible solutions, including government development and technological development. Does this chapter's scientific principles provide comprehensive knowledge of the role of CA in promoting sustainable agriculture for food production encouraging long-term achievement by blending real-world experiences.

2. Principles of conservation agriculture

Three basic concepts support conservation agriculture (CA): crop rotation, permanent land cover, and minimal or no land disturbance. By mimicking natural processes, these processes work together to create complex and robust soil systems (**Figure 1**).

2.1 Minimal soil disturbance

Reduction or elimination of soil disturbance through passive or mitigation methods helps to preserve soil structure and soil biodiversity. According to research, unplanned soil organic matter significantly increases (SOC) levels, and microbial activity improves effectiveness [14].

2.1.1 No-till farming

In this method, crops are grown without first tilling the soil, directly into the residues of previous crops. This method reduces soil erosion, increases drainage, and maintains soil structure [15]. Studies show that no-tillage farming can increase SOC stocks by improving soil accumulation and controlling soil organic matter degradation [16]. For example, it shows that no-till farming improves soil fertility and structure by increasing soil organic matter and microbial bioavailability increase. Furthermore, no-till farming can save on labor and fuel costs and will benefit farmers economically [17].

2.1.2 Reduced tillage practices

Reducing the depth and frequency of cultivation increases the accumulation potential of the soil and conserves soil resources, further improving soil health and

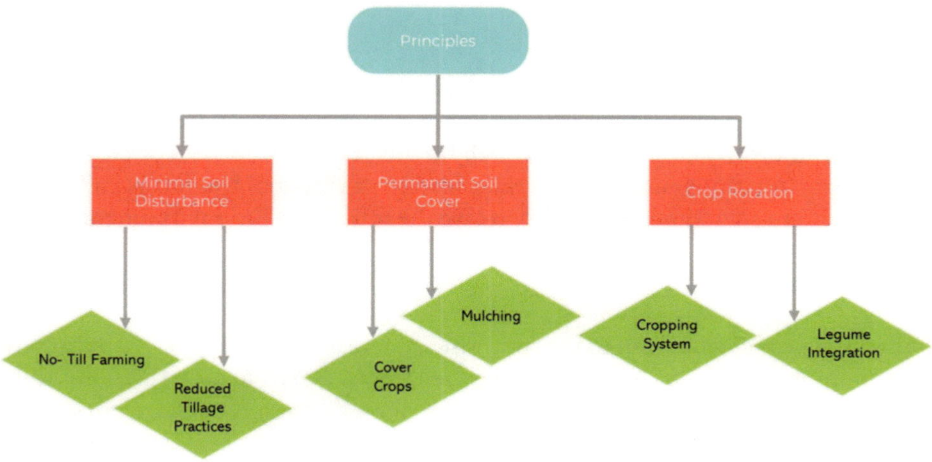

Figure 1.
Principles of conservation agriculture.

crop yields [18]. A diverse and functional soil microbiome is another benefit of reduced tillage systems, and this microbiota is important for organic waste breakdown and nutrient cycling [19].

2.2 Permanent soil cover

Using mulch or cover crops to cover the soil regularly can prevent erosion, prevent weed growth, and release soil water retention effectively. Cover crops enhance nitrogen cycling and improve soil organic matter [20].

2.2.1 Cover crops

Cover crops such as vetch, rye, and clover are grown seasonally to reduce soil erosion and increase productivity by fixing atmospheric nitrogen [21]. In addition, covering crops increases organic matter and soil structure, both of which are important for soil health [22]. For example, cover crops can reduce nitrate leaching, improve water quality, and support resilient agricultural conditions.

2.2.2 Mulching

The application of organic mulches, such as wood chips or straw, to the soil's surface, helps to improve moisture conservation, regulation of temperature, and the addition of organic matter through decomposition [23]. Compost can increase plant growth and production as it has been shown to retain soil moisture and temperature increase consumption [24]. Also, pollution can be controlled to prevent weed growth, reduce the need for pesticides, and provide agricultural incentives for environmental control [25].

2.3 Crop rotations

Multi-level crop rotations disrupt pest and disease cycles, increase soil fertility, and support ecosystems. Corn can cycle atmospheric nitrogen; therefore, it is important to use low levels of fertilizers [26, 27].

2.3.1 Cropping systems

Rotational cropping disrupts pest and disease cycles, improves soil structure, and increases nutrient use efficiency [28]. Research studies result in a reduction of pest and disease attacks by crop rotations which further increase the yield of crops. For example, crop rotations can reduce greater storage of essential chemicals and pesticides, to save construction costs and decrease the impact on the environment.

2.3.2 Legume integration

Legumes provide nitrogen to the soil, which decreases the usage of chemical fertilizers and improves the fertility of the soil when they are used in crop rotations [29]. Legumes, including beans and peas, work together with bacteria that fix nitrogen from the atmosphere. This process increases the amount of nutrients in the soil by converting atmospheric nitrogen into a form that can be used by plants [30]. Besides contributing to successful crops, this natural cropping method also increases the overall sustainability of the agricultural system [31].

3. Permanent raised beds in conservation agriculture

Under the basic principles of conservation agriculture minimal soil erosion, constant land cover, crop rotation and diversity, and permanently raised beds are a cornerstone strategy. By building these beds, you can maintain soil structure, promote healthy ecosystems, and reduce the need for regular tillage. In addition, soil erosion in crop areas is reduced by restricting walking to designated paths, thereby increasing root access and overall plant health. Raised beds are softened with organic matter such as grass clippings or grass mulch. Mulching provides organic matter during soil decomposition, which inhibits weed growth and maintains soil moisture. Planting cover crops in the off-season improves soil structure, increases soil fertility, and reduces soil erosion.

Raised beds are usually 3–4 feet wide in plan and style, making it easy to get out from either side without walking upstairs. Soil erosion can be reduced by providing channels between beds, which can be covered with gravel or dirt [32]. Prepare the soil by filling the beds with mixtures, native soils, and other organic matter to create a favorable growing environment [33]. The fertility and structure of the soil are maintained by compost and organic matter regularly [34]. Raised beds also improve water efficiency. In tropical areas, better drainage reduces the possibility of flooding, while in drier climates, raised beds can be created to improve water retention, especially when mulching and other organic matter are added [35]. Mulching suppresses weeds and reduces the chemical demand for pesticides. Insect traps or row covers can be used as physical barriers to protect crops from pests [36].

Many smallholder agricultural systems, particularly in developing countries, have used sustainable raised beds effectively to increase crop yields, conserve water, and improve soil health [37]. They are also well-loved in urban agriculture due to their efficient use of space and ability to produce large quantities of food in small areas [38]. Properly raised beds support resilient farming systems, encourage more efficient land use, and help farmers meet long-term sustainability goals while maintaining or increasing yields.

4. Impact of conservation agriculture on soil health

The use of conservation agriculture (CA) techniques is associated with many benefits in terms of soil physical, chemical, and biological properties. These improvements are important for sustainable production, environment protection, and climate change mitigation in sustainable agricultural systems (**Figure 2**).

4.1 Physical properties

Conservation agriculture (CA) significantly improves soil fertility, improves soil structure, overall soil stability, drainage, storage, and erosion management. Agriculture and the long-term future of land management depends on these developments.

4.1.1 Soil aggregation

The concept of "soil aggregates" describes how soil particles combine to form aggregates that increase soil porosity and structure. Under the strategies of CA, agricultural land reduction and permanent land cover grow and strengthen these aggregates:

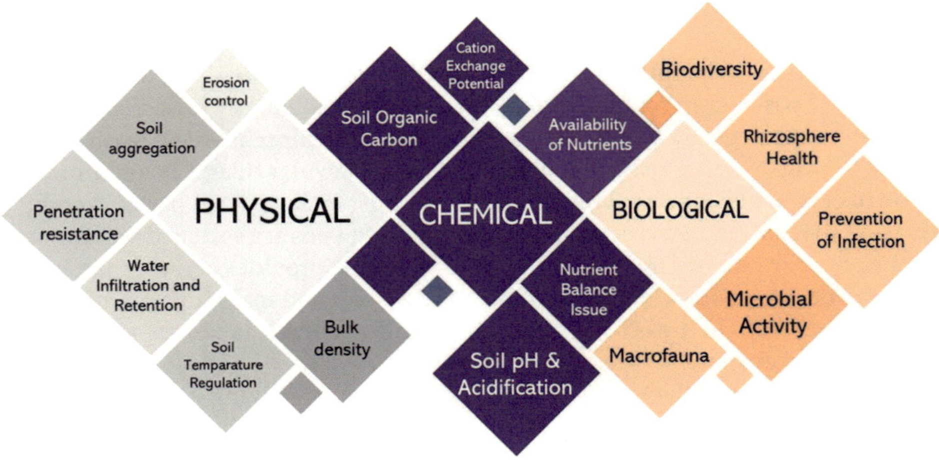

Figure 2.
Impacts of conservation agriculture on soil health.

4.1.1.1 Macro-aggregate stability

No-till system results in greater macro-aggregate stability. Comparing no-till farming to conventional tillage shows the production of larger aggregates of soil, which is useful for preserving soil organic matter and improving soil structure in the state of a no-till tillage system [39]. Long-term no-till methods improve soil macro-aggregate stability, which improves soil health [40].

4.1.1.2 Aggregate formation

Cover crops and Mulch provide permanent cover for the soil, which leads to the production of organic matter and encourages the production of aggregates. The increment of organic carbon in the soil leads to the binding of soil particles together and improves soil aggregation [41]. Research conducted found how soil aggregate stability was increased by different cover cropping strategies [42].

4.1.1.3 Soil porosity

Greater aggregation of soil particles increases the porosity of the soil, which promotes higher root and water penetration. Greater soil porosity in CA systems in India has been linked to increased development of root and crop production [43]. The techniques under CA can increase the porosity of soil, which improves soil aeration and root growth [44].

4.1.2 Water infiltration and retention

Under CA practices, improved soil structure increases water infiltration and storage capacity, which is important for crop health and drought tolerance:

4.1.2.1 Soil drainage

CA methods significantly increase soil drainage, especially no-till farming. Since no-till soils have better soil structure and water content than conventional soils because no-till soils have higher permeability. This upgrade improves groundwater recharge while lowering runoff. CA techniques can improve water infiltration rates, which is necessary for reducing the risk of flooding [45].

4.1.2.2 Water conservation

Moisture content is increased by mulch and cover crops which improves soil organic matter. CA sites in semiarid regions of India have more water, which improves drought resistance and meets water requirements during dry seasons [46].

4.1.2.3 Reduced runoff

By reducing surface runoff, CA systems enhanced infiltration and retention capacities to stop soil erosion and nutrient loss. Soil covers significantly reduce runoff and protect soil and water resources. CA methods can effectively reduce runoff, thereby improving water conservation [47].

4.1.3 Erosion control

Permanent soil conservation and erosion reduction reduce soil erosion in many ways:

4.1.3.1 Ground cover

It helps prevent soil erosion by protecting the surface from rainfall energy and by planting mulch and perennials. The protective layer created by CA techniques can prevent soil displacement, resulting in a 90% reduction in soil erosion when compared to conventional tillage [48]. These results are supported by a meta-analysis which demonstrates a significant decrease in erosion caused by CA methods [49].

4.1.3.2 Root system

Cover crops and crop rotations increase soil particle and root binding. Cover crops have large root systems that contribute to soil stability and reduce soil erosion. The CA method improved soil stability, and reduced soil loss in the Deccan Plateau of India [50]. According to the most recent study found that cover crops with deep roots are very useful in reducing soil erosion [51].

4.1.3.3 Surface roughening

CA pavements covered with organic debris increase surface roughness, slow flow and increase sediment deposition. This roughness contributes to the reduction of surface water velocity, resulting in a decrease in sedimentation. CA can improve water retention and reduce erosion by increasing surface roughness [52].

4.1.4 Bulk density

Over time, CA methods tend to deplete more soil organic matter, making the soil softer and more conducive to root growth and water circulation. Systematically do not work and tend to have lower soil volumes than tilled systems, resulting in more plant growth. Reducing the cultivated area and incorporating more organic matter into CA systems can significantly reduce soil volume and improve root growth [53].

4.1.5 Soil temperature regulation

The application of water to the soil makes the soil surface impermeable, thus causing evaporation. Organic mulches protect plant roots from strong temperature fluctuations by helping to maintain a more constant soil temperature. Mulching in CA systems can effectively regulate soil temperature, and enhance crop development and yield [54].

4.1.6 Penetration resistance

Increased organic matter and less tillage reduce soil vulnerability and infiltration resistance. As a result, the roots can grow deeper and absorb more water and nutrients. Root systems under CA showed higher resistance to water infiltration than conventional soils [55]. CA processes can improve root development and overall plant health by improving soil penetration.

4.2 Chemical properties

The use of conservation tillage (CA) significantly improves soil chemical properties of soil, resulting in increased soil organic carbon (SOC), improved nutrient availability, reduced soil acidity, and other changes. A good use of crop cover and organic amendments supports more nutrient balance and overall soil health [56, 57].

4.2.1 Soil organic carbon (SOC)

SOC levels are significantly increased by CA practices, especially no-till farming and cover crops. In general, SOC is important for soil health, soil structure, and nutrient availability. This increases the soil's ability to hold nutrients and water, both of which are essential for microbial activity and plant growth. No-till methods can sequestrate more carbon in the soil than conventional methods and improve soil fertility and structure. CA processes significantly increased SOC concentration and nutrient availability in the Indian Sub-Himalaya, increasing soil fertility and agricultural productivity [58]. Higher SOC under CA has also been observed in studies in India in rice and wheat systems, that improve soil health. It also supports growth and resistance to climatic stresses.

4.2.2 Availability of nutrients

4.2.2.1 Enhanced nutrient cycling

CA improves nutrient availability by increasing biological activity and organic matter from cover crops and residues. When these organic matter are degraded, essential nutrients such as nitrogen, phosphorus, and potassium are leached into the soil, resulting in increased plant biomass. Soil products from crop coatings increase nutrient retention and reduce fertilizer requirements. Increased nitrogen availability in CA-managed areas of the Indian Indo-Gangetic plain, resulting in enhanced crop growth and yield [59].

4.2.2.2 Biological activity

Nutrient cycling is greatly contributed by the increased microbial activity observed in CA systems. As organic matter decays, microorganisms release the right nutrients for plants to absorb. Over time, this process improves soil fertility in addition to increasing nutrient levels. Nutrient cycling and availability are far better in CA systems with high microbial and activity levels, reducing the need for chemical fertilizers.

4.2.3 Soil pH and acidification

4.2.3.1 Soil pH control

Cover crops and organic amendments in CA systems help stabilize soil pH and reduce acidification. By adding organic matter, these activities make the soil less acidic and improve plant growth conditions. Especially cover crops of legumes can reduce soil acidity by supplementing organic residues with buffering pH levels [60]. This buffering effect encourages good nutrient uptake and microbial activity and helps maintain a constant pH in the soil.

4.2.3.2 Reduced acidification

Continuous addition of organic matter through the CA process neutralizes soil acidity and reduces the negative effects of acidity on crop yield and soil health. In the Western Ghats of India, CA processes improve soil pH and reduce the need for the application of lime, and other pH modifying agents were shown to be organic the more frequent deposition and increased biological activity in CA systems are due to this pH gradient [61].

4.2.4 Nutrient balance issues

CA methods encourage continuous reapplication of nutrients through microbial activity and organic matter breakdown, resulting in greater nutrient balance. This balance is important for long-term soil fertility and yield. The atmospheric nitrogen can be fixed with cover crops such as maize, enriching the soil and reducing the usage of nitrogen fertilizers. This practice increases the importance of sustainable soil management and maintains nutrient balance.

4.2.5 Cation exchange potential (CEC)

The ability of SOC and organic matter to store positively charged ions (cations) such as calcium, magnesium, and potassium for increased CEC increases the availability and retention of nutrients for the plant spike [62]. Increased SOC concentration in CA systems improves CEC, enhances nutrient retention, and reduces leaching losses.

4.3 Biological properties

The application of conservation agriculture (CA) practices significantly improves soil ecosystem health. Higher organic matter and less disturbance to improve the habitat for ketch flies, earthworms, and other beneficial organisms. These biological functions are important for soil structure and nutrient cycling [63].

4.3.1 Microbial activity

Higher SOC and lower disturbance increase microbial biomass and activity, which is important for biodegradation and nutrient cycling. The no-till system increased microbial biomass and their activity increased dramatically. Reducing soil tillage increases soil microbial and enzymatic activity, thus increasing yield and nutrient cycling. The microbial activity increased in soils treated with CA in India [64]. CA processes in the Indo-Gangetic plain increase soil microbial populations, thereby increasing soil health and agricultural productivity [65].

4.3.1.1 Soil enzyme activity

Enzymes that are necessary for breaking down organic compounds and liberating nutrients for plant absorption are released by soil microbes. Effective CA methods facilitate nitrogen cycling, which increases the activity of soil enzymes. Nonagricultural conditions significantly increase the activity of enzymes important for soil fertility and health, such as urease, phosphatase, and dehydrogenase [66]. Due to higher enzyme activity in CA systems, high soil fertility and nutrient cycling support crop production efficiency. Studies in Indian agricultural systems have shown a similar increase in enzyme activity under CA, which enhances soil health and nutrient availability.

4.3.1.2 Mycorrhizal fungi

These symbiotic fungi attach to plant roots to help plants absorb water and nutrients. When soil disturbance decreases, as reflected by the CA method, mycorrhizal fungi grow. CA fields have higher rates of mycorrhizal colonization, improving plant growth and phosphate uptake [67]. Mycorrhizal relationships enhance plant health and nutrient utilization, increasing yields, and keeping the soil in a CA system. In semiarid areas of Rajasthan, the CA method has been shown to increase mycorrhizal availability, resulting in improved drought tolerance of plants.

4.3.1.3 Soil respiration

Soil respiration is improved by using healthy soil and CA methods, indicating microbial activity. Increased soil respiration means greater biological activity, which is important for organic matter decomposition and nutrient cycling. Significantly higher soil respiration rates in the CA method compared to conventional agriculture [68]. Increased soil respiration in the CA system indicates greater productivity of healthy soil bacteria, improving soil fertility and productivity.

4.3.2 Macrofauna

Earthworms and other soil microorganisms thrive in CA systems, improving soil fertility, structure, and nutrient availability. Compared to conventional farming, informal systems significantly increase earthworm levels. Non-tillage methods increase earthworm levels, enhancing soil structure and nutrient availability. Similar results were found in the study in Northern India, where the density of earthworms was higher in CA areas [69]. More earthworm activity was observed in the CA areas which further increased soil porosity and drainage [70].

4.3.3 Biodiversity

Crop rotations and cover crops provide a wealth of soil ecological knowledge, enhancing ecosystem services, including nutrient cycling and pest management. The diverse agroecosystems increase biodiversity, increasing soil resilience and productivity. By increasing soil microbial production, crop diversity in rotation promotes ecological stability and productivity. In the Indian state of Punjab, greater biodiversity and better management of pests in CA systems [71]. CA methods improved soil health and crop productivity by increasing soil fungal populations and microbial activity in rice and wheat cropping systems in India.

4.3.4 Rhizosphere health

The soil layer around plant roots, known as the rhizosphere, benefits from CA by improving root growth and productivity. In CA sites, dense root systems provide beneficial microbial interactions that improve plant health and nutrient utilization. By promoting a robust and nutrient-rich rhizome, CA pathways enhance plant-bacterial symbiosis [72]. Good rhizome round health in CA systems promotes robust plant growth and stress tolerance.

4.3.5 Prevention of infection

By promoting a high and well-balanced soil microbial population that interacts with harmful microorganisms, CA methods can reduce soil-borne diseases. Natural modification of CA systems and reduced cultivation create beneficial microbial populations competing with diseases pesticides required to maintain healthy crops also need to be reduced also on infection control [73].

5. Mitigating climate change and global warming

Conservation agriculture (CA) is key to mitigating climate change by reducing greenhouse gas emissions, improving the resilience of agricultural ecosystems to extreme weather events, and sequestering carbon on the ground (**Figure 3**):

5.1 Carbon storage

By significantly increasing soil organic carbon (SOC), CA practices especially cover crops and no-till crops—act as carbon sinks and reduce global warming calculation enables maximum utilization of CA receiving CO_2 concentration well atmosphere. This method of removing CO_2 by adding organic matter to the soil and stabilizing it with soil aggregates prevents CO_2 from returning to the atmosphere. The CA systems in India capture more carbon than conventional methods, contributing to climate change mitigation. Their study showed that CA practices in the southern Himalayan Indian region significantly increase SOC concentrations, improve soil fertility, and contribute to climate change mitigation.

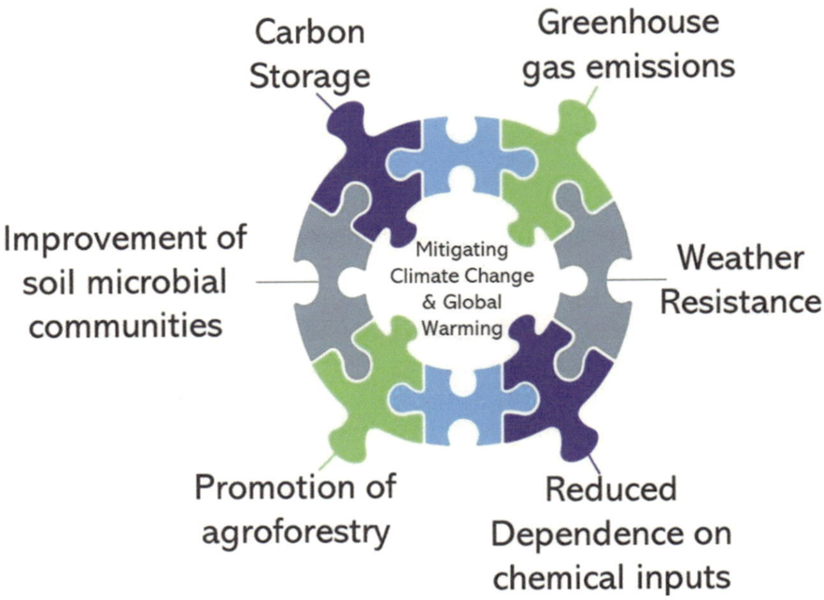

Figure 3.
Mitigating climate change and global warming.

5.2 Greenhouse gas emissions

CO_2, methane (CH_4), and nitrous oxide (N_2O) emissions in CA systems are reduced through tillage reduction and better soil health. Because CA systems have better soil structure and organic matter management than cultivated land, they emit less N_2O [74]. The main reason is that CA processes improve soil aeration and reduce anaerobic conditions, which are favorable for N_2O generation. In India, farms using CA released fewer greenhouse gases [75]. N_2O and CH_4 emissions were reported to be reduced under CA processes due to residue efficiency and reduced soil disturbance, indicating the environmental benefits of CA in tropical settings in agroecosystems.

5.3 Weather resistance

Improved soil structure and water-holding capacity in CA systems enhance flood and drought resistance. CA processes help the soil to retain moisture effectively and resist extreme weather conditions. Although the soil structure is good for runoff and sedimentation during periods of heavy rainfall, the high soil moisture content helps in crop survival during drought. In semiarid India shows that CA methods increase crop resistance to drought conditions. They found that even under adverse weather conditions, CA-managed fields had improved water infiltration and retention, resulting in more robust crop yields.

5.4 Reduced dependence on chemical inputs

Conservation agriculture (CA) strategies to reduce the demand for fertilizer crops by improving soil fertility and nutrient cycling, a key contributor to greenhouse gas emissions. The inclusion of legumes and organic matter in crop rotations encourages natural nitrogen fixation, which reduces nitrogen feed requirements and reduces N_2O, a greenhouse gas temperature, and the carbon footprint associated with fertilizer production and consumption. CA processes reduce the need for pesticides and fertilizers and reduce agriculture costs and the impact of agricultural activities on the environment. The reduced reliance on chemicals used in CA systems results in dramatic economic and ecological benefits, which are a factor of importance in sustainable agriculture and the protection of natural resources.

5.5 Promotion of agroforestry

It is a common practice in California to add trees and shrubs to farms to further enhance carbon capture and biomass sequestration. Trees capture carbon in their biomass in addition to adding carbon to the soil on their root leaves. In addition, agroforestry systems insulate crops from wind and shade and provide microclimate and crop resilience to extreme temperatures [76]. Agroforestry practices in India generate soil biomass carbon maximize conservation, mitigate climate change, and provide additional income for farmers. It has also been shown to be beneficial [77].

5.6 Improvement of soil microbial communities

Carbon sequestration and mitigation of greenhouse gases in soil are greatly contributed by an active and diverse soil microbial community promoted by CA pathways. Microbial bacteria participate in pathways species that reduce N_2O emissions break down organic matter and fix carbon in soil. In addition to improving soil health, increased microbial activity through CA contributes to the reduction of greenhouse gas emissions from all agricultural systems.

6. Tailoring practices to local conditions

To achieve successful results, conservation agriculture (CA) strategies must be adapted to local conditions. This means that factors such as soil structure, climate, and special agricultural challenges must be carefully considered. Utilizing the full potential of CA methods requires careful planning of planting times and the selection of appropriate local cover crops.

6.1 Site-specific management

Effective CA techniques must be adapted to the specific needs of the farm, accounting for factors such as soil properties and climate change. The importance of tailoring strategies to specific areas, emphasizes the importance of developing sector-specific CA strategies, particularly within the agroecological sectors [78].

6.2 Investing in the right technology

The modern equipment needed to use conservation agriculture (CA) to improve agricultural productivity and simplify operational methods, such as precision and no-till seeding, is reduced employees are required to manage and use resources effectively, making it important for the implementation of CA strategies. The important role of such tools in the adoption of CA [79]. Based on empirical evidence of how technological advances significantly enhance farmers' capacity to adopt CA strategies [80]. With appropriate technological investments, farmers can solve problems associated with traditional farming methods, such as resource waste, soil erosion, and precision mechanization to control yield consumption accuracy, resulting in better crop placement and results. In addition, the use of the latest technologies is consistent with many goals of sustainable agriculture through conservation practices that prioritize ecological resilience and soil health. Modern technologies can help farmers fully reap the benefits of CA by increasing agricultural productivity and environmental impact.

6.3 To provide training and support to farmers

Education and support for farmers are essential to encourage the adoption of conservation agriculture (CA). Its long-term sustainability and effective implementation depend largely on farmers' extension services and training programs that will provide them with the information and skills they need. To use the CA methods potentially, farmers need effective knowledge and technical support. Extension services and farmer field schools are important forums for developing new farmer training and support programs to provide farmers with the necessary skills. Programs such as the National Mission for Sustainable Agriculture (NMSA) in India, are crucial in providing farmers with inputs and training and highlight the key role of extension services in adopting CA emphasis.

7. In practical applications

See **Table 1**

S. No	Practices	Results
1	Transition to no-till and reduced crop systems	The implementation of transition to reduced or no-till agriculture requires careful planning and purchase of specialized equipment to minimize soil erosion and increase productivity must be accompanied by a comprehensive program incorporating state-of-the-art techniques such as integrated pest management (IPM).
2	Equipment	An important feature of no-till and reduced systems is specialized equipment such as precision and informal seeding, these technological improvements improve soil health and productivity.
3	Weed management	Weed control in no-till systems combines rotational cropping, cover crops, and specific pesticides. This combined approach handles weed control effectively with minimal soil disturbance.
4	Cover crop management and mulching	Careful selection and maintenance of cover crops based on specific agricultural objectives is essential for soil health and moisture drainage effectiveness. In comparison, applying appropriate mulching techniques improves soil organic matter and contributes to its conservation.
5	Cover crop selection	Cover crops should be carefully selected to improve soil fertility and prevent erosion, considering their specific characteristics and benefits.
6	Nursery methods	Sustainable agriculture is supported by the use of appropriate materials and techniques, such as the addition of organic matter or crop residues, to preserve soil integrity and to increase moisture content.

S. No	Practices	Results
7	Efficient crop rotation	An important aspect of CA is a crop rotation system that balances agricultural needs with environmental sustainability. Integrated crop rotations, based on characteristics including soil nutrient requirements and market demand, improve soil resilience and health.
8	Crop compatibility	Long-term agricultural sustainability depends on crop rotation that reduces soil nutrient depletion and pests through careful crop selection in rotation.
9	Market and Environmental Factors	Crop rotations that are in line with environmental factors and market dynamics guarantee financial success while fostering ecological resilience, highlighting the comprehensive approach that is fundamental to CA principles.

Table 1.
In-practical applications of conservation agriculture.

8. Toward sustainable agriculture

An important aspect of sustainable agriculture is conservation agriculture (CA), which provides a comprehensive approach to agricultural management that ensures longevity for farmers and the environment In-depth research examines how California supports sustainable agriculture economically, environmentally, and humanely.

8.1 Economic benefits

There are significant economic benefits in the use of CA techniques as they reduce input costs and increase farm profitability. CA helps to save farmers' costs by reducing the use of expensive inputs such as pesticides and fertilizers to maintain or increase yields over time. This reduction in input costs directly contributes to increased economic growth for farmers. Additionally, due to the sustainability of CA processes, farmers benefit from stronger crops year after year, increasing the economic stability of agribusiness. In India highlighting the economic benefits of CA by showing that it increases gross farm income and reduces investment costs [81].

8.2 Environmental benefits

CA practices are necessary to protect the environment by reducing soil erosion, improving water quality, and enhancing biodiversity. CA reduces soil erosion by wind and rainfall through the amount of agricultural residue that occurs each year or covers crops in the soil. This protects important topsoil and reduces erosion. In addition, the improved soil structure and organic matter created by CA processes increase infiltration and storage capacity and reduce the potential for nutrient leaching into nearby waters. Cover crops show that it also improves water quality and ecosystem health. In addition, CA supports aboveground and belowground ecosystems that provide habitats and food for a variety of organisms, thereby creating resilient ecosystems that are resilient to environmental hazards well established.

8.3 Social benefits

By enhancing nutrition and improving rural livelihoods, CA strategies have significant positive social impacts in addition to economic conditions. CA contributes to sustainable and predictable agricultural production, contributing to food security locally and globally by increasing soil health and combating climate change. This is especially important in areas where severe weather events are more likely to be encountered, as the implementation of CA practices can protect populations from floods, droughts, and other climate-related hazards. CA increases food production and reduces sensitivity to climate shocks, thus strengthening community resilience. In addition, CA methods can increase overall agricultural productivity and reduce the need for expensive inputs, giving farmers greater economic empowerment, and thus improving rural livelihoods. In promoting sustainable agricultural systems and offers empirical evidence of the beneficial effects of CA adoption on family food security and resilience [82].

9. Challenges and future directions

9.1 Challenges

Conservation agriculture (CA) has many advantages, but several barriers prevent its widespread adoption. These challenges include the need for specialized equipment, high labor costs, and loss of initial production. The long-term viability and widespread use of CA depends on solving these problems.

9.1.1 Reducing initial crop yield

Reducing primary yield is a major issue in agricultural production. Yields may temporarily decline during CA transitions as soil structures change in new practices. For example, yields may take years to solidify, which poses a significant adoption challenge. Research found comparable challenges faced by farmers in India, corroborating these findings and highlighting the importance of providing adaptive support to mitigate the effects of artificial basic losses at the bottom of the table [83].

9.1.2 Increase labor

The additional effort required to implement CA techniques such as memory management creates additional problems. These practices require considerable effort, especially residue management and cover crop abandonment. The increased labor demand from Indian farmers, and highlighted the need for labor-saving solutions to make CA more effective for farmers with limited labor resources the emphasis.

9.1.3 Specialized equipment

The need for specialized equipment, such as seed drilling and no-till planters, which can be expensive, is another barrier. The importance of appropriate tools to ensure the effective adoption of CA. The high cost of such devices has been a major obstacle in India. This highlights the need for accessible and affordable technological solutions to facilitate the widespread adoption of CA practices. Labor-saving technology, making special tools available at reasonable prices, and financing and technical assistance to farmers throughout the transition are among the many ways they need just to address a few of those issues. These obstacles may be removed to greatly increase the adoption of conservation agriculture and result in more sustainable and successful farming methods worldwide.

9.2 Future directions

9.2.1 Research and innovation

Ongoing research and innovation are essential to improve conservation agriculture (CA) methods and develop new technologies. Precision agriculture, improving cover crops, and soil health management systems are some areas of focus that can help farmers adapt CA practices.

9.2.1.1 Precision agriculture

Precision agriculture technologies, such as soil sensors and GPS-guided equipment, provide viable opportunities to increase the efficiency of CA. With the use of these instruments, farmers may maximize the utilization of resources, accurately control inputs, and raise farm production. Research highlights how precision input control through precision farming can transform California. CA production has improved significantly in places like India where precision agricultural technology has been implemented.

9.2.1.2 Cover crop breeding

Cover crop breeding programs have the potential to produce effective traits particularly suitable for CA systems. The potential flexibility and efficiency of CA methods can be enhanced by various characteristics including high pest density, fast growth rate, and high biomass. The importance of creating cover crop varieties specific to California. In India to create cover crops that adapt well to their environment, which would increase the efficacy of CA in a variety of agricultural environments.

9.2.2 Policy and incentives

The implementation of supportive legislation and incentives has a significant impact on the adoption of agricultural conservation (CA) by policymakers. From grants and subsidies to technical assistance programs, these programs aim to reduce the financial burden and risk of changes in CA programs.

9.2.2.1 Financial incentives

Subsidies and subsidies play an important role in alleviating financial problems faced by farmers considering adopting CA. Policymakers can facilitate investment in CA by providing financial support to farmers so that they can invest in the necessary equipment and cover crops. Financial incentives play an important role in encouraging adoption by reducing the initial costs for CAs. Government initiatives like the National Mission for Sustainable Agriculture (NMSA) by providing adequate financial support for CA processes greatly support farmer rescue efforts in India.

9.2.2.2 Technical assistance

Extension services and technical assistance programs play an important role in encouraging the uptake of CA by providing farmers with the information and resources needed for the effective adoption of CA. The need for technical support to facilitate CA implementation. The extension program in India was successful in educating farmers about CA and providing support, increasing farmer adoption rates.

10. Conclusion

A sustainable approach to improving soil health, reducing climate change, and increasing agricultural production is conservation agriculture (CA). Farmer-friendly practices such as minimizing soil disturbance, preserving permanent soil cover, and varying crop rotations can yield durable advantages for both their soils and the communities they serve. Adopting CA helps farms protect biodiversity, lessen erosion, save water resources, and maintain soil fertility and structure. By promoting resilience in the face of climatic unpredictability and guaranteeing food security for future generations, these techniques are in line with the larger objectives of sustainable agriculture. Although switching to CA systems has presented some difficulties, there remains hope for the future of sustainable agriculture. Research is still being done to improve CA procedures and create innovative devices that increase their efficacy. Incentives and regulations that promote universal adoption, such as financial aid and technical help, are also very important. Going forward, maintaining momentum in pushing the use of CAs worldwide is essential. This calls for increased funding for projects aimed at capacity-building, innovation, and research. We can enable farmers with the tools and resources they need to become more resilient and sustainable farmers in the future by encouraging cooperation between stakeholders. Conservation agriculture may flourish going forward with our combined efforts, benefiting ecosystems and societies alike.

Author details

Dwaram Mrudula[1*], Banoth Bixapathi Nayak[1], Chebrolu Sravani[1], Thulisegari Prasanna[1], Mude Ramesh Naik[2] and Mohammad Afrose[3]

1 SR University, Warangal, Telangana, India

2 Senior Scientist Research Systems Management Division, ICAR-National Academy of Agricultural Research Management, Hyderabad, India

3 Chaitanya University, India

*Address all correspondence to: mrudula1816@gmail.com

References

[1] Lal R. Restoring soil quality to mitigate soil degradation. Sustainability. 2015;7(5):5875-5895

[2] Kassam A, Friedrich T, Derpsch R. Global spread of conservation agriculture. International Journal of Environmental Studies. 2019;**76**(3):369-380

[3] Friedrich T, Derpsch R, Kassam A. Adoption of conservation agriculture and the role of policy and institutional support. Plant Production Science. 2009;**12**(1):3-14

[4] Pretty J. Agricultural sustainability: Concepts, principles and evidence. Philosophical Transactions of the Royal Society B: Biological Sciences. 2008;**363**(1491):447-465

[5] Smith RG, Atwood LW, Warren ND. Crop rotations and tillage system effects on weed dynamics, soil quality, and crop performance. Agronomy Journal. 2008;**100**(3):636-644

[6] Lal R. Sequestering carbon and increasing productivity by conservation agriculture. Journal of Soil and Water Conservation. 2015;**70**(3):55A-62A

[7] Hobbs PR, Sayre K, Gupta R. The role of conservation agriculture in sustainable agriculture. Philosophical Transactions of the Royal Society B: Biological Sciences. 2008;**363**(1491):543-555

[8] Derpsch R, Franzluebbers AJ, Duiker SW, Reicosky DC, Koeller K, Friedrich T, et al. Why do we need to standardize no-tillage research? Soil and Tillage Research. 2014;**137**:16-22

[9] Giller KE, Andersson JA, Corbeels M, Kirkegaard J, Mortensen D, Erenstein O, et al. Beyond conservation agriculture. Frontiers in Plant Science. 2015;**6**:870

[10] Montgomery DR. Soil erosion and agricultural sustainability. Proceedings of the National Academy of Sciences. 2007;**104**(33):13268-13272

[11] Derpsch R, Friedrich T, Kassam A, Li H. Current status of adoption of no-till farming in the world and some of its main benefits. International Journal of Agricultural and Biological Engineering. 2010;**3**(1):1-25

[12] Jat RA, Jinger D, Kumar K, Singh R, Jat SL, Dinesh D, et al. Scaling-up of conservation agriculture for climate change resilient agriculture in South Asia. In: Scaling-up Solutions for Farmers: Technology, Partnerships and Convergence. 2021. pp. 351-380

[13] Sidhu HS, Singh M, Singh Y, Blackwell J, Lohan SK, Humphreys E, et al. Development and evaluation of the turbo happy seeder for sowing wheat into heavy rice residues in NW India. Field Crops Research. 2015;**184**:201-212

[14] Six J, Ogle SM, Breidt FJ, Conant RT, Mosier AR, Paustian K. The potential to mitigate global warming with no-tillage management is only realized when practised in the long term. Global Change Biology. 2004;**10**(2):155-160

[15] Triplett GB, Dick WA. No-tillage crop production: A revolution in agriculture. Agronomy Journal. 2008;**100**(Suppl. 3):S-153

[16] Lal R. Soil carbon sequestration impacts on global climate change and food security. Science. 2004;**304**(5677):1623-1627

[17] Hobbs PR. Conservation agriculture: What is it and why is it important for future sustainable food production? Journal of Agricultural Science. 2007;**145**(2):127-137

[18] Franzluebbers AJ. Water infiltration and soil structure related to organic matter and its stratification with depth. Soil and Tillage Research. 2002;**66**(2):197-205

[19] Kladivko EJ. Tillage systems and soil ecology. Soil & Tillage Research. 2001;**61**(1-2):61-76

[20] Schipanski ME, Barbercheck M, Douglas MR, Finney DM, Haider K, Kaye JP, et al. A framework for evaluating ecosystem services provided by cover crops in agroecosystems. Agricultural Systems. 2014;**125**:12-22

[21] Clark A, editor. Managing Cover Crops Profitably. Diane Publishing; 2008

[22] Snapp SS, Swinton SM, Labarta R, Mutch D, Black JR, Leep R, et al. Evaluating cover crops for benefits, costs and performance within cropping system niches. Agronomy Journal. 2005;**97**(1):322-332

[23] Dabney SM, Delgado JA, Reeves DW. Using winter cover crops to improve soil and water quality. Communications in Soil Science and Plant Analysis. 2001;**32**(7-8):1221-1250

[24] Teasdale JR, Mohler CL. The quantitative relationship between weed emergence and the physical properties of mulches. Weed Science. 2000;**48**(3):385-392

[25] Bond W, Grundy AC. Non-chemical weed management in organic farming systems. Weed Research. 2001;**41**(5):383-405

[26] Smith RG, Gross KL, Robertson GP. Effects of crop diversity on agroecosystem function: Crop yield response. Ecosystems. 2008;**11**(3):355-366

[27] Liebman M, Dyck E. Crop rotation and intercropping strategies for weed management. Ecological Applications. 1993;**3**(1):92-122

[28] Altieri MA. The ecological role of biodiversity in agroecosystems. Agriculture, Ecosystems & Environment. 1999;**74**(1-3):19-31

[29] Bowman G, Shirley C. Managing cover crops profitably. 3rd ed. Beltsville, MD: Sustainable Agriculture Research and Education (SARE) Program; 2000

[30] Peoples MB, Brockwell J, Herridge DF, Rochester IJ, Alves BJR, Urquiaga S, et al. The contributions of nitrogen-fixing crop legumes to the productivity of agricultural systems. Symbiosis. 2009;**48**:1-17

[31] Crews TE, Peoples MB. Legume versus fertilizer sources of nitrogen: Ecological tradeoffs and human needs. Agriculture, Ecosystems & Environment. 2004;**102**(3):279-297

[32] Jeavons J. How to Grow More Vegetables: (and Fruits, Nuts, Berries, Grains, and Other Crops) Than You Ever Thought Possible on Less Land with Less Water Than You Can Imagine. Ten Speed Press; 2017

[33] Martin DL, Gershuny G, editors. The Rodale Book of Composting: Easy Methods for Every Gardener. Rodale; 1992

[34] Magdoff F, Van Es H. Building Soils for Better Crops. Vol. 4. Beltsville: Sustainable Agriculture Network; 2000

[35] Bubenzer GD, Meyer LD. The simulation of rainfall and its application to soil erosion research. Transactions of the ASAE. 1965;**8**(1):73-75

[36] Nair SS, Ngouajio M, Biernbaum JA. Integrating row covers and compost-based organic mulches for sustainable pest management in organic tomato production. HortScience. 2016;**51**(4):424-431

[37] Pretty J, Toulmin C, Williams S. Sustainable intensification in African agriculture. International Journal of Agricultural Sustainability. 2006;**9**(1):5-24

[38] Gittleman M, Jordan K, Brelsford E. Using citizen science to quantify community garden crop yields. Cities and the Environment (CATE). 2012;**5**(1):4

[39] Six J, Bossuyt H, Degryze S, Denef K. A history of research on the link between (micro) aggregates, soil biota, and soil organic matter dynamics. Soil and Tillage Research. 2004;**79**(1):7-31

[40] Zhang Q et al. Effects of long-term no-tillage on soil aggregate stability and organic carbon distribution in wetland paddy fields. Catena. 2020;**189**:104471

[41] Blanco-Canqui H, Lal R. Soil structure and organic carbon relationships following 10 years of wheat straw management in no-till. Soil and Tillage Research. 2008;**95**(1-2):240-254

[42] Tautges NE et al. Long-term diversified crop rotations support higher soil microbial diversity and organic matter than monocultures. Science. Advances. 2019;**5**(11):eaaw9750

[43] Ghosh PK, Mohanty M, Bandyopadhyay KK, Painuli DK, Misra AK. Growth, competition, yield advantage and economics in soybean/pigeonpea intercropping system in semi-arid tropics of India: I. Effect of subsoiling. Field Crops Research. 2010;**96**(1):80-89

[44] Pires LF et al. Soil bulk density and porosity about soil organic matter under no-till and conventional tillage. Soil Science Society of America Journal. 2021;**85**(1):123-131

[45] Basche AD, DeLonge MS. Comparing infiltration rates in soils managed with conventional and alternative farming methods: A meta-analysis. PLoS One. 2019;**14**(9):e0215702

[46] Sharma P, Abrol V, Sharma RK. Impact of tillage and mulch management on economics, energy requirement and crop performance in maize–wheat rotation in rainfed sub-humid inceptisols. Indian Journal of Soil Conservation. 2011;**39**(3):264-271

[47] Blanco-Canqui H et al. Cover crops and soil health indicators: A meta-analysis. Soil Science Society of America Journal. 2020;**84**(5):1385-1400

[48] Blanco H, Rattan L. Principles of Soil Conservation and Management. Vol. 167169. New York: Springer; 2008

[49] Palm CA et al. Conservation agriculture and ecosystem services in east and southern Africa. Agriculture, Ecosystems & Environment. 2020;**295**:106920

[50] Sharda VN, Dogra P, Prasher SO. Estimation of soil erosion by USLE and GIS techniques in Deccan plateau region of India. Agricultural Research. 2017;**6**(2):146-157

[51] Dabney SM et al. Cover crops for the reduction of soil erosion and sedimentation in the Mississippi

River Basin. Journal of Soil and Water Conservation. 2019;**74**(5):487-498

[52] Abid M, Asad SA, Raza H, Zulfiqar F, Thapa GB, Ahmad I, et al. Exploring Farmers' Intentions to Adopt Climate-Smart Agricultural Practices (CSA) in High Altitude Farming Regions of Hindu Kush Himalaya: An Application of Structural Equation Modeling. 2021. SSRN 3990571

[53] Pulido-Moncada M et al. Soil bulk density and penetration resistance under conservation agriculture: A review. Soil and Tillage Research. 2021;**212**:105039

[54] Wang J et al. Soil temperature and moisture regulation by different mulching practices in a rainfed cropping system on the loess plateau of China. Soil and Tillage Research. 2020;**197**:104492

[55] Bot A, Benites J. The Importance of Soil Organic Matter: Key to Drought-Resistant Soil and Sustained Food Production (No. 80). Food & Agriculture Organisations of United Nations; 2005

[56] Mikha MM, Rice CW. Tillage and manure effects on soil and aggregate-associated carbon and nitrogen. Soil Science Society of America Journal. 2004;**68**(3):809-816

[57] Teasdale JR, Coffman CB, Mangum RW. Potential long-term benefits of no-tillage and organic cropping systems for grain production and soil improvement. Agronomy Journal. 2007;**99**(5):1297-1305

[58] Bhattacharyya R, Kundu S, Pandey SC, Singh KP, Gupta HS. Soil quality changes under conservation agriculture with drip irrigation in maize (*Zea mays* L.)–mustard (*Brassica juncea* L.) cropping system in Indian sub-Himalayas. Soil & Tillage Research. 2015;**150**:31-42

[59] Verma SR, Sharma A, Singh B. Effect of conservation agriculture practices on nutrient availability and rice-wheat productivity in Indo-Gangetic Plains. Journal of Plant Nutrition and Soil Science. 2018;**181**(4):499-508

[60] Teasdale JR, Cavigelli MA, Heckman JR. Benefits of cover crops in no-till corn production. Agronomy Journal. 2007;**99**(5):1502-1507

[61] Singh RK, Kumar S, Yadav B. Evaluation of soil physical properties under different cropping systems in Western Ghats of India. Soil Science Society of America Journal. 2019;**83**(4):1032-1042

[62] Brady NC. The Nature and Properties of Soils. 1984. 750 p

[63] Brussaard L, de Ruiter PC, Brown GG. Soil biodiversity for agricultural sustainability. Agriculture, Ecosystems & Environment. 2007;**121**(3):233-244

[64] Bhaduri D, Purakayastha TJ, Patra AK. Soil biological activity, carbon dynamics and productivity of chickpea (*Cicer arietinum* L.) grown under different crop residues and phosphorus management in a typic haplustept. Agriculture, Ecosystems & Environment. 2014;**184**:111-123

[65] Verma S, Sharma RK, Timsina J, Jat ML. Conservation agriculture and climate resilience in Indo-Gangetic Plains of India: A review. International Journal of Current Microbiology and Applied Sciences. 2018;7(8):1193-1215

[66] Acosta-Martinez V, Zobeck TM, Gill TE, Kennedy AC. Enzyme activities and microbial community structure in semiarid agricultural soils. Biology and Fertility of Soils. 2003;**38**:216-227

[67] Miller RM. Arbuscular mycorrhizal colonization of switchgrass in remnant and reconstructed tallgrass prairie. Agriculture, Ecosystems & Environment. 2000;**77**(1-2):117-127

[68] Jacinthe PA, Lal R. Tillage effects on carbon sequestration and microbial biomass in reclaimed farmland soils of southwestern Indiana. Soil Science Society of America Journal. 2009;**73**(2):605-613

[69] Yadav RS, Yadav PK, Kumar A, Jat HS, Meena BR. Earthworm population dynamics and soil biological activity under different tillage and residue management practices in North-Western India. Soil and Tillage Research. 2016;**157**:32-42

[70] Singh G, Sharma RK, Soni ML, Kumar S. Conservation agriculture effects on soil health and crop productivity in an arid region of India. Soil Use and Management. 2019;**35**(1):19-29

[71] Kaur T, Sidhu HS, Jat ML. Conservation agriculture impacts on soil health and sustainability: Lessons from the South Asia experience. Journal of Soil and Water Conservation. 2020;**75**(4):86-98

[72] Phillips RE, Blevins RL, Thomas GW, Frye WW, Phillips SH. No-tillage agriculture. Science. 1980;**208**(4448): 1108-1113

[73] Hoitink HAJ, Boehm MJ. Biocontrol within the context of soil microbial communities: A substrate-dependent phenomenon. Annual Review of Phytopathology. 1999;**37**:427-446

[74] Six J, Conant RT, Paul EA, Paustian K. Stabilization mechanisms of soil organic matter: Implications for C-saturation of soils. Plant and Soil. 2004;**241**(2):155-176

[75] Pathak H, Greenhouse gas emission from Indian agriculture: Trends, drivers and mitigation strategies. Proceedings of the Indian National Science Academy. 2015;**81**(5):1133-1149

[76] Nair PKR, Nair VD, Kumar BM, Showalter JM. Carbon sequestration in agroforestry systems. Advances in Agronomy. 2009;**108**:237-307

[77] Pandey DN. Multifunctional agroforestry for ecological services and environmental benefits: An overview. Journal of Tropical Forestry. 2007;**23**(1):1-10

[78] Jat RK, Jat ML, Sidhu HS. Development and adoption of precision conservation agricultural technologies in the Indo-Gangetic Plains of India. Indian Journal of Agronomy. 2019;**64**(4):317-326

[79] Friedrich T, Kassam A, Shaxson F. Conservation agriculture. In: Agriculture for Developing Countries. Science and Technology Options Assessment (STOA) Project. Vol. 1, No. 3. Karlsruhe, Germany: European Technology Assessment Group; 2009. pp. 1-13

[80] Erenstein O, Sayre K, Wall P, Hellin J, Dixon J. Conservation agriculture in maize- and wheat-based systems in the (sub)tropics: Lessons from adaptation initiatives in South Asia, Mexico, and Southern Africa. Journal of Sustainable Agriculture. 2012;**36**(2):180-206

[81] Powlson DS, Stirling CM, Thierfelder C, White RP, Jat ML. Does conservation agriculture deliver climate change mitigation through soil carbon sequestration in tropical agro-ecosystems? Agriculture, Ecosystems & Environment. 2016;**220**:164-174

[82] Das S, Jat ML, Choudhary M, Jat RK, Sidhu HS. Conservation agriculture and sustainable intensification for enhancing resilience and ecosystem services in South Asia. Frontiers in Sustainable Food Systems. 2020;**4**:70

[83] Erenstein O, Hellin J, Chandna P. Poverty mapping based on livelihood assets: A meso-level application in the Indo-Gangetic Plains, India. Applied Geography. 2012;**34**:385-392

Chapter 2

Nutrient Management for Sustainable Soil Fertility

Isidora Radulov and Adina Berbecea

Abstract

The sustainable use of agricultural land and the maintenance of soil fertility require a permanent exchange of nutrients between organic matter, mineral colloids, and the soil solution. Soil fertility is influenced by the soil's physical, chemical, and biological properties. Nutrient availability to plant roots is facilitated by nutrient ion transport in soil solution, influenced by factors such as soil water content, pH, cation exchange capacity (CEC), redox potential, organic matter content, microbial activity, and fertilizer application. Adequate soil moisture enhances nutrient absorption and organic matter decomposition; soil structure, texture, and drainage impact water and nutrient movement, by facilitating nutrient retention or causing runoff and nutrient losses; CEC, determined by soil colloids, is crucial for nutrient retention and availability, affecting soil pH and buffering capacity; organic matter improves soil structure, water retention, and nutrient supply, while also acting as a chelating agent for metal ions; microbial activity drives nutrient cycling, with soil pH playing a significant role in nutrient availability. Redox processes, influenced by soil aeration, impact nutrient forms and availability, with anaerobic conditions leading to nutrient losses and potential toxicity. Proper nutrient management, considering soil properties and plant requirements, is essential for maintaining soil fertility and preventing environmental pollution.

Keywords: soil, fertility, nutrient management, soil organic matter, fertilizers

1. Introduction

The sustainable use of agricultural land and the maintenance of soil fertility involve a continuous exchange of nutrients between organic matter, mineral colloids, and soil solution. In all agricultural systems, significant amounts of nutrients are removed over time with the crops. When the rate of nutrient removal exceeds the nutrient input, soil nutrient reserves are depleted, and eventually, crop yields decrease. When nutrient input exceeds crop requirements, the nutrient levels in the soil increase. Excess nutrients can lead to the pollution of groundwater and surface water.

Soil is the primary medium for plant nutrition as well as for the application of fertilizers and amendments. A soil's ability to continuously and simultaneously supply plants with adequate nutrients and water, while meeting other vegetation needs, represents soil fertility. Fertile soil is defined by the combination of physical properties (texture, structure, profile, water retention capacity, etc.) and physicochemical properties (pH, nutrient content, organic matter content, cation and anion exchange capacity, and the sum of exchangeable bases) [1].

Most soils are not naturally fertile, requiring human intervention through various practices (application of fertilizers or amendments) to achieve the desired physicochemical conditions. To assess the actual contribution of soil to the chemical elements needed by plants and to determine the necessity of nutrient supplementation through fertilizers, it is essential to understand the soil's constitutive components and its content of various nutrient forms [2].

Fertilizer resources have different physicochemical characteristics and can affect soil ecosystems as well as its chemical, physical, and biological properties. Therefore, it is important to understand their impact on soil properties to determine the potential benefits or adverse effects of fertilizers. Each crop has a specific nutrient consumption influenced by the obtained or estimated yield. In addition to specific consumption, other factors to consider when calculating the required fertilizer resources include soil nutrient supply status, nutrient loss through fixation or leaching, soil type, precipitation, and irrigation.

Fertilization plans and nutrient quantities determined based on soil properties and plant requirements achieve desired effects if the factors influencing fertilizer use positively act within the ecosystem. The wide range of mineral and organic fertilizers provides more or less efficient fertilization options, so solutions should be adapted to the soil and plant needs. Nutrient management must ensure, through assortments, doses, and application methods, an increase in soil fertility, high and quality yields, and environmental protection [3].

Understanding the chemical composition of plants helps establish the necessary agrochemical measures, determining the quantity and quality of crops. The chemical elements involved in plant nutrition and fertilization have specific and essential roles in plant growth and development processes, and the deficiency of one or more nutrients disrupts the vital cycle.

2. Essential nutritive elements

Depending on their role in metabolism, nutrients are classified as follows:

- Essential elements indispensable for plant growth and development: C, O, H, N, P, K, S, Ca, Mg, Fe, Mn, B, Zn, Cu, Co, Mo.

- Non-essential elements: their presence can favor plant growth and development, but their absence has no noticeable effects: Rb, Na, Si, Br, F, I, Ni, Sn, Sr., W, V, etc.

The term "essential element" was proposed by Arnon and Stout in 1939. According to their criteria, for a nutrient element to be considered essential, it must meet three conditions:

1. Its absence or insufficiency in the nutrient medium must prevent normal plant growth and the completion of the life cycle;

2. The element must have a specific and direct role in the plant's nutrition process;

3. The element cannot be replaced by any other element in its role.

Based on the quantity in which they are found in plants, elements are classified as follows (relative to dry matter):

1. Macronutrients: found in concentrations greater than 10^{-2}%. These can be primary (N, P, K) or secondary (S, Ca, Mg).

2. Micronutrients: found in concentrations of 10^{-2}–10^{-5}%. These include Fe, Mn, B, Zn, Cu, Co, Mo, Cl.

3. Trace nutrients: found in concentrations less than 10^{-6}%. These include some radioactive elements (U, Th, Ac) whose presence in the nutrient medium can stimulate plant development (**Table 1**).

Plants require 16 chemical elements for normal growth and completion of their life cycle. The soil provides 13 of these 16 essential elements needed for the nutrition of higher plants. These elements must be continuously available and in balanced quantities to ensure the process of photosynthesis and other metabolic processes in the plant.

In general, soil contains high amounts of these nutrient elements, but only a small fraction of them is available to plants. For example, although the total iron content in soil can exceed 50,000 ppm, less than 5 ppm is accessible to plants. Nutrient availability to plants is influenced by: the form and chemical properties of the nutrient elements, soil pH, interactions with soil colloids, microbial activity, and the physical properties of the soil (aeration, compaction, temperature, and moisture).

The chemical form of an element strongly influences how it interacts with other elements and compounds in the soil. Nutrient elements can exist in various forms, with the most common ones presented in **Table 2**.

Soil nitrogen is the most challenging element to characterize: it occurs in both inorganic and organic forms (as both an anion and cation), in soil solution, and in gaseous form. Plants absorb it only in its inorganic form. The nitrogen forms contained in organic and mineral fertilizers are ammonia, urea, ammonium ions, and

Nutrients from air and water	Nutrients from soil, fertilizers, and soil amendments		
	Main nutrients	Secondary nutrients	Micronutrients
Carbon (C)	Nitrogen (N)	Sulfur (S)	Iron (Fe)
Hydrogen (H)	Phosphorus (P)	Calcium (Ca)	Boron (B)
Oxygen (O)	Potassium (K)	Magnesium (Mg)	Manganese (Mn)
			Zinc (Zn)
			Copper (Cu)
			Molybdenum (Mo)
			Cobalt (Co)
			Chlorine (Cl)

Table 1.
Essential nutrients for plant growth.

Nutrient	The main form used by the plant
Carbon	CO_2 (g)
Hydrogen	H_2O (liq), H^+
Oxygen	H_2O (liq), O_2 (g)
Nitrogen	NH^{4+}, NO^{3-}
Phosphorus	$H_2PO_4^-$, HPO_4^{2-}
Potassium	K^+
Calcium	Ca^{2+}
Magnesium	Mg^{2+}
Sulfur	SO_4^{2-}
Iron	Fe^{2+}, Fe^{3+}
Manganese	Mn^{2+}
Zinc	Zn^{2+}
Copper	Cu^{2+}
Boron	$B(OH)_3$
Molybdenum	MoO_4^{2-}
Chlorine	Cl^-

Table 2.
The forms of essential nutrients absorbed by plants.

nitrate ions. Ammonia (NH_3) reacts rapidly with soil water to form the ammonium cation (NH_4^+). Urea $(NH_2)_2CO$ is quickly converted by urease into ammonia. When urea or fresh manure is applied to the soil surface, nitrogen is lost as ammonia, especially in dry soils with high temperatures and pH values. After incorporation and decomposition in the soil, urea quickly transitions into ammonia and then ammonium. The ammonium cation, due to its positive charge, is retained by the negative charges of the colloidal complex. This prevents nitrogen loss through leaching, except in soils with low cation exchange capacity.

Potassium, calcium, magnesium, manganese, zinc, iron, and copper also appear in cationic form. Except for potassium and ammonium, which have a single positive charge (K^+ and NH_4^+), the remaining elements possess two or more positive charges. The greater the cation's charge, the more strongly it will be retained by the negative charge of the colloidal complex. When the sum of the cations' charges exceeds the soil's capacity to retain the nutrients, potassium and ammonium will be released before the cations with two or more positive charges, making them available to plants. Leaching of K^+ generally occurs in sandy soils with low cation exchange capacity. Due to their high positive charges, micronutrients form very strong bonds (coordination bonds) in the soil. Under normal conditions, they cannot be leached (**Table 3**).

Soil microorganisms (Nitrosomonas, Nitrobacter) convert the ammonium ion into the nitrate ion (NO_3^-). The chloride ion (Cl^-) behaves similarly to the nitrate ion. The negative charge of the nitrate ion is repelled by the negative charges of the colloidal complex, leading to its leaching. Sulfur is found in the soil both in plant residues and

Nutrient (symbol)	Nutrient status	Main visual symptoms
Nitrogen (N)	Deficiency	Yellowing of the leaves starting from the tip, especially of the old ones, slow growth, poor fruit development.
	Excess	The luxuriant growth of the plant, the decrease in resistance to the attack of diseases and pests, the decrease in the taste qualities of vegetables and fruits, the accumulation of nitrates occurs.
Phosphorus (P)	Deficiency	The leaves acquire a reddish color, slowing down the growth and development of the plants.
	Excess	Excess phosphorus can cause deficiencies of microelements, especially Fe and Zn.
Potassium (K)	Deficiency	Marginal necrosis of leaves, irregular fruit shape and color.
	Excess	Excess potassium can cause magnesium and calcium deficiencies.
Sulfur (S)	Deficiency	Yellowing of new leaves, the older leaves remain green, reduced growth.
	Excess	Damaging of root system and foliage.
Calcium (Ca)	Deficiency	Reduced growth or death of growth tips, poor fruit development.
	Excess	Excess calcium can cause magnesium or potassium deficiencies.
Magnesium (Mg)	Deficiency	Interveinal chlorosis of older leaves, some plants may display red, purple, or brown tints, slow plant growth, and poor crop production.
	Excess	Excess magnesium can cause calcium or potassium deficiencies.
Iron (Fe)	Deficiency	Interveinal chlorosis of younger leaves.
	Excess	Stunted growth and discolored bronzing foliage.
Boron (B)	Deficiency	Brown and necrotic spots, cracking and corky areas in fruit and tubers.
	Excess	Marginal leaf chlorosis and necrosis, scorching, burning, and premature drop of interior foliage.
Zinc (Zn)	Deficiency	the shortening of the internodes and the reduction of the size of the leaves, the arrangement of the leaves and terminal branches in a rosette, and the appearance of yellow spots on the leaves.
	Excess	reduced growth and development, slowing down of seed germination, and elongation of roots and stems.
Manganese (Mn)	Deficiency	Interveinal chlorosis with a relatively wide area of green associated with the veins, but leaves do not become cream color. Leaves, shoots, fruits, or stems may be contorted, and plant parts may be reduced in size or have necrotic spots.
	Excess	Necrotic spots on the older leaves.
Copper (Cu)	Deficiency	Bleaching of leaf tips, withering and drying of upper tier leaves, internodes remain small, flowers and reproductive organs atrophy.
	Excess	Reduction of plant growth, symptoms of iron deficiency.
Molybdenum (Mo)	Deficiency	The falling and twisting of the leaves, brown spots appear on the leaves, in the form of sticks.
	Excess	Coloring the stems in reddish or golden yellow.
Chlorine (Cl)	Deficiency	Chlorotic and necrotic leaves with sharp demarcations between dead and living tissue, wilting of leaf margins and highly branched roots.
	Excess	Premature yellowing of the older leaves, burning of the leaves margins and tips.

Table 3.
Visual symptoms of plant nutrient deficiency and excess.

in organic matter, but is taken up by plants as the sulfate ion (SO_4^{2-}). Microorganisms (*Thiobacteria*) convert elemental sulfur and organic sulfur in the soil into the sulfate ion. Like the sulfate ion, the molybdate ion (MoO_4^{2-}) is easily leached from the soil. These two ions can form weak bonds with the positively charged colloids in the soil, such as iron and aluminum hydroxides. However, they are not strongly retained in the soil and are leached more slowly than nitrate and chloride ions.

Phosphorus is found in the soil as primary and secondary phosphate ions, depending on the soil pH. Unlike nitrate and chloride anions, phosphate anions are not leached. Phosphorus strongly binds to Fe^{3+}, Al^{3+}, and Ca^{2+} in the soil solution, as well as to iron oxides, iron and aluminum hydroxides, the positive charges of clays, and limestone. These strong bonds limit the mobility of phosphorus in the soil. Boron is found in the form of boric acid (H_3BO_3), which is leachable and weakly bound in the soil.

The primary macroelements (N, P, and K) are frequently used as fertilizers, usually annually. The secondary macroelements (S, Ca, Mg) are applied only to specific deficient soils or to certain species with high demands for these elements. Micronutrients are used only on soils with low available content or for species that have high nutritional requirements for these elements.

Except for carbon in the form of CO_2, hydrogen as H_2O, and oxygen as O_2, plants absorb essential nutrients from the soil.

Carbon (C) is a fundamental element in soil, primarily present in organic forms. It is one of the most critical factors contributing to soil fertility, protecting against degradation, erosion, desertification, and leaching [4].

In soil, carbon exists in the following forms:

- Soil Organic Carbon (SOC): SOC is the primary component of soil organic matter, encompassing living organisms (microorganisms, plant roots, and soil fauna), fresh plant and animal residues (decomposing organic matter), and humus (organic matter formed through the microbial decomposition of plant and animal material). SOC is a crucial determinant of soil quality, fertility, and atmospheric CO_2 sequestration. It comprises a complex mixture of organic compounds derived from various stages of organic matter decomposition, facilitated by interactions with soil biota, water, and nutrient cycling processes. SOC exhibits dynamic behavior, continuously cycling into and out of the soil. It plays an essential role in forming soil organic acids, which facilitate the dissolution of soil minerals and the availability of nutrients to plants, as well as in nutrient leaching processes. SOC is recognized as a vital indicator of soil health. The dissolved organic carbon fraction represents the most mobile and reactive component of soil carbon, capable of being adsorbed by soil minerals or exported to aquatic systems [5].

- Soil Inorganic Carbon (SIC): SIC is predominantly found as carbonates (e.g., calcium carbonate, $CaCO_3$) in arid and semi-arid regions. The SOC to SIC ratio varies widely depending on the location. Approximately 40% of the Earth's surface consists of arid and semi-arid zones, where SIC is the predominant carbon form. In these regions, the rate of inorganic carbon accumulation is approximately 1.4 times higher than that of organic carbon [6].

Nitrogen is a crucial nutrient in soil, playing a vital role in plant growth and overall soil health. Being a key component of amino acids, proteins, and chlorophyll essential for plant growth and development, the nitrogen level in soil is an indicator of fertility (**Figure 1**).

Approximately 95–99% of the nitrogen in soil is in the form of plant and animal residues, soil microorganisms, and relatively stable soil organic matter (SOM). This nitrogen is not directly available to plants, but a part can be converted into an available form by microorganisms. However, a very small fraction of organic nitrogen may exist in the form of soluble organic compounds easily accessible to plants.

The form accessible to plants consists of ammonia (NH_4^+) and nitrates (NO_3^-). Mineral nitrogen is introduced into the soil either as a product of microbial fixation, atmospheric deposition of compounds resulting from the burning of fossil fuels or emissions from agriculture, or inputs from rock weathering [7].

In soil, nitrogen undergoes several processes that result in the increase of forms accessible to plants such as:

- fixation: conversion of atmospheric N_2 into NH_3 by bacteria and plants, making it accessible to plants;

- mineralization: decomposition of organic matter by microorganisms, converting organic nitrogen into mineral nitrogen (NH_3) accessible to plants.

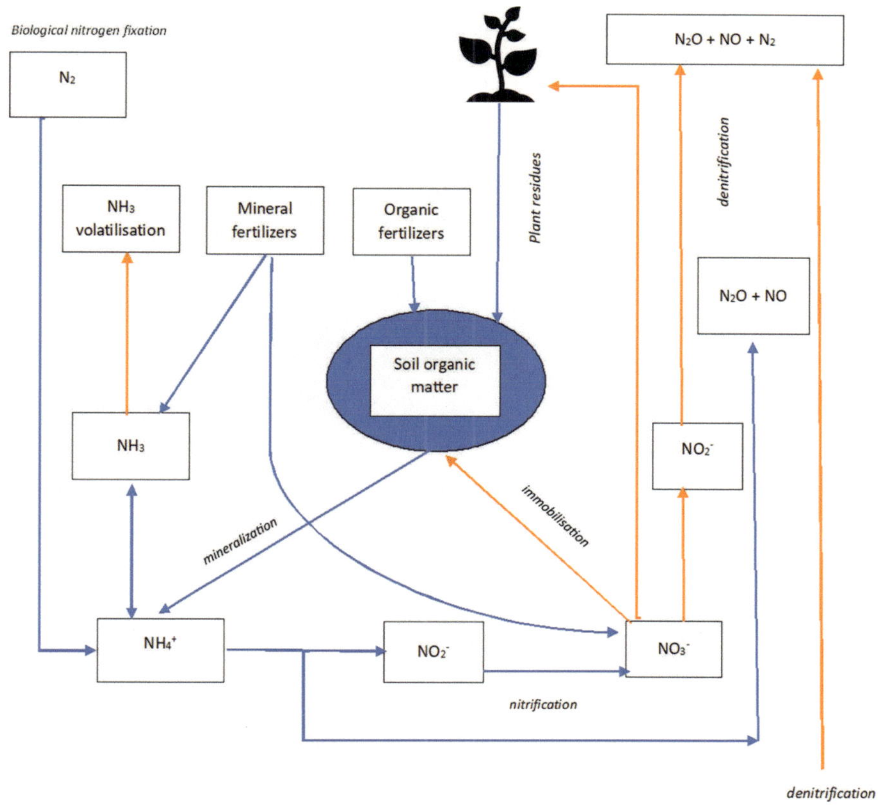

Figure 1.
Nitrogen cycle.

- nitrification: oxidation of NH_3 by nitrifying bacteria in two stages: $NH_3 \rightarrow NO_2^-$ (Nitrosomonas) and $NO_2^- \rightarrow NO_3^-$ (Nitrobacter).

Also, in the soil, processes are taking place that lead to the decrease of the nitrogen content:

- Denitrification: reduction of NO_3^- to N_2 by denitrifying bacteria, releasing it into the atmosphere.

- Leaching: loss of nitrates from the soil due to leaching with water, causing groundwater pollution [8, 9].

Phosphorus is the second most essential nutrient for crops, a major component of soil organic matter in natural terrestrial ecosystems, primarily resulting from the weathering of phosphate minerals in parent rock. In soil, phosphorus exists both in organic and mineral forms, differing in behavior, presence, and accessibility to plants (**Figure 2**) [10]. Organic phosphorus originates from the decomposition of plant and animal fragments and soil microorganisms, directly or through biochemical transformations of organic matter. It exists in both stabilized (inositol phosphates and phosphonates) and active forms (orthophosphate monoester and organic polyphosphates). This form results from decomposition and mineralization mediated by soil microorganisms in two stages: biochemical mineralization (inorganic P is released from organic compounds by phosphatase enzymes) and biological mineralization (inorganic P is released from organic matter during its oxidation by soil microorganisms). These pro-

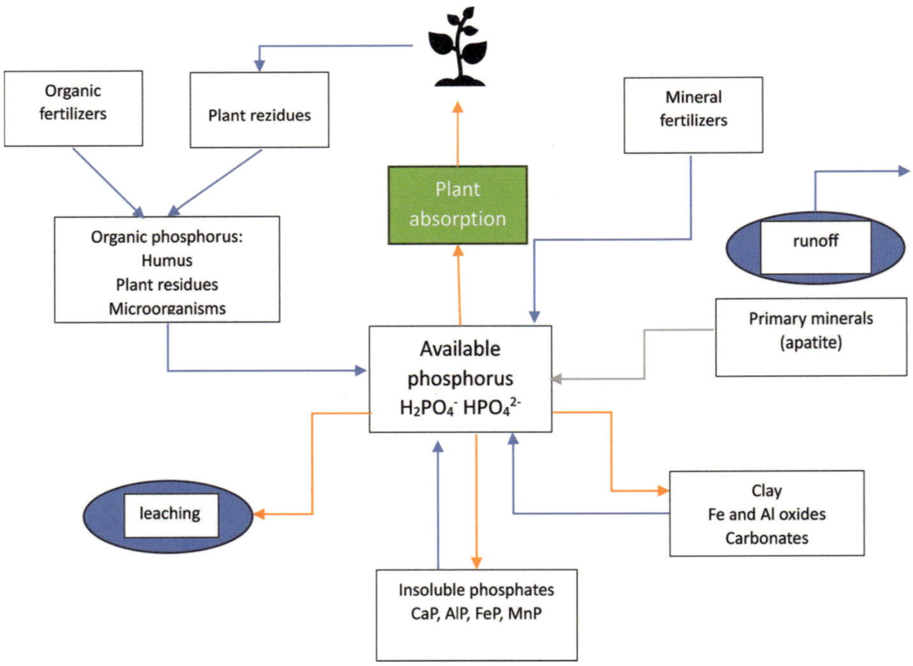

Figure 2.
Phosphorus cycle.

cesses are strongly influenced by soil moisture content, temperature regime, and the properties of the soil surface layer.

Phosphorus bound to organic matter is a momentarily unavailable reserve but is gradually released. It can serve plant nutrition as organic matter mineralizes. Phosphorus bound to humus through adsorption mechanisms is easily mobilizable, contributing to soil supply through anion exchange processes.

Inorganic phosphorus exists precipitated as simple calcium, iron, and aluminum phosphates, free ions in soil solution, or adsorbed states. Complex calcium phosphates (hydroxyapatite, fluorapatite, chlorapatite), tricalcium phosphate, and complex iron and aluminum phosphates with advanced crystallization are insoluble or poorly soluble forms of phosphorus, inaccessible or hardly accessible to plants, known as non-labile phosphorus. Phosphate ions adsorbed on the surface of clay minerals or iron and aluminum hydroxides, which can be released back into solution, represent mobile phosphorus. It constitutes the main reserve that replenishes phosphorus in soil solution consumed by plants [11].

Accessible phosphorus soluble in soil solution is represented by primary orthophosphate ions ($H_2PO_4^-$) and secondary orthophosphate ions (HPO_4^{2-}). The concentration of these two species depends on soil pH. At pH values <7, $H_2PO_4^-$ predominates, while at pH >7, HPO_4^{2-} predominates. The total concentration of these two ions in soil solution decreases significantly with increasing pH above 7.5 and with decreasing pH below 5.5, being maximal between 6.3 and 6.5.

Soil pH influences the conversion of phosphorus into forms accessible to plants. In acidic soils, compounds formed from the reaction between applied phosphorus fertilizers and $Al(OH)_3$ and $Fe(OH)_3$ either precipitate or are adsorbed by clay particles, becoming inaccessible to plants. The maximum accessibility of phosphorus for plants is at pH = 5.5–7.0. Soils rich in clay retain large amounts of phosphorus depending on the type of clay (montmorillonite fixes more phosphorus). Soil structure indirectly influences phosphorus accessibility through aeration effects, as increased compaction hinders root penetration, preventing adequate phosphorus supply to plants.

The presence of organic matter increases phosphorus availability for plants, with certain organic ions able to release phosphate ions fixed on soil particles. Soils rich in aluminum and iron minerals, and low in calcium and magnesium, are poor in phosphorus and strongly fix it. The phosphorus fixation capacity decreases when calcium and magnesium are in normal amounts in soil composition.

Soil moisture favors phosphate ion absorption by plants and the diffusion process. Excessive drying reduces phosphate solubility. Increasing soil moisture increases the amount of solubilized phosphates. As soil temperature increases within certain favorable limits for plant growth, phosphate solubility improves, enhancing mineral phosphorus nutrition [12].

In soils, potassium exists both in forms that are poorly soluble and poorly accessible to plants, as well as in forms easily soluble in water and readily accessible. The ratio between these forms varies depending on the nature of the clay mineral, clay percentage, and pH [13].

Native potassium exists as a constituent element in the crystal lattice of primary and secondary minerals, representing the largest part of soil potassium. Following the alteration processes of primary minerals, part of the potassium is retained in the crystal lattice of secondary silicates (clay minerals), part is retained and adsorbed in the clay-humic complex, and part is removed through leaching. The retention and fixation capacity of potassium depends on the mineralogical nature of the clay and the quantitative ratio in which it is found.

Fixed potassium is the form of potassium retained in the spaces between two packets of aluminosilicates forming the crystal lattice of clay minerals, in the pores of zeolites, and permutites. Fixation occurs through alternative wetting and drying of the soil when the aluminosilicate packets forming clay particles move apart or come together, and potassium ions penetrate these spaces along with water molecules, satisfying the negative valences resulting from isomorphic substitutions. A stronger fixation of potassium becomes possible if isomorphic substitution occurs at the octahedral aluminum sheet where the Al^{3+} ion is replaced by divalent Mg^{2+} ions.

This form of potassium represents an important potential reserve for plant supply with this ion, as well as a transition form between exchangeable potassium and potassium in the crystal lattice (**Figure 3**).

The fixation and release of potassium from soils occur through a dynamic and reversible process controlled by the type and size of mineral and secondary particles present in soils. The process is also affected by soil structure, soil pH, temperature, drying and wetting, freezing and thawing.

Exchangeable potassium is the potassium retained ionically by the negative charges of the adsorbing complex, which can be easily displaced (through exchange) with weak saline solutions. Therefore, it refers to potassium ions adsorbed by soil colloids, quantitatively representing less than 1% of the total potassium in mineral soils.

The importance of exchangeable potassium for plants lies in the fact that it can easily pass into solution through exchange with other cations. This potassium can be absorbed by root hairs through exchange with hydrogen ions permanently adsorbed on the root surface.

The accessibility of exchangeable potassium for plants depends on the physical-chemical and physical properties of soils, adsorption capacity, potassium saturation, composition, and quantity of accompanying ions. Exchangeable potassium is more

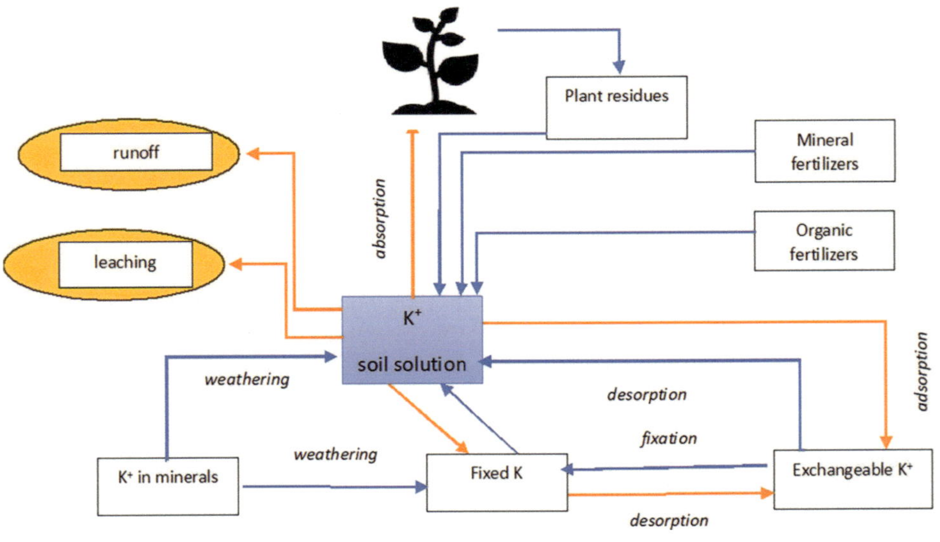

Figure 3.
Potassium cycle.

mobile as the soil's adsorption capacity is lower and its potassium content is higher. Soil saturation with calcium reduces potassium mobility [14].

Soluble potassium is the most accessible form for plants. Soluble potassium is the potassium found in soluble salts in soil solution and that which is in solution in ionizable form in equilibrium with ions adsorbed by the colloidal complex. The appearance of soluble potassium in water is the result of a series of processes:

- Hydrolysis of potassium minerals;
- Degradation of minerals by plant roots;
- Action of nitrifying bacteria's nitric acid and other vital activity products of microorganisms;
- Replacement of exchangeable potassium by salts reaching the soil with fertilizers and root secretion products.

In soil, sulfur is found both in organic and mineral forms. Organic sulfur is mostly present in the upper horizon, increasing with the soil's humus content. Mineral sulfur can be found in the form of sulfates and sulfides, as adsorbed SO_4^{2-} ions by soil colloids (especially hydrated iron and aluminum oxides) or as soluble SO_4^{2-} ions dissociated in soil solution. The retention of SO_4^{2-} ions in soil is influenced by several factors, such as:

Nature of cations: Ca^{2+}, Mg^{2+}, Ba^{2+}, and Fe^{2+} intensify SO_4^{2-} ion fixation, while K^+ and Na^+ reduce it;

Soil pH: In acidic soils, SO_4^{2-} ions are retained due to the presence of iron and aluminum hydroxides, while in soils with pH > 6, SO_4^{2-} ions are leached together with basic cations, leading to pH decrease.

Organic sulfur undergoes mineralization in soil under the action of microorganisms, transforming into accessible sulfates for plants. It has been determined that 30–70% of the sulfur consumed by plants comes from the mineralization of sulfur-containing organic matter.

The formation of SO_4^{2-} ions in soil is accompanied by a decrease in pH value, intensifying the solubilization of PO_4^{3-} ions and increasing their accessibility to plants. SO_4^{2-} ions in soil solution and adsorbed SO_4^{2-} ions are the main sulfur sources for plants. The sulfate ion is absorbed by plant roots through diffusion and drainage. Most plants' sulfur needs are satisfied in soils containing 3–5 ppm SO_4^{2-}. However, some plants, such as alfalfa, require higher concentrations of SO_4^{2-} ions in the soil.

Due to its negative charge and high solubility, SO_4^{2-} ions can be leached from the soil surface. SO_4^{2-} ions can be adsorbed in acidic soils due to the presence of aluminum and iron oxides, but adsorption becomes negligible at pH values greater than 6.5. Adsorbed SO_4^{2-} ions are in equilibrium with SO_4^{2-} ions in soil solution; as the concentration of SO_4^{2-} ions in solution increases, the amount of adsorbed SO_4^{2-} ions also increases.

Under anaerobic conditions, SO_4^{2-} ions are reduced by bacteria from the *Desulfovibrio*, *Chlorobium*, *Chromatium*, and *Sporovibra* genera, resulting in H_2S (hydrogen sulfide), which is toxic to plants. In soils with mobile iron, H_2S does not form; instead, FeS and FeS_2 (insoluble) form. Reduction of sulfates in submerged soils, such as rice paddies and swamps, leads to increased pH and the amount of ferrous ions. When these soils are drained, the pH value decreases due to the oxidation of sulfides to sulfates. After draining some swamps and peat bogs, the pH can drop significantly to 2–3, which is difficult to correct by applying amendments.

Calcium, an essential mineral for plants, is an abundant element in soil where it functions as a binding agent in the aggregation of organic and inorganic soil particles, forming microaggregates important for developing good soil structure [15]. On the other hand, it maintains balance among adsorbed nutrients, occupying spaces that would otherwise be taken by acidic elements. The fraction of calcium present in soil solution is quickly accessible to plants. Besides being absorbed by plants as a nutrient, calcium can be leached, retained by soil particles, or precipitated as secondary minerals due to its high mobility in soil.

Exchangeable calcium represents the fraction retained on the surface of clay particles or organic soil colloids. Due to its charge (+2) and small hydrated radius, calcium is retained much more strongly compared to other cations. Being the dominant cation in cation exchange capacity (CEC), it can easily pass into soil solution.

In most soil types, in addition to soil solution and exchangeable forms, calcium becomes accessible to plants through the weathering of secondary minerals such as $CaCO_3$—calcite.

The average magnesium content in the Earth's crust is 219 g/kg, while in soil, it ranges from 3 to 25 g/kg [16]. Although magnesium is needed by plants in smaller amounts compared to calcium, it is not less important, activating more enzymes necessary for energy production than any other mineral nutrient. Magnesium behavior in soil is similar to calcium, although its content is lower. This is because Mg^{2+} is not as strongly retained by clay particles and organic colloids, and the magnesium content in the parent material is lower than that of calcium.

In neutral and slightly acidic soils, magnesium is the second most abundant exchangeable cation after calcium. Soils with higher magnesium content than calcium show clay dispersion and surface sealing because, despite being a divalent cation, it has a larger hydrated radius and does not contribute to particle aggregation.

Magnesium sources in soil include silicates and carbonates ($MgCO_3$—magnesite, $MgCO_3xCaCO_3$—dolomite, and $CaCO_3$ with 1–3% Mg—calcite). Magnesium trapped in silicate interlayers is immobile but can be released into the mobile fraction through long weathering processes.

Iron, the fourth most common element, represents 5% of the Earth's crust (by mass percentage) and is present in all soils, found in the crystal lattice of numerous minerals (augite, biotite, hornblende, olivine). Weathering these minerals forms Fe^{3+} oxides with very low solubility, accumulating in the clay fraction as hydrous oxides. The concentration of soluble iron in soil is extremely low. Soluble inorganic forms include Fe^{3+}, $Fe(OH)^{2+}$, and Fe^{2+}, depending on the oxic/anoxic conditions in the soil. Soluble Fe inorganic solubility is influenced by pH; acidic soils have higher soluble Fe inorganic content than alkaline soils [17]. Accessibility of iron to plants is influenced by soil pH, organic matter content, aeration, and the presence of other macro- and micronutrients. Under waterlogged conditions and acidic pH, anaerobic bacteria reduce Fe^{3+} to Fe^{2+}, which can be absorbed by plants, leading to toxicity. Soil organic matter can immobilize iron in the form of complexes accessible to plants as it decomposes, representing the primary source of accessible iron [18].

Manganese concentration, a redox-sensitive element, represents approximately 0.1% of the Earth's crust (by mass percentage), reflecting its level in parent material. During rock weathering, Mn^{2+} is released and quickly oxidized to Mn^{3+}/Mn^{4+}, forming over 30 oxides and hydroxides [19]. Due to its omnipresence, high redox potential, and large surface area, it controls the distribution, transport, and transformations of other elements in the soil. The primary mode of solubilizing manganese oxides is the microbial reduction of Mn^{4+} under waterlogged conditions, attributed mainly to bacteria.

Manganese accessibility depends on soil properties, increasing with decreasing pH and becoming toxic in acidic soils (pH <5). A small portion of the soil's manganese content is found in soil solution, quickly accessible to plants. On the other hand, manganese in soil solution can easily bind to organic matter, becoming inaccessible to plants.

Most copper in soil is found in the crystal lattice of primary and secondary minerals: CuO—cuprite; CuS—covellite; CuS_2—chalcocite; $CuFeS_2$—chalcopyrite; $CuFe_2O_4$—cupricpyrite; $Cu_2CO_3(OH)_2$—malachite; $Cu_3(OH)_2(CO3)_2$—azurite. Additionally, a significant amount of soil copper is bound to soil organic matter (SOM).

Copper ions in soil are either retained by the negative charges of inorganic (carbonates, phyllosilicates, and hydrous oxides of Al, Fe, and Mn) or organic groups (carboxyl groups in humic and fulvic acids) present in soil organic matter or are found in soil solution, generally in very low concentration [20]. The concentration of Cu^{2+} in soil solution dramatically decreases with increasing pH, while the concentration of organic copper complexes is less dependent on pH. The concentration of Cu^{2+} in soil solution is governed by the adsorption of Cu to soil particles of an organic or mineral nature, making copper very immobile and not easily leached into depth. Copper content decreases along the soil profile. Copper deficiency mainly occurs in soils rich in humus, where certain organic compounds strongly bind Cu^{2+}, immobilizing it.

Plants absorb chlorine in the form of the Cl^- ion dissolved in the soil solution, being necessary for certain photochemical reactions within the photosynthesis process. In alkaline soils, chlorine accumulates in the form of salts. Chlorine availability to plants is influenced to a small extent by soil acidity, soil aeration, or organic matter content. Chlorine absorption affects the degree of hydration of plant cells and balances positive ion charges in cation transport. Chlorine deficiency causes rapid withering of plants and excessive development of lateral roots, and the leaves acquire a "bronze" color [21]. Saline soils can contain high amounts of chlorine, and to prevent toxicity caused by this nutrient, it is recommended to increase the anion exchange capacity in the root zone by incorporating natural or chemical fertilizers with high anion exchange capacity; leaching the excess of soluble Cl^- from the root zone by waterlogging, or providing readily available resources of nutrient anions to replace Cl^- [22].

3. Influence of soil properties on fertility

To meet plant nutrient requirements, these nutrients must reach the root surface, possible through the transport of nutrient ions in soil solution. The concentration of nutrient ions in soil solution is influenced by soil water content, pH, cation exchange capacity, redox potential, organic matter content, microbial activity, and fertilizer application.

Soil moisture is important for root growth and nutrient absorption. Adequate moisture improves nutrient absorption in roots through diffusion and intensifies organic matter decomposition, releasing N, P, and S. Low moisture can lead to nutrient retention in insoluble compounds. Soil moisture and its water retention capacity are influenced by texture, structure, aeration, and drainage. Good soil structure is essential for water and nutrient movement, penetration, and retention. Large spaces between soil aggregates allow free movement of water and dissolved nutrients, leading to leaching losses. Small spaces or lack of spaces between aggregates, especially due to compaction, impede water movement through the soil profile, causing runoff. Drainage and aeration affect nutrient loss and solubility. Poorly drained soils are poorly aerated as they flood easily, promoting nitrogen loss through denitrification. Excessively drained soils (sandy) favor leaching losses. Flooded or very wet soils increase the solubility of minerals such as iron and manganese.

Cation exchange capacity (CEC) is given by the concentration of all cations adsorbed by the soil's colloidal complex and depends on the nature of colloids. The more clay or organic matter a soil contains, the higher its cation exchange capacity. Organic colloids have a higher cation exchange capacity than clays. The total cation exchange capacity of clays depends on their nature.

The total cation exchange capacity (CEC) is important for the following reasons:

- some exchangeable cations, such as calcium, magnesium, and potassium, can be directly taken up by plants;

- cations retained through adsorption cannot be lost through leaching.

Pesticides and other chemicals with positively charged groups can be adsorbed, being removed from the soil solution, thereby preventing their loss and soil pollution.

There is a close relationship between CEC and soil pH: the lower the CEC value, the more rapidly the soil pH will decrease. Soils with low CEC values are susceptible to developing deficiencies in K, Mg, and other cations [23]. Soils with high CEC values can retain a larger amount of H^+, thus having a greater buffering capacity against acidification (**Table 4**) [24].

High clay content increases CEC and, therefore, the capacity to retain nutrients, while high sand content decreases CEC and nutrient retention capacity (Kawaguchi et al.). The higher the total cation exchange capacity of a soil, the more nutrients it can retain. Although CEC is only one of the factors influencing soil fertility, the higher its value, the lower the need to replace nutrients through the application of mineral and organic fertilizers.

Redox processes in soil are mainly due to the activity of microorganisms that, using the energy released from the decomposition of plants and animals, carry out oxidation through electron transfer from organic matter (reducing agent) to other

Colloid	CEC (me/100 g soil)
Kaolinite	3–10
Montmorillonite	80–150
Illite	10–30
Humus	200
Smectite	100
Iron and aluminum oxides	4

Table 4.
The cationic exchange capacity of some soil colloids.

electron-accepting substances (oxidizing agents). Oxygen is the strongest electron acceptor, releasing a large amount of energy in the oxidation processes of organic matter, being the only electron acceptor that plants can use. The most important chemical elements participating in soil redox reactions are C, O, N, Mn, Fe, and S. In aerated soil conditions (presence of oxygen), nutrients are oxidized to stable forms: NO_3^-, SO_4^{2-}, Fe^{3+}, Mn^{4+}, etc. In the absence of oxygen in soil (anaerobic conditions), nutrients are found in reduced forms (H_2S, Fe^{2+}, Mn^{2+}), which can accumulate in toxic amounts for plants. Nitrates can be reduced to lower nitrogen oxides and even to molecular nitrogen, leading to nitrogen losses from the soil.

For optimal plant growth and development conditions, oxidation processes must predominate over reduction processes in the soil. In total anaerobic conditions (due to water stagnation on the soil), the soil's redox potential has negative values, resulting in the total reduction of iron and sulfates to elemental sulfur and sulfides (H_2S—toxic for plants). Aerobic microorganisms' activity is blocked, and anaerobic fermentation occurs, forming methane and other gases toxic to plants.

Soil organic matter, a heterogeneous mixture of substances in various stages of decomposition, of organic nature and very rich in carbon (C), is a key component of soil that influences its physical, chemical, and biological properties. It enhances soil quality by increasing water retention and improving soil structure, thereby reducing erosion [25].

Organic matter is an important component of the soil, having the following roles:

- improves the physical properties of the soil, contributes to the formation of structure, porosity, and water retention;
- reduces nutrient losses through erosion;
- improves nutrient supply (N, P, and S);
- increases the soil's cation exchange capacity;
- contributes to the solubilization of nutrients;
- protects some nutrients from retrogradation (fixation in non-exchangeable forms) or leaching losses;
- has the capacity to chelate metal ions (Fe, Zn, B, Cu, Mo, Co).
- Depending on the origin and chemistry, soil organic substances are classified into two categories:
- non-humified and intermediate organic substances;
- humified organic substances specific to soil (humus).

Humus increases the soil's water retention capacity. The most important property of humic substances is their ability to retain water. They participate in forming soil structures, allowing water infiltration and retention in the root zone. Due to their large specific surface area and electric charges, humic substances can retain up to seven times their volume in water, having a higher retention capacity than clays.

The retained water can be used as a medium for nutrient movement to soil organisms and plant roots when needed.

Humic substances contribute to forming a friable soil structure. Humic substances form complex colloidal aggregates with the soil's mineral fraction (humus-clay). These aggregates formed by electrostatic retention help create a friable surface soil structure. Soils with friable, adequate structures are easier to work and more aerated, favoring gas exchange between soil gas phase and atmosphere and water infiltration.

Humic substances act as a pH buffer system. Numerous field experiments on acidic or alkaline soils have shown that adding humic substances neutralized the soil pH. Following pH neutralization, nutrients bound in insoluble and inaccessible compounds due to acidic or alkaline conditions became accessible to plants. Humic substances release CO_2 from calcium carbonates present in the soil, which can be absorbed by plants or transformed into carbonic acid. Carbonic acid contributes to releasing nutrients from soil minerals for plants.

Humic substances stabilize and inactivate soil enzymes. Humic substances bind to soil enzymes through covalent bonds, contributing to their stabilization and making them less accessible to microbial degradation. Once stabilized and bound to humic substances, enzyme activity is reduced or ceases. Since many of these bonds are weak, enzymes can be released during soil pH modification. When certain components of humic substances react with soil enzymes, they are more strongly bound. This enzyme stabilization process restricts potential pathogenic agents' activity. As pathogenic potential releases enzymes to destroy plant defenses, they will bind to humic substances, preventing the pathogen from invading the potential host plant.

Soil microorganisms decompose organic matter, releasing nutrients in forms that plants can absorb. Healthy microbial activity is a sign of fertile soil [26]. Enzymes produced by soil microbes play a role in nutrient cycling and organic matter decomposition, influencing soil fertility [27]. Overall soil fertility is a complex interplay of these properties, and managing them effectively is key to maintaining productive agricultural systems.

Soil texture determines the reaction surface area of soil particles; the coarser the soil, the smaller the reaction surface area. Heavy clay soils have particles with a large reaction surface area. The nature of soil colloids will determine the reactivity level between colloidal particles and fertilizer particles. As soil organic matter content increases, the reactive potential between soil and fertilizer particles increases [28].

Soil acidity and alkalinity influence the accessibility of essential nutrients. Soil pH affects nutrient availability, with most nutrients being readily available in soils with a pH between 6.5 and 7.5. While the availability of nitrogen (N), potassium (K), and sulfur (S) is less influenced by pH, phosphorus (P) availability is directly affected. Phosphates become immobilized as poorly soluble compounds with iron (Fe) and aluminum (Al) in acidic soils, and with calcium (Ca) and magnesium (Mg) in alkaline soils. Most micronutrients tend to be less available at pH values exceeding 7.5.

Soil pH is often linked to microorganism populations; acidification of soil generally leads to a reduction in the number and diversity of microorganisms. Specifically, bacteria, epiphytes, and actinomycetes decline as soil pH decreases. This effect is more pronounced in surface soil layers compared to deeper ones [29]. However, direct evidence of soil acidification on soil biodiversity is limited [30]. Conversely, the literature indicates that in soils with a slightly alkaline pH, there is greater diversity in microbial communities. Among all soil properties, pH has the most significant impact on the diversity of microbial communities [31]. Aerobic nitrogen-fixing bacteria, both symbiotic (*Rhizobium*) and non-symbiotic (*Azotobacter*), have an optimal pH range of 6.5–7.5. Anaerobic nitrogen-fixing bacteria (*Clostridium*) are more tolerant to acidity, with an optimal pH range of 5.7–7. Nitrifying bacteria have a pH range of 6.6–8, so slightly acidic to slightly alkaline. Acidic reaction strongly inhibits nitrifying microorganism activity [32]. Denitrifying bacteria have an optimal pH range of 7–8, with denitrification occurring at a pH between 3.9 and 9 [27, 28]. Fungi, unlike bacteria, prefer an acidic environment.

Most cultivated plants cannot tolerate pH < 4.5 or pH > 8.3 and large variations in it. Most plants grow well on neutral soils (pH = 6.8–7.2) or slightly acidic (pH = 6.3–6.8). Generally, plants tolerate acidic environments better than alkaline ones. The highest sensitivity to acidity and alkalinity is observed at the beginning of the vegetation period, during the early stages, particularly at the seedling stage. Plants' sensitivity to extreme pH conditions increases under water and nutrient stress [33]. Based on their sensitivity to soil reaction, cultivated plants can be grouped as follows:

Very sensitive to acidity: sugar beet, fodder beet, cauliflower, garden beans, quince, sour cherry, which thrive on soils with pH = 7–8;

Sensitive to acidic reaction: sunflower, beans, wheat, corn, peas, hemp, pumpkins; these develop well on soils with slightly acidic reaction pH = 6–7;

Tolerant to acidity: rye, oats, tomatoes, carrots, vine, apple, optimal pH is between 5 and 6;

Very tolerant to acidity (prefer acidic reaction): potatoes, lupine, gooseberry, raspberry.

If many plant species develop poorly or die on highly acidic soils, it is not only due to the high concentration of H^+ ions but also the entire soil complex, especially mobile Al^{3+} and Mn^{2+} in soil solution above certain limits, deficiencies of some nutrients, and imbalances produced by acidity in their accessibility to plants.

The nitrogen source for plants is organic matter, whose decomposition is reduced at pH values below 7. Phosphorus has greater mobility in soils with pH = 5.5–7.0. At pH < 5.5, iron and aluminum phosphates form, which are insoluble and inaccessible to plants. At pH values >7.0 and up to 8.5, secondary and tertiary calcium phosphates form, which are poorly soluble and inaccessible to plants. At pH > 8.5, sodium phosphates, which are easily soluble and accessible to plants, form. In acidic soils, potassium is easily leached compared to neutral and alkaline soils. In calcium-amended soils, potassium mobility decreases. Ca2+ ions cause significant desorption of K^+ from the colloidal complex and antagonize its absorption by plants.

Calcium and magnesium are more soluble at pH values above 6.0. Iron, copper, and zinc have reduced mobility at pH > 7.5. Boron is very mobile at low pH, while molybdenum is at neutral pH (6.7–7) [34].

In acidic soils, aluminum partially transitions into soluble forms. High concentrations of Al^{3+} in the nutrient solution have harmful effects on roots, causing morphological changes, blackening, and alterations in cation absorption and retention capacity. Al^{3+} occupies exchange sites on root hairs, preventing the absorption of other nutrients: K^+, Ca^{2+}, Mg^{2+}, Fe^{2+}, etc. In the presence of Al^{3+}, phosphorus introduced into the soil through fertilizers is inaccessible to plants due to the formation of insoluble $AlPO_4$, making it inaccessible to plants. Therefore, in highly acidic soils with high aluminum content, large quantities of phosphate fertilizers are necessary to prevent phosphorus deficiency and the harmful effects of aluminum by precipitating it as tertiary phosphate. The negative effect of aluminum starts at 15–50 ppm of exchangeable Al, and at 0.3 me/100 g soil (27 ppm Al), lime amendments are required [35].

In addition to aluminum, manganese also causes toxicity effects at very variable concentrations, depending on plant sensitivity to this element. While some plants exhibit significant disturbances at concentrations of 1–4 ppm Mn, corn can tolerate concentrations above 15 ppm Mn without growth inhibition. Manganese accumulates more in the aerial part of the plant, causing protein metabolism disturbances. Common toxicity symptoms include the appearance of brown spots on leaves. Excess manganese in soil solution sometimes causes symptoms indicating false iron deficiency.

Acidic soils are generally poor in nutrients, have low base saturation, and low biological activity. As a result, the basic nutrient elements in the adsorptive complex and soil solution are low. The complex-solution exchanges occur unfavorably for plant nutrition, and nutrient reserve replenishment is slow [36, 37].

In salt-free soils, the acidic reaction is due to:

- hydrogen released by various organic acids or carbonic acid reacting with soil colloids, penetrating their diffuse layer, replacing various basic cations either leached or precipitated;
- presence of hydrogen, aluminum, or iron ions in the colloidal complex. Bivalent and trivalent metal cations exhibit the strongest acidic properties in water.

Therefore, an acidic reaction is due to the presence of hydronium ions (H_3O^+) in solution. Acidic reaction in soils can also result from applying fertilizers with a physiological acidic reaction. Correcting acidic reaction is done by applying materials called lime amendments. Lime amendments are considered those materials containing basic cations that can replace H^+ and Al^{3+} ions in the adsorptive complex and accompanying anions or those resulting from amendment transformation in soil, neutralizing H^+ ions or forming weakly dissociated acids (carbonic or silicic acid).

Soil alkalinity is determined by weak acid anions passing from the solid phase into soil solution and acting as bases, with their base role increasing with the higher Kb constant. The main anions determining soil alkalinity are carbonate (CO_3^{2-}) and bicarbonate (HCO_3^-). The presence of salts that hydrolyze alkaline, such as Na_2CO_3, K_2CO_3, $NaHCO_3$, $KHCO_3$, $MgCO_3$, and $CaCO_3$, also contributes to soil alkalinity.

Alkaline soils are formed due to complex interactions between natural (geomorphological, hydrogeological, climatic, hydrochemical) and anthropic factors (irrigation, excessive fertilization with mineral fertilizers on soils containing mineral-rich groundwater). Also, due to reduced leaching in areas with low precipitation, salts in the soil profile cannot be removed, concentrating carbonate or bicarbonate ions through evapotranspiration processes. As aridity increases, these alkaline materials and associated salts of sodium and gypsum accumulate closer to the soil surface. Large and excessive accumulations of soluble salts such as Na_2CO_3, $NaHCO_3$, Na_2SO_4, $NaCl$, $MgCl_2$, $MgSO_4$, and $CaSO_4$ have negative effects on plant growth and development, becoming more harmful as the soil's adsorptive complex becomes increasingly saturated with Na^+ ions at the expense of other cations, especially Ca^{2+} and Mg^{2+}. The negative influence of these salts on plant development is due to the specific toxicity of different cations and anions forming these salts. Among cations, Na^+ is prominent, while among anions, CO_3^{2-} and HCO_3^- are notable. Na_2CO_3 is more toxic compared to Na_2SO_4, $NaCl$, $CaCl_2$, or $MgCl_2$.

Due to sodium toxicity causing reduced plant growth, excessive sodium concentration in tissues, extremely low Ca^{2+}/Na^+ and K^+/Na^+ ratios, calcium deficiency, and inhibited root system growth, replacing sodium with calcium in the soil's adsorptive complex is necessary.

4. Fertilizer resource application

The physical form of the fertilizer resource will determine the application method. Solubility in soil solution can be a positive or negative factor in root absorption, determining the nutrient's accessibility for plants. High solubility can reduce nutrient availability due to rapid reactions between them and the soil's colloidal complex, although it enhances nutrient ion movement through diffusion in soil solution.

Using complex fertilizer resources ensures plant supply with multiple nutrients in a single application, which is advantageous. However, applying complex fertilizers also has disadvantages: the fixed ratio between elements only matches certain soil conditions, plant requirements, and vegetation stages. The chemical reactivity of compounds in a fertilizer resource can create a solution concentration imbalance due to precipitation or cation exchange phenomena involving nutrients. It can also modify soil reaction due to chemical reactions in soil, releasing $H+$ ions [38–40].

The application method of fertilizer resources significantly influences nutrient accessibility for root absorption and nutrient distribution in the root hair zone. Surface spreading and incorporation of fertilizer resources will disperse fertilizer particles, affecting nutrient accessibility and favoring nutrient fixation processes at soil component levels. Applying in bands reduces soil-fertilizer particle contact, reducing retention capacity in soil and increasing accessibility for plants by enhancing absorption potential at root contact. Repeated band application creates zones with high or low nutrient accessibility for plants, impacting production. Fertilizer application along rows during planting or vegetation maximizes root absorption and minimizes nutrient fixation in soil. Fertigation is effective for crops where the timing and rate of essential nutrient application need to be precise during critical nutrition periods. The timing of fertilizer resource application will significantly influence plant response: the shorter the time between the critical nutrition phase and fertilizer application, the more positive the yield effect [41].

Since acidic soils have a low content of organic matter and alkaline cations, it is recommended that the correction of the pH be accompanied by organic fertilization, and that this be repeated periodically for the regeneration of humic substances. The semi-fermented manure, applied before plowing and incorporated immediately, favors the rapid solubilization of the amendment in the soil. Manure applied to land where calcareous substances have been applied without immediate incorporation leads to nitrogen losses in the form of ammonia. The application of ammonium nitrogen fertilizers ($N\text{-}NH^{4+}$) together with calcareous amendments should be avoided, because there are losses of nitrogen through volatilization, as well as of superphosphate, in order to avoid the retrogradation of phosphorus into insoluble compounds that are not accessible to plants. The following can be mixed with calcareous amendments before incorporation: potassium fertilizers, natural phosphates, and nitrogen fertilizers [42].

Nutrient concentration in soil will change following fertilizer resource application. The balance between different nutrient forms in soil will change depending on the nutrient's nature. For example, applying potassium fertilizers will alter the balance between potassium ions in soil solution, those retained by the colloidal complex, and those fixed. The rate of balance change depends on soil's physical-chemical properties, moisture level, and the presence or absence of a crop [43].

Improper use of fertilizer resources, without considering the soil's physical-chemical characteristics, nutrient supply level, plant nutrient requirements, and expected yields, can lead to environmental pollution, especially surface water. Nitrate ions applied as highly soluble nitrogen fertilizers in soil water will undergo leaching processes under certain circumstances, concentrating over time in groundwater and surface water. Uncontrolled decomposition of animal waste and wastewater from the livestock sector, as well as uncontrolled mineralization of sewage sludge and municipal wastewater, constitute sources of nitrates.

Phosphate water pollution can occur if phosphate fertilizers are improperly used on highly permeable sandy soils, allowing fertilizer particles to pass through without being adsorbed. Phosphates and nitrates applied as fertilizers intensify eutrophication, ultimately degrading these waters and partially or totally destroying fauna through oxygen elimination and the formation of harmful chemicals [44].

Although each main macronutrient has a direct role in increasing yield, its high values are, most of the time, the result of the positive interaction between nitrogen and phosphorus. Various long-term researches on the effect of nutrients show the differences between the roles of N, P, and K on quality and production of different crops (**Table 5**).

Through chlorophyll, nitrogen contributes to photosynthetic activity, the basis of all plant processes, including growth and development. As a component of auxin, nitrogen favors the division of meristematic cells, stimulates the absorption of water and mineral substances, and through cytokinetins stimulates the multiplication of non-meristematic cells by division. It can be stated that nitrogen is the main element that determines crop production. From **Table 5**, it can be seen that, although the application of phosphorus and potassium causes a yield increase, nitrogen fertilization brings the highest yield increases both in the case of sunflower and in the case of wheat.

Along with nitrogen, phosphorus is necessary for cell division and cell growth after division. It stimulates root development, flowering, fruiting, the formation of reproductive organs, the storage of reserve substances, and the ripening of fruits and seeds. Favoring the processes of nutrition and growth, phosphorus reduces the specific water consumption of plants. A good supply of plants with phosphorus, along with potassium, magnesium, and calcium, increases the frost resistance of plants. Intensive nutrition, through the administration of phosphorus fertilizers, stimulates twinning in cereals and accelerates the maturation of plants.

Potassium is a main nutrient that contributes to obtaining high quality yields. Although crop plants need potassium and nitrogen in approximately equal amounts, in agricultural practice, there is an imbalance in the supply of these nutrients to the detriment of potassium. Potassium is one of the main elements that regulates the transpiration of plants and increases their resistance to drought and frost, as well as to the attack of pests and diseases. It determines the quality of vegetable products and increases their shelf life. It is well known effect of potassium fertilization to promote normal development and elasticity of stem tissue and to increase the breaking strength of the plant. By reducing the incidence of leaf and stem diseases, potassium fertilizers had a notable contribution to metabolite accumulation, especially in grain protein substances.

Nitrogen fertilizers determine both the variation in protein content and its composition. From **Figure 4**, an increase in the crude protein content can be observed, as the doses of nitrogen fertilizers increase. Although the influence of phosphorus on protein content is not as great as that of nitrogen, P will support the effect of nitrogen and mitigate the negative influence of high doses of nitrogen on protein quantity and quality. Fertilization with potassium, especially on soils with a moderate supply of this element and under conditions of multi-year use of nitrogen and phosphate

Fertilization		Sunflower		Wheat	
P and K kg\ha	N Kg\ha	kg/ha	%	kg/ha	%
0	0	1212,50	100.00	2112	100.00
	50	1403.75	115.77	2652	124.64
	100	1711,25	141.13	3087	145.09
	150	1905,00	157.11	3372	158.48
	200	2123,75	175.15	3465	162.86
50	0	1526,25	125.88	2449	115.10
	50	1911,25	157.63	3331	156.56
	100	2093,75	172.68	3863	181.56
	150	2081,25	171.65	4342	204.07
	200	2228,25	183.77	4724	222.03
100	0	1783,00	147.05	2769	130.14
	50	2238,75	184.64	2667	125.35
	100	2425,00	200.00	4395	206.57
	150	2671,25	220.31	5154	242.24
	200	2787,50	229.90	4962	233.21
150	0	1960,00	161.65	3002	141.09
	50	2330,00	192.16	4028	189.32
	100	2793,75	230.41	5242	246.37
	150	3120,00	257.32	5870	275.89
	200	2887,50	238.14	5893	276.97

Table 5.
Sunflower and wheat yield after N, P, and K fertilization [45, 46].

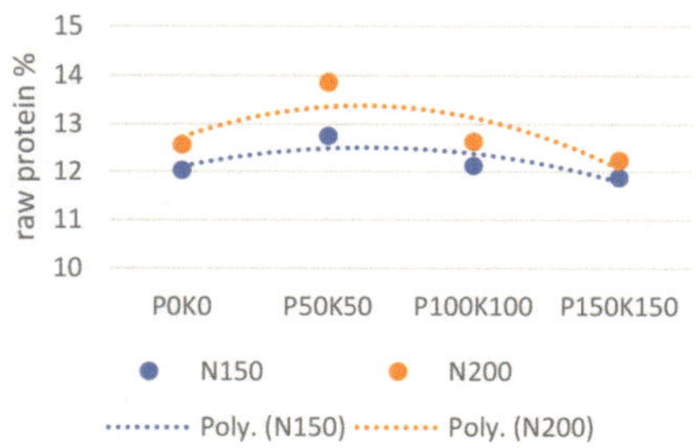

Figure 4.
Nitrogen, phosphorus, and potassium influence on raw protein content of corn grains [47].

fertilizers, can lead to an increase in protein production. The decrease in the percentage of raw protein content, at the application of high doses of potassium fertilizers, can be explained by the fact that while potassium is needed in large quantities for plant growth and protein biosynthesis, it stimulates more intense carbohydrate synthesis and transport than synthesis of nitrogen substances. Therefore, when applying large amounts of potassium fertilizers, carbohydrate synthesis and their transportation from other organs to grains increase and protein content decreases [47].

Although sulfur and zinc favor the formation of protein in the wheat grain, the determining role is played by nitrogen fertilization.

5. Nutrient use efficiency

Nitrogen fertilizers primarily contribute to stimulating the mineralization of organic matter in the soil, which may undergo the following transformations:

- 50–60% are taken up by plants,
- 15% are lost through denitrification;
- 5–15% (sometimes even 30%, on light soils) are lost through leaching;
- 10–20% are immobilized in the soil.

When reducing nitrogen losses and increasing the efficiency of the fertilizer, it is very important to apply them at the right time, in optimal doses and the correlation with the other fertilizers, in accordance with a rational tillage system, with the rotation and with the correct irrigation regime.

A determining factor in choosing the type of fertilizer is the reaction of the soil. On acidic soils, $pH < 6.3$, fertilizers with low physiological acidity or equal to 0 are used: urea, sodium nitrate, calcium nitrate, nitro-lime. On soils with a neutral reaction, $pH = 6.4–7.3$, ammonium nitrate, urea, ammoniacal water, and ammonium sulfate are used with good results. In the case of alkaline soils, $pH > 7.5$, the best results were given by ammonium sulfate, together with urea [48].

A neutral to slightly basic soil reaction favors the absorption of ammonia fertilizers, while an acidic reaction favors the absorption of nitric fertilizers [49].

Along with the reaction of the soil, the presence of other ions in its solution can influence the efficiency of nitrogen fertilizers. Thus, the presence in the soil solution of Ca^{2+} and Mg^{2+} simultaneously with the presence of Cl^- and SO_4^{2-} favors the penetration of NH^{4+} and inhibits the penetration of NO_3^-. On the other hand, the presence of K^+ and Na^+, together with $H_2PO_4^-$ and HPO_4^{2-}, inhibits the penetration of NH_4^+ and intensifies the penetration of NO_3^- [50].

If plant residues are incorporated into the soil, it is necessary to raise the nitrogen dose by approximately 6–10 kgN/t plant residue.

Foliar fertilization increases the coefficient of use of nitrogen from fertilizers, being recommended for this type of fertilization are nitric and amide forms, as they do not cause burns on plants.

The application of nitrogen fertilizers together with the irrigation water presents the advantage of administering nitrogen during the periods of maximum consumption when the plants are in an advanced state of vegetation.

The fractional application of nitrogen, one part for basic fertilization, one part for sowing or planting, and one part for vegetation, leads to the reduction of nitrogen losses through leaching, volatilization, denitrification, increasing the coefficient of use of nitrogen in the fertilizer. In autumn, with the basic fertilization, it is recommended to apply ammoniacal nitrogen fertilizers, not nitric, to avoid leaching losses.

Incorporating ammoniacal nitrogen fertilizers and urea into the soil increases their efficiency as a result of reducing volatilization losses [49].

Phosphorus use coefficient from fertilizers in the first year depends on the type of soil, the plant species, and the culture technology, being between 5 and 25%. The efficiency of the use of phosphorus fertilizers is influenced by several factors: pH, temperature, humidity, soil texture, its content in mobile phosphorus, the species and age of the plant, the type of fertilizer, the method of application, and the culture technology.

Plants require P especially in the first phase of vegetation, when the root system is poorly developed and has a reduced capacity to take phosphorus from less accessible forms. That is why it is necessary that a large amount of the P dose be applied in autumn under basic plowing.

The species of plants with a short vegetation period and those with a poorly developed root system utilize more effectively P from the easily soluble fertilizers in water. Such plants, which respond well to fertilization with easily soluble P fertilizers, are vegetables (potatoes, onions, beans, cabbage, tomatoes, cucumbers). Also, grasses use less P from the soil and from hardly soluble fertilizers, responding well to fertilization with easily soluble P.

Unlike grasses, legumes strongly solubilize P from poorly soluble compounds. Plants that make good use of P from the soil reserve and less soluble fertilizers are those with a long vegetation period and those with deep and rapid root system growth. Due to the low solubility and poor mobility of P in the soil, simple P fertilizers are basic fertilizers applied in autumn, before sowing.

It is recommended that fertilizers with water-soluble P be applied in granulated form so that the reduced contact surface between the granule and the soil reduces the amount of retrograded P. Fertilizers with hardly soluble P will be applied in powder form, to ensure a more efficient mixing with the soil and a better mobilization of P [51].

In order to increase the efficiency of P fertilizers, it is recommended to apply them locally, in strips or in rows, compared to spreading them over the entire surface. When the fertilizer is in contact with a smaller volume of soil and the adsorption-retrogradation processes are limited, small and moderate doses of P will reach an effective distance from the plant's root.

On soils with an extreme reaction (acidic or alkaline), the fixation of phosphorus takes place, and the application efficiency is reduced, which is why it is recommended to amend these soils before fertilizing with phosphorus. On such soils, and especially on those with a high phosphorus fixing capacity, it is recommended to apply granulated superphosphates in rows or strips, in order to reduce the amount of phosphorus adsorbed.

In order to avoid the negative effects of overfertilization with phosphorus on the environment and to establish necessary doses of phosphate fertilizer to obtain optimal yields, it is necessary to evaluate the losses of this element in the respective environment (leaching, infiltrations, runoff\erosion). The recommendations must take into account the risk of pollution of water resources in which eutrophication phenomena may occur [52].

Potassium fertilizers are applied as basic fertilizers, in the fall before sowing and rarely simultaneously with sowing. The coefficient of use of potassium in fertilizers is 40–60%. On soils with reduced cationic exchange capacity, such as sandy soils and those poor in organic matter, potash fertilizers are applied annually, fractionally: a part as close as possible to sowing and a part in vegetation.

On soils with medium cation exchange capacity, it is recommended to apply fertilizers during autumn plowing. On soils with a high cationic capacity, such as clay soils, the efficiency of potassium fertilizers is reduced and only localized application is recommended. Reserve fertilization can also be done on these soils, once every 2–3 years, under deep plowing in the fall, incorporating a higher dose than those applied normally.

At the same dose and the same method of application, the efficiency of using potassium from fertilizers decreases as the state of supply of the soil with available potassium increases [53].

Depending on the potassium requirement, plants can be classified as follows:

- plants that consume a lot of potassium are, in general, the crops that synthesize and deposit carbohydrates. Ex.: potato, sunflower, cabbage, cauliflower, carrot, celery, tomatoes, vine;

- plants that moderately consume potassium are mainly legumes, which synthesize proteins: beans, soybeans, clover, alfalfa, but also early cabbage, strawberry, apple, hair;

- low potassium consuming crops: lupine, rye, oats, red beet.

Dicotyledons respond better to potassium fertilization than monocotyledons. In the case of species with a long vegetation period, the fractional application of potassium can be effective.

When choosing the type of *sulfur* fertilizer, the pH of the soil and the need to apply the accompanying nutrient must be taken into account. Sulfur fertilizers that contain nitrogen will be applied on alkaline and neutral soils, and those that contain calcium will be applied on saline and alkaline soils. As a source of sulfur on acid soils, simple superphosphate is preferable. Plants absorb sulfur more easily in the form of soluble sulfate than in the form of gypsum, which dissolves slowly [54].

Sulfur deficiency occurs on sandy and poor soils in organic matter, as well as on soils with low phosphorus content and intensively fertilized with N and P fertilizers, without sulfur input. Sulfur deficiency can occur on irrigated soils where sulfur is leached into the soil depth and on those that are not fertilized with natural organic fertilizers. In general, sulfur deficiency occurs on soils with mobile sulfur content <5–10 ppm.

The effectiveness of sulfur fertilizers is all the greater as the plants are more sensitive to the lack of this nutrient. Cruciferous plants (mustard, rapeseed), onions, garlic, and legumes (alfalfa, clover) are high sulfur-consuming plants.

The application of *iron* fertilizers to the soil is done before sowing, but it is less effective because the iron from inorganic fertilizers is quickly insoluble in the form of oxides. On carbonate soils, precipitation of Fe^{2+} ions takes place in the form of $FeCO_3$. Foliar application is more effective than soil application, but due to the low mobility of iron in the plant, foliar spraying must be repeated several times. Iron chelates are recommended for the preparation of solutions, but ferrous sulfate can also be used [55].

Boron fertilizers are applied to the soil in the fall by spreading evenly and incorporating them under plowing. Foliar fertilization is practiced especially in perennial crops and on soils where a strong fixation of boron can occur if applied to the soil.

The application of boron fertilizers is done once every 4–5 years for the most demanding crop in the rotation, except for sandy and acidic soils and in areas with intense precipitation where it is necessary to return more often.

Currently, digital solutions and technologies based on data and knowledge will allow the calculation of nutrient quantities related to local needs in a more precise way, making them accessible to as many users as possible. The new tools and sensors for diagnosing soil and crops and monitoring environmental factors that allow real-time prediction models, but also decision support based on artificial intelligence will play an increasing role in responsible plant nutrition. Personalized plant nutrition is required that must minimize the compromises between productivity and the environment [56].

6. Conclusion

Fertile soil contains all the nutrients necessary for plant growth and development, major nutrients (e.g., nitrogen, phosphorus, and potassium), secondary nutrients (e.g., calcium, magnesium, sulfur) as well as micronutrients needed in smaller quantities (e.g., iron, zinc, copper, manganese, boron, molybdenum, chlorine). Different crops have varying nutrient requirements and proper management ensures the right balance of nutrients, preventing deficiencies or excesses that can harm plant health. These nutrients are critical for plant growth, flowering, and fruiting. Adequate nutrient supply leads to higher crop yields and continuous cropping can deplete soil nutrients over time. Knowing how these nutrients behave in the soil and what are the factors that influence their availability for plants ensures their correct management and helps replenish these nutrients, preventing soil exhaustion and maintaining sustainable soil fertility. Also, a sustainable nutrient management contributes to soil organic matter content which improves soil structure, water retention, and nutrient availability. Efficient nutrient use minimizes nutrient runoff into water bodies, which can cause pollution.

Conflict of interest

The authors declare no conflict of interest.

Author details

Isidora Radulov* and Adina Berbecea
University of Life Sciences "King Mihai I" from Timisoara, Timisoara, Romania

*Address all correspondence to: isidora_radulov@usvt.ro

References

[1] Benton, Jones J Jr. Plant Nutrition and Soil Fertility Manual. 2nd ed. Boca Raton: CRC Press; 2012, ISBN: 978-1-4398-1609-7

[2] Brady NC, Weil RR. Elements of the Nature and Properties of Soils. 2nd ed. Upper saddle River, NJ: Prentice-Hall, Inc.; 2003

[3] Rusu M, Marghitas M, Tania Mihaiescu I, Oroian AD. Tratat de Agrochimie. Romania: Ceres Bucuresti; 2005. pp. 231-123

[4] Jannatul F, Varun P. A review on the possible factors influencing soil inorganic carbon under elevated CO_2. Catena. 2021;**204**:105434. DOI: 10.1016/j.catena.2021.105434, ISSN 0341-8162

[5] Kothawala DN, Moore TR, Hendershot WH. Soil properties controlling the adsorption of dissolved organic carbon to mineral soils. Soil Science Society of America Journal. 2009;**73**:1831-1842. DOI: 10.2136/sssaj2008.0254

[6] Zamanian K, Zhou J, Kuzyakov Y. Soil carbonates: The unaccounted, irrecoverable carbon source. Geoderma. 2021;**384**:114817, ISSN 0016-7061. DOI: 10.1016/j.geoderma.2020.114817.], 10.1016/j.geoderma.2020.114817.]

[7] Bingham AH, Cotrufo FM. Organic nitrogen storage in mineral soil: Implications for policy and management. Science of the Total Environment. 2016;**551-552**:116-126. DOI: 10.1016/j.scitotenv.2016.02.020, ISSN 0048-9697

[8] Lamb JL, Fernandez FG, Kaiser DE. Understanding Nitrogen in Soils, AG-FO-3770-B (Revised 2014). United States: Nutrient Management, University of Minnesota; 2014

[9] Shafreen M, Vishwakarma K, Shrivastava N, Kumar N. Physiology and Distribution of Nitrogen in Soils. Cham: Springer; 2021. DOI: 10.1007/978-3-030-71206-8_1

[10] Muindi E. Understanding: Soil phosphorus. International Journal of Plant & Soil Science. 2019;**3**(1):1-18. DOI: 10.9734/ijpss/2019/v31i230208

[11] Mengel K, Kirkby EA, Kosegarten H, Appel T. Phosphorus. In: Principles of Plant Nutrition. Dordrecht: Springer; 2001. DOI: 10.1007/978-94-010-1009-2_9

[12] Isidora R. Soil Chemistry and Management of Nutrients and Fertilizers. Timișoara: Eurostampa; 2007, ISBN 978-973-687-537-3

[13] Harpreet K. Forms of potassium in soil and their relationship with soil properties - a review. International Journal of Current Microbiology and Applied Sciences. 2019;**8**(10):1580-1586. DOI: 10.20546/ijcmas.2019.810.184

[14] Isidora R, Goian M. Potassium in Agriculture and Nutrition. Timisoara: Solness; 2004, ISBN 973-729-006-2

[15] Xiaolu D, Zhang C, Zhang J, Ma D, Chen L, Zhou G, et al. Relationship between calcium forms and organic carbon content in aggregates of calcareous soils in northern China. Soil and Tillage Research. 2024;**244**:106210, ISSN 0167-1987. DOI: 10.1016/j.still.2024.106210

[16] Yan B, Hou Y. Effect of soil magnesium on plants: A review.

IOP Conference Series: Earth and Environmental Science. 2018;**170**:022168. DOI: 10.1088/1755-1315/170/2/022168

[17] Petra M. Marschne's Mineral Nutrition of Higher Plants. 3rd ed. USA: Elsevier; 2012. ISBN 978-0-12-384905-2

[18] Jones JD. Iron availability and management considerations: A 4R approach. Crops & Soils. 2020;**53**:32-37. DOI: 10.1002/crso.20019

[19] Li H, Santos F, Butler K, Herndon E. A critical review on the multiple roles of manganese in stabilizing and destabilizing soil organic matter. Environmental Science & Technology. 2021;**55**(18):12136-12152. DOI: 10.1021/acs.est.1c00299

[20] Mengel K, Kirkby EA, Kosegarten H, Appel T. Soil copper. In: Mengel K, Kirkby EA, Kosegarten H, Appel T, editors. Principles of Plant Nutrition. Dordrecht: Springer; 2001. DOI: 10.1007/978-94-010-1009-2_16

[21] Schulte EE. Soil and Applied Chlorine A3556. USA: University of Winsconsine Extension; 1999

[22] Geilfus C-M. Chloride in soil: From nutrient to soil pollutant. Environmental and Experimental Botany. 2019;**157**:299-309. DOI: 10.1016/j.envexpbot.2018.10.035, ISSN 0098-8472

[23] Cornell University Agronomy Fact Sheet #22: Ction Exchange Capacity (CEC). 2007. Available from: http://nmsp.css.cornell.edu/publications/factsheets.asp

[24] Li K-W, Lu H-l, Nkoh Nkoh J, Hong Z-n, Xu R-K. Aluminum mobilization as influenced by soil organic matter during soil and mineral acidification: A constant pH study. Geoderma. 2022;**418**:115853. DOI: 10.1016/j.geoderma.2022.115853, ISSN 0016-7061

[25] Ontl TA, Schulte LA. Soil carbon storage. Nature Education Knowledge. 2012;**3**(10):35

[26] Shawver SE, Brady JA, McGahan DG. Variation of prokaryotic and fungal soil communities across a vegetative transect. Agricultural Research & Technology Open Access Journal. 2018;**15**(3):555953. DOI: 10.19080/ARTOAJ.2018.15.555953

[27] Li T, Meng L, Li N, Zhao Y, Uwizeyimana H. Study on microorganism distribution characteristics of three types of typical acidified soils. In: Proceedings of the 2015 AASRI International Conference on Circuits and Systems. France: Atlantis Press; 2015. DOI: 10.2991/cas-15.2015.19, ISBN 10.2991/cas-15.2015.19

[28] Cresser J, Killham N, Edwards J. Soil Chemistry and its Application. Cambridge: Cambridge University Press; 1993. pp. 65-68

[29] Tibbett M, Gil-Martínez M, Fraser T, Green ID, Duddigan S, De Oliveira VH, et al. Long-term acidification of pH neutral grasslands affects soil biodiversity, fertility and function in a heathland restoration. Catena. 2019;**180**:401-415. DOI: 10.1016/j.catena.2019.03.013, ISSN 0341-8162

[30] Neina D. The role of soil pH in plant nutrition and soil remediation. Applied and Environmental Soil Science. 2019;**2019**:5794869. DOI: 10.1155/2019/5794869, 9 pages

[31] Saleem S, Mushtaq NU, Rasool A, Shah WH, Tahir I, Rehman RU. Chapter two - plant nutrition and soil fertility: Physiological and molecular avenues for crop improvement. In: Sustainable Plant Nutrition. London, UK: Academic Press; 2023. pp. 23-49. DOI: 10.1016/B978-0-443-18675-2.00009-2. ISBN 9780443186752

[32] Barrow NJ, Hartemink AE. The effects of pH on nutrient availability depend on both soils and plants. Plant and Soil. 2023;**487**:21-37. DOI: 10.1007/s11104-023-05960-5

[33] Bennett JA, Klironomos J. Mechanisms of plant-soil feedback: Interactions among biotic and abiotic drivers. The New Phytologist. 2019;**222**:91-96. DOI: 10.1111/nph.15603

[34] Jones C, Jacobsen J. Plant nutrition and soil fertility. In: Nutrition Management & Modification. Vol. 2. USA: Montana State University; 2005. p. 1. Available from: https://citeseerx.ist.psu.edu/viewdoc/download?doi=10.1.1.577.4503&rep=rep1&type=pdf

[35] Rahman MA, Lee SH, Ji HC, Kabir AH, Jones CS, Lee KW. Importance of mineral nutrition for mitigating aluminum toxicity in plants on acidic soils: Current status and opportunities. International Journal of Molecular Sciences. 2018;**19**(10):3073. DOI: 10.3390/ijms19103073

[36] Shahne AA, Shivay YS. Soil health and its improvement through novel agronomic and innovative approaches. Frontiers in Agronomy. 2021;**3**:680456

[37] Adnane B, Karim L, Mohamed C, Youssef Z, Driss D. Soil microbial resources for improving fertilizers efficiency in an integrated plant nutrient management system. Frontiers in Microbiology. 2018;**9**:1606. DOI: 10.3389/fmicb.2018.01606

[38] Katharin H, Felix E, Paula R, Beatrice G. Achieving the nutrient reduction objective of the farm to fork strategy. An assessment of CAP subsidies for precision fertilization and sustainable agricultural practices in Germany. Frontiers in Sustainable Food Systems. 2023;**7**:1088640. DOI: 10.3389/fsufs.2023.1088640

[39] Beckman J, Ivanic M, Jelliffe J. Market impacts of farm to fork: Reducing agricultural input usage. Applied Economic Perspectives and Policy. 2022;**44**:1995-2013. DOI: 10.1002/aepp.13176

[40] Kim H-J, Sudduth KA, Hummel JW. Soil macronutrient sensing for precision agriculture. Journal of Environmental Monitoring. 2009;**11**:1810. DOI: 10.1039/b906634a

[41] Hasler K. Environmental impact of mineral fertilizers: Possible improvements through the adoption of eco-innovations [thesis]. Netherlands: Wageningen University; 2017. DOI: 10.18174/422865

[42] Gosling P, Shepherd M. Long-term changes in soil fertility in organic arable farming systems in England, with particular reference to phosphorus and potassium. Agriculture, Ecosystems & Environment. 2005;**105**:425-432. DOI: 10.1016/J.AGEE.2004.03.007

[43] Alloway BJ. Micronutrients and crop production: An introduction. In: Alloway BJ et al., editors. Micronutrient Deficiencies in Global Crop Production. The Netherlands: Springer; 2008. pp. 1-39

[44] Drinkwater LE, Snapp SS. Nutrients in agroecosystems: Rethinking the management paradigm. Advances in Agronomy. 2007;**92**:163-186

[45] Sala F, Radulov I, Crista F, Berbecea A. The correlation between the fertilization system and yield for the sunflower crop. Research Journal of Agricultural Science. 2010;**42**(3):296-301, ISSN 2066-1843

[46] Crista F, Radulov I, Crista L, Lato A, Sala F, Berbecea A, et al. Changing quality indicators of wheat crops following fertilizers application. Research Journal of Agricultural Science. 2013;**45**(1):3-12, ISSN 2066-1843

[47] Florin C, Isidora R, Laura C, Adina B, Alina L. Influence of mineral fertilization on the amino acid content and raw protein of wheat grain. Journal of Food Agriculture & Environment. 2012;**10**(3-4):47-50, Part: 1, ISSN 1459-0255

[48] Vitousek PM, Naylor R, Crews T, David MB, Drinkwater LE, Holland E, et al. Nutrient imbalances in agricultural development. Science. 2009;**324**:1519-1520

[49] Dawson JC, Huggins DR, Jones SS. Characterizing nitrogen use efficiency in natural and agricultural ecosystems to improve the performance of cereal crops in low-input and organic agricultural systems. Field Crops Research. 2008;**107**:89-101. DOI: 10.1016/J. FCR.2008.01.001

[50] Congreves KA, Otchere O, Ferland D, Farzadfar S, Williams S, Arcand MM. Nitrogen use efficiency definitions of today and tomorrow. Frontiers in Plant Science. 2021;**12**:637108. DOI: 10.3389/ fpls.2021.637108

[51] Djodjic F, Bergström L, Grant C. Phosphorus management in balanced agricultural systems. Soil Use and Management. 2005;**21**:94-101

[52] Fixen P, Brentrup F, Bruulsema T, Garcia F, Norton R, Zingore S. Nutrient/fertilizer use efficiency: Measurement, current situation and trends. In: Drechsel P, Heffer P, Magen H, Mikkelsen R, Wichelns D, editors. Managing Water and Fertilizer for Sustainable Agricultural Intensification. Paris: IFA; 2015. pp. 8-38

[53] Manuel N-C, Francisco R, Santa-María GE. Editorial: Nutrient use-efficiency in plants: An integrative approach. Frontiers in Plant Science. 2020;**11**:623976. DOI: 10.3389/ fpls.2020.623976

[54] Baligar VC, Fageria NK, He ZL. Nutrient use efficiency in plants. Communications in Soil Science and Plant Analysis. 2001;**32**(7-8):921-950. DOI: 10.1081/CSS-100104098

[55] Fageria NK, Baligar VC, Li YC. The role of nutrient efficient plants in improving crop yields in the twenty first century. Journal of Plant Nutrition. 2008;**31**(6):1121-1157. DOI: 10.1080/01904160802116068

[56] Dobermann A, Bruulsema T, Cakmak I, Gerard B, Majumdar K, McLaughlin M, et al. Responsible plant nutrition: A new paradigm to support food system transformation. Global Food Security. 2022;**33**:100636. DOI: 10.1016/j. gfs.2022.100636

Chapter 3

Organic Fertilizers and Biopesticides: Eco-Innovations, Circular Economy, and Agribusiness Developed by Organic Farmers in Uganda

Bienvenu A. Dagoudo, Charles Ssekyewa and Silvère D. Tovignan

Abstract

Agriculture production now faces the challenge of reducing soil fertility and biodiversity components as a result of climate change and the use of large amounts of chemical inputs per unit of production. To reduce the negative consequences of using chemicals on human beings and ecosystems, farmers eco-innovate through agricultural practices, recycling residues and waste, and producing organic fertilizers and bio-pesticides. This study assesses the contribution of locally made organic fertilizer and biopesticides through recycling and eco-innovation to organic farms and agribusiness in Wakiso, Massaka, and Jinja regions of Uganda. A multiple-stage stratified random sampling procedure was used to gather data from 150 organic farmers. The results revealed that organic farmers surveyed, a majority of whom were women (64%) with a farm size average equal to 0.79 ha (±0.67), produced organic fertilizers and biopesticides from locally available materials, such as wood ash, rice, milk, and plant leaves. It was observed that the farmers could recycle maize and sweet potato residues for feeding animals and domestic food waste for organic fertilizing soil. Some organic farmers eco-innovate by creating small- and medium-sized businesses that produce organic fertilizers and biopesticides for the local market. It is highlighted that endogenous knowledge underpins the majority of biopesticides and organic fertilizers produced by the organic farmers surveyed.

Keywords: biopesticides, circular economy, organic fertilizer, eco-innovations, agribusiness

1. Introduction

Agricultural production systems face the challenge of using chemicals to manage fertilization, pests, diseases, and weeds, which reduce soil fertility, contaminate water and food, and devastate the environment and ecosystem. These negative effects result in crop yield losses, which expose farmers to food insecurity. Farmers, in their instinct to survive, rely on accessible resources to avoid a catastrophic situation, eco-innovate through production, regenerate biodiversity, and conserve ecosystems. Eco-innovations have long been a focal point for preserving agrobiodiversity in agricultural landscapes by recycling waste and residues and promoting a circular economy. How do organic farmers eco-innovate in Uganda?

Eco-innovation refers to new products, processes, ideas, or behaviors that are aimed at reducing the environmental impacts of resource utilization [1, 2]. Eco-innovation is focused on innovative activities that improve the ecological environment for sustainable development [3]. As a result of increasing mass production, eco-innovation incorporates the economy, ecology, and societal aspects to limit the overuse of the current environmental resource [4]. The reuse of derived agricultural products (waste, residue, manure, etc.) as organic fertilizers and biopesticides initiates a cycle in which farmers design a circular economy based on their capacity. Circularity is a concept developed for using natural resources in productive processes, which implies the reduction or avoidance of waste and trade-offs [5, 6]. The circularity of reusing waste contributes to the independence of organic farmers from external inputs and reduces the cost of production. Organic production generates a circular economy that participates in the conservation of soil fertility and biodiversity restoration. For Geissdoerfer et al. [7], the circular economy promotes a regenerative system in which the waste of resources, emission of gases, and leakage of energy are minimized by lowering downs, closing, and narrowing material and energy loops. Murray et al. [8] draws on the human dimension to define the circular economy as a transformation of production systems that prioritizes ecosystem resilience and human well-being. However, the circular economy through organic inputs emphasizes agriculture resilience and biodiversity restoration by generating business and economic opportunities for farmers and providing environmental and societal benefits. Organic agriculture has been recognized as the agriculture practice that guarantees the health and sustainability of agroecosystems [9, 10]. As a holistic practice, organic agriculture revealed the logical association and/or synchronization of systemic elements [11], such as water, soils, crops, and livestock. Organic agriculture aims to establish integrated agroecosystems that are socially, environmentally, and economically viable [12]. However, banning synthetic agrochemicals is the watchword for organic production in order to achieve biodiversity conservation and agricultural sustainability. Therefore, using organic inputs (organic fertilizers and biopesticides) derived from plants and microorganisms allows for the conservation of biodiversity and ecosystems while producing food. It is highlighted that organic fertilizers and biopesticides have a positive impact on the socioeconomic and ecological environment, particularly in soil nutrients, which will significantly contribute to food production through quality and safety, human health, and environmental preservation, [13, 14], decreasing the effects of groundwater contamination [15]. Thus, through a circular economy that is less reliant on depleting natural resources than traditional economies, organic fertilizer, and biopesticide production have opened the door to business by implementing a spiral-loop system that minimizes the flow of materials and energy. This will maintain the potential for economic growth and development [16, 17], despite the resources engaged. Organic fertilizer and biopesticide production provide agribusiness opportunities because they rely on natural internal resources, such as plant sources, residues, animal and food waste, and so on, resulting in an integrated system of production, agroforestry, and livestock. This article examines the production of organic fertilizers and biopesticides as eco-innovations for organic farmers to improve production through recycling and the circular economy, as well as agribusiness for organic production development.

2. Methodology

2.1 Study area

The research was carried out in these three Ugandan districts: Wakiso, Masaka, and Jinja. The area studied is characterized by a tropical climate where annual rainfall

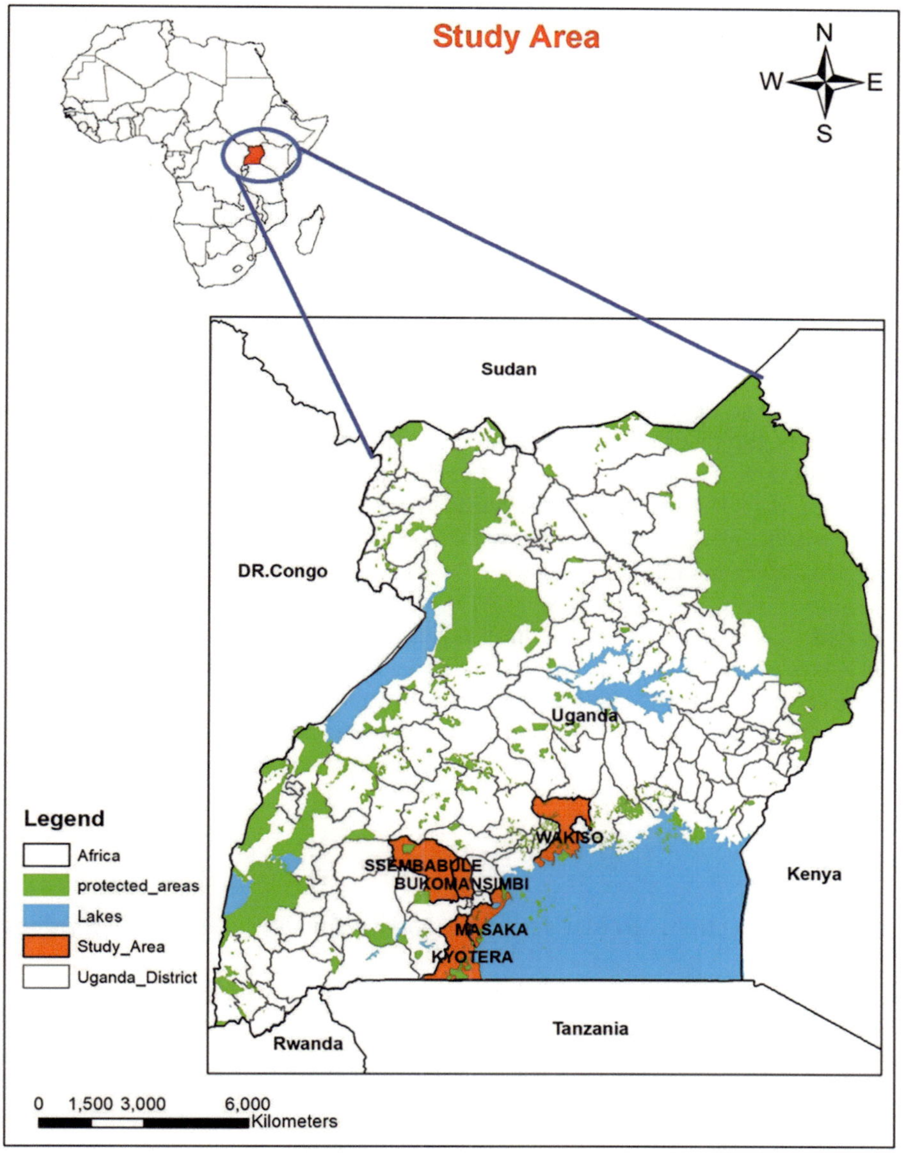

Figure 1.
Area study.

Institution	District	County	Number of organic farmers
Agency For Integrated Rural Development	Wakiso	Busiro	50
St. Jude Family Project	Masaka	Bukoto	50
KSAM Farm Limited	Jinja (Buikwe)	Kizigo	50
Total			150

Table 1.
Number of organic farmers interviewed.

averages around 1614 mm and is bimodal with two seasons. The first season lasted from March to May, while the second ran from October to December. The average temperature ranged from 18.14 to 26.53°C, with a minimum of 18.5°C and a maximum of 28.1°C [18, 19]. **Figure 1** depicts the maps of the study area.

2.2 Data collection

The data were collected through field research, which focuses on farmer visits and participatory observation to identify organic fertilizers and biopesticides. This data collection procedure follows two steps, resulting in a sample of 150 organic farmers.

- The first step was to purposefully select members of the institution for the National Organic Agriculture Movement of Uganda (NOGAMU) to host data collection and provide the location of organic farmers with experience in organic fertilizers and biopesticides. This step was taken to select from the Agency for Integrated Rural Development and the St. Jude Family Project, KSM Farm Limited as NOGAMU members. These institutions have at least 5 years of experience teaching organic farmers how to make organic fertilizer and biopesticides from locally available materials.

- The second step involved randomly selecting organic farmers to observe the effective use of organic fertilizer and biopesticides. Therefore, 150 organic farmers were selected, 50 from each institution (**Table 1**). The visit to a farm is critical for data collection in terms of residue and waste valorization, as well as the innovativeness of organic fertilizer and biopesticides.

3. Results

3.1 System of recycling, circular economy, and agribusiness in organic production

The visit to organic farms reveals their production system that emphasizes the following three components: recycling, circular economy, and agribusiness. These three interconnected components form the foundation of eco-innovation. Recycling covers the value chain of organic production. Organic farmers highlight that they recycle the waste from agriculture production (maize residue, sweet potato residue, mulch of maize, etc.), agroforestry (plant leaves, herbs), and livestock (cattle, sheep, and, pig manure). Recycling extends to residues from crop processing (coffee husk) and consumption. The circular economy is defined as the reuse of organic value chain waste as organic fertilizer and biopesticides to control diseases and pests on the same farm. Agribusiness emerges when organic farms operate as small- and medium-sized businesses, producing organic fertilizer and biopesticides for local markets. **Figure 2** depicts the interconnect recycling, circular economy, and agribusiness in organic production.

3.2 Sample description

The findings revealed that organic farmers' respondents are women.
(64%). Organic farming combines multiple agricultural activities, that are better suited to women than men. Producing organic fertilizer and biopesticides, and using readily available materials is not a one-person job. Organic households surveyed have an average of six people, with a minimum of one and a maximum of 20 (**Table 2**). Organic farmers participating in this survey have an average age of 53.1 years. Organic farmers faced the challenges of accessing resources that justify the small size of their farms (average 0.79 ha ±0.67). **Table 2** depicts the descriptive analysis of organic farmer respondents.

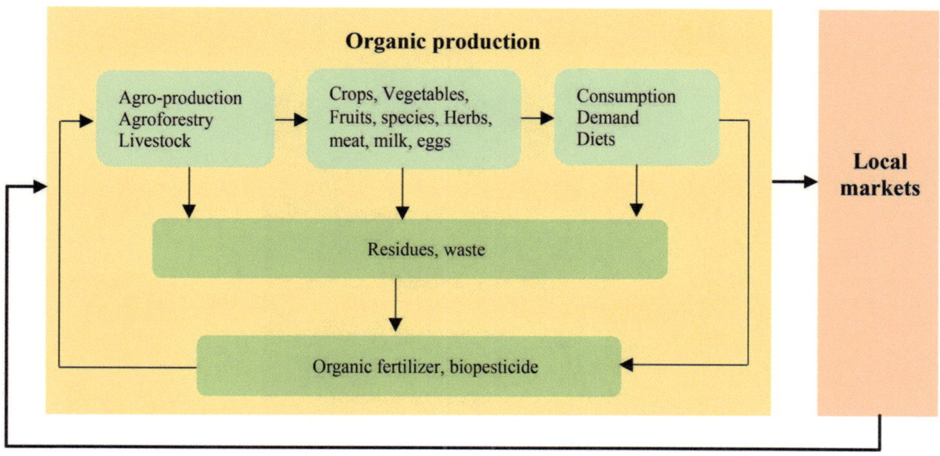

Figure 2.
Organic production through recycling, circular economy, and agribusiness.

Variable		Frequency		Percentage (%)	
Sex	Female	96		64	
	Male	54		36	
		Mean	*Std. dev.*	*Min*	*Max*
Age		53.1	14.16	20	90
Family size		6.34	2.96	1	20
Male actives		1.97	1.22	1	9
Female actives		2.36	1.54	1	14
Organic farm size (ha)		0.79	0.67	0.04	4.86

Table 2.
Descriptive analysis of organic farmers.

3.3 Organic fertilizers and biopesticides availability

Organic farming prohibits the use of chemicals for fertilization, disease control, pest control, and weed control. Organic farmers, on the other hand, emphasized on organic fertilizer to increase production yield and biopesticides to control disease or pests (**Figure 3**). Organic fertilizers and biopesticides were not widely manufactured in Uganda, resulting in high market prices and limited availability for small farmers. Small farmers created organic fertilizers and biopesticides from plant leaves, herbs, residues, or waste using their knowledge and local training. The majority of farmer respondents (93.33%) stated that they know how to make organic fertilizers or biopesticides, such as manure, vermicompost, Ash NPK, native, Indigenous Micro-Organisms (IMO), Lacto, green tea, and ash biopesticides, among others. At least two of their organic fertilizers or biopesticides were used on the farm by 80.66% of the organic farmers surveyed. Some organic products (IMO) can serve as both fertilizer and pesticide. **Figure 4** shows some materials for organic fertilizers or biopesticides in the area studied.

Figure 3.
Vermicompost from coffee husks and maize corn.

Figure 4.
Organic fertilizer NKP from wood ash.

3.4 Recycling of residues and waste for agriculture production

The small size of organic farms in the area studied (0.79 ha, ±0.67) provides small-quality residues and waste. In order to optimize the recycling of residues and waste, organic farmers used an innovation called "urban farming." Urban farming innovation contributes to optimizing the small size of organic farms by intercropping and increasing the yield of production. **Figure 5** is an example of urban farming innovation.

Organic farming is the integration of crop production, livestock, and agroforestry. However, the majority of farm visits recycle the residues from crop production to feed animals, such as cattle, sheep, and pigs. The manure from livestock is used as fertilizer in crop production. **Figure 6** depicts the goat fed with corn and sweet potato residues and cattle fed with sweet potato residues.

3.5 Organic agribusiness

Organic agribusiness focuses on the production of organic production inputs such as organic fertilizer and biopesticide. According to organic farmers, organic production lacks organic fertilizer and biopesticide which affect negatively the yield production. However, some organic farmers initiate the production of organic fertilizer and biopesticide based on the experience from endogenous knowledge and local organization support. As small and medium enterprises, they produce for the local market. In the majority of cases, the enterprise is an individual enterprise, and organic farmers entrepreneurs are called "innovators." Some organic fertilizers are certified by the Ministry of Agriculture, Animal Industry, and Fisheries (e.g., *worders foliar fertilizer*), and some organic fertilizers and biopesticides are still in the process of certification (e.g., *Jimusa foliar NPK fertilizer*, *Kulika biopesticide*). Ministry of Agriculture, Animal Industry, and Fisheries certification gives the right of property of the innovation to be sold at the national market. **Figures 7–9** show *worders foliar fertilizer, Jimusa foliar NPK fertilizer*, and *Kulika biopesticide,* respectively.

Figure 5.
Urban farming innovation through the recycling of residues and waste.

Figure 6.
Sheep fed with corn and sweet potato residues and cattle fed with sweet potato residues.

Figure 7.
Organic fertilizer.

Figure 8.
Organic fertilizer.

DOI: http://dx.doi.org/10.5772/intechopen.1005854

Figure 9.
Biopesticide.

4. Discussion

Organic production in Uganda is a well-organized system that registers 139,191 agricultural producers who harness 16,376 ha of land in the country for organic production [20]. Organic farming necessitates the use of organic fertilizers and biopesticides. Despite the existence of organic certification institutions, organic farmers continue to face difficulties in obtaining organic fertilizers and biopesticides. However, organic farmers use internal resources to promote organic production, such as recycling residues and waste to fertilize the soil and/or sourcing organic fertilizers and biopesticides from small-to-medium enterprises. Some authors point out that animal-based organic waste (manure), compost (plant sources and food waste), and urban waste (sewage sludge and household waste) are all types of organic waste that can be recycled as agricultural fertilizer [21]. The results revealed that organic farmers use sheep, cattle, and pig manure as organic fertilizers. Manures are known to contain beneficial nutrients such as nitrogen and phosphorus, which are essential components of fertilizers for good crop quality and high productivity yields [22]. However, applying sheep manure helps to phytostabilize soils, which is beneficial to plant growth and strengthens plants' ability to manage pollutants in the soil [22]. The conversion of manure to energy fertilizers has minimal environmental impact and allows for plant nutrient recycling efficiency [23, 24]. Recycling plant sources and housing food waste for compost or fertilizer, the results highlight the vermicompost from the coffee husk and maize corn, Organic fertilizer NKP from wood ash, Indigenous Micro-Organisms, etc. in the area study. However, composting is a sustainable technique for converting organic and biodegradable food waste into a stable form of organic matter and fertilizers that can be used as soil amendments in agriculture [25]. Vermicompost has an excellent structure, aeration, porosity, and increased moisture-holding capacity to promote plant growth [26–28]. Vermicompost improves soil structure and nutrient availability by increasing vegetative growth and yield while having no negative effects on human or environmental health [29], reducing pathogens, and providing microbiological quality [30]. Reusing residues and waste as fertilizer and biopesticide in farms reduces input production costs while also promoting a circular economy. Recycle-reduce-reuse allows used resources and materials to remain in the production and consumption processes for a long time until they are physically degraded [3]. Recycling-reduce-reuse is sometimes referred to as a circular process that integrates materials and economies. A circular economy prioritizes resource utilization, particularly waste recycling, throughout the value chain, thereby reducing environmental pressures and reliance on nonrenewable resources [31, 32]. Organic farmers, on the other hand, recycle plant source residues and food waste, as well as reuse animal organic manure, to fertilize soil, regenerate soil's natural properties, and conserve agrobiodiversity through eco-innovation. Eco-innovation encompasses the production, process of production, and assimilation, emphasizing the life cycle of resources and materials used as a result of reducing environmental pollution, risk, and other negative impacts [1]. Therefore, product eco-innovations are the most marketable type of innovation, but to be a tool for implementing circular economy concepts, they must be designed to be fully reusable through repairing, maintaining, remanufacturing, recycling, and cascading the use of components and materials [33]. Organic farmers in Uganda, through eco-innovation and circular economy, develop an agribusiness that focuses on the production of organic fertilizer and biopesticides. The results revealed in the local market the existence of some organic fertilizers and biopesticides, such as "*Worders foliar*" fertilizer, "*Jimusa foliar NPK*" fertilizer, and "*Kulika*" biopesticide. It is highlighted that endogenous knowledge underpins the majority of biopesticides and organic fertilizers produced by the organic farmers surveyed. Rising demand for organic products has created domestic and international agribusiness opportunities in the food market (over 106 billion euros in 2019), which has influenced many farmers to shift to organic farming across countries [34]. Organic farmers who eco-innovate by creating small- and medium-sized enterprises that produce organic fertilizers and biopesticides revealed the lack of support from scientific researchers in organic production in creating standards using (quantity per area size) biopesticides and organic fertilizers through experimentation. They point out the high cost of some laboratory analyses, which they cannot support through business.

5. Conclusion

The chapter revealed that organic fertilizers and biopesticides are essential for eco-innovation and recycling residues and waste, as well as the development of agribusiness and the circular economy in organic agriculture. Organic fertilizers and biopesticides are pointed out as inputs for regenerating microorganisms in soils that participate in the conservation of agrobiodiversity and the preservation of sustainable ecosystems. Through an integrated system, organic farmers combine agricultural production, agroforestry, and livestock, then recycle the residues and waste to produce organic fertilizers and biopesticides. The reuse of organic residues and waste contributes to a circular economy, which reduces input costs and boosts organic farmers' incomes. However, organic farmers eco-innovate by using local recycling materials to make fertilizers and biopesticides, which represent potential agribusiness. Local demand for organic fertilizers and biopesticides revealed opportunities for adding value, creating jobs, and increasing wealth along the organic production value chain. Based on endogenous knowledge and local training, organic farmers strive to create small- and medium-sized enterprises for producing organic fertilizer and biopesticides. Without scientific proof, the certification of organic fertilizers and biopesticides remains a challenge, even for organic farmers known as "innovators." The certification and the manufacturing of organic fertilizers and biopesticides are critical for developing standards for organic inputs in order to reverse the conventional agriculture trend because endogenous knowledge underpins the majority of biopesticides and organic fertilizers produced by organic farmers. However, we suggested that:

- Scientific researchers should take advantage of this opportunity to test biopesticides and organic fertilizers, developing standards (quantity per area size);
- The government should promote biopesticides and organic fertilizers that protect biodiversity while also facilitating local economies and employment for agricultural independence.
- NGOs and other food system stakeholders should focus their efforts on strengthening business partnerships in organic systems production to ensure healthy and safe food for the population.

Conflict of interest

The authors declare no conflict of interest.

Author details

Bienvenu A. Dagoudo[1,3]*, Charles Ssekyewa[2] and Silvère D. Tovignan[3]

1 Africa Center for Agroecology and Livelihood Systems, Faculty of Agriculture, Uganda Martyrs University, Kampala, Uganda

2 St. Lawrence University, Kampala, Uganda

3 Faculty of Agronomy, Department of Rural Economics and Sociology, University of Parakou, Parakou, Benin

*Address all correspondence to: akowedahobienvenu@gmail.com

References

[1] Kemp R, Pearson P. Final report MEI project about measuring eco-innovation. UM Merit, Maastricht. 2007;**10**:1-120. Available from: https://www.oecd.org/environment/consumption-innovation/43960830.pdf

[2] Rennings K. Redefining innovation—Eco-innovation research and the contribution from ecological economics. Ecological Economics. 2000;**32**:319-332. Available from: https://www.sciencedirect.com/science/article/pii/S0921800999001123

[3] Shih D-H, Lu C-M, Lee C-H, Cai S-Y, Wu K-J, Tseng M-L. Eco-innovation in circular agri-business. Sustainability. 2018;**10**:1140. Available from: https://www.mdpi.com/2071-1050/10/4/1140

[4] Xavier AF, Naveiro RM, Aoussat A, Reyes T. Systematic literature review of eco-innovation models: Opportunities and recommendations for future research. Journal of Cleaner Production. 2017;**149**:1278-1302. Available from: https://www.sciencedirect.com/science/article/pii/S0959652617303712

[5] Fidelis M, de Moura C, Kabbas Junior T, Pap N, Mattila P, Mäkinen S, et al. Fruit seeds as sources of bioactive compounds: Sustainable production of high value-added ingredients from by-products within circular economy. Molecules. 2019;**24**:3854. Available from: https://www.mdpi.com/1420-3049/24/21/3854

[6] Sachs JD, Schmidt-Traub G, Mazzucato M, Messner D, Nakicenovic N, Rockström J. Six transformations to achieve the sustainable development goals. Nature Sustainability. 2019;**2**:805-814. Available from: https://www.nature.com/articles/s41893-019-0352-9

[7] Geissdoerfer M, Savaget P, Bocken NM, Hultink EJ. The circular economy–A new sustainability paradigm? Journal of Cleaner Production. 2017;**143**:757-768. Available from: https://www.sciencedirect.com/science/article/pii/S0959652616321023

[8] Murray A, Skene K, Haynes K. The circular economy: An interdisciplinary exploration of the concept and application in a global context. Journal of Business Ethics. 2017;**140**:369-380. Available from: http://link.springer.com/10.1007/s10551-015-2693-2

[9] Giovannucci D, Ponte S. Standards as a new form of social contract? Sustainability initiatives in the coffee industry. Food Policy. 2005;**30**:284-301. Available from: https://www.sciencedirect.com/science/article/pii/S030691920500031X

[10] Pretty J. Agricultural sustainability: Concepts, principles and evidence. Philosophical Transactions of the Royal Society B: Biological Sciences. 2008;**363**:447-465

[11] Scofield AM. Organic farming—The origin of the name. Biological Agriculture and Horticulture. 1986;**4**(1):1-5. Available from: https://www.tandfonline.com/doi/abs/10.1080/01448765.1986.9754481

[12] Lampkin NH. Organic farming: Sustainable agriculture in practice. In: Lampkin NH, Padel S, editors. The Economics of Organic Farming–An International Perspective. Wallingford, UK: CAB; 1994. pp. 3-9

[13] Alfa MI, Adie DB, Igboro SB, Oranusi US, Dahunsi SO, Akali DM. Assessment of biofertilizer quality

and health implications of anaerobic digestion effluent of cow dung and chicken droppings. Renewable Energy. 2014;**63**:681-686. Available from: https://linkinghub.elsevier.com/retrieve/pii/S0960148113005284

[14] Silva WO, Stamford NP, Silva EVN, Santos CERS, Freitas ADS, Silva MV. The impact of biofertilizers with diazotrophic bacteria and fungi chitosan on melon characteristics and nutrient uptake as an alternative for conventional fertilizers. Scientia Horticulturae. 2016;**209**:236-240. Available from: https://www.sciencedirect.com/science/article/pii/S0304423816303028

[15] Guo Z, Zhang J, Fan J, Yang X, Yi Y, Han X, et al. Does animal manure application improve soil aggregation? Insights from nine long-term fertilization experiments. Science of the Total Environment. 2019;**660**:1029-1037. Available from: https://www.sciencedirect.com/science/article/pii/S0048969719300579

[16] Linder M, Williander M. Circular business model innovation: Inherent uncertainties. Business Strategy and the Environment. 2017;**26**:182-196. Available from: https://onlinelibrary.wiley.com/doi/10.1002/bse.1906

[17] Liu JY-S. Circular economy and environmental efficiency–the case of traditional Hakka living system. Procedia-Social and Behavioral Sciences. 2012;**57**:255-260. Available from: https://www.sciencedirect.com/science/article/pii/S1877042812046459

[18] Majaliwa JGM, Tenywa MM, Bamanya D, Majugu W, Isabirye P, Nandozi C, et al. Characterization of historical seasonal and annual rainfall and temperature trends in selected climatological homogenous rainfall zones of Uganda. Global Journal of Science Research. 2015;**15**:21-40. Available from: https://oar.icrisat.org/id/eprint/10818

[19] UBOS. Uganda Annual Agricultural Survey 2018. 2020. Available from: https://www.ubos.org/explore-statistics/20/

[20] IFOAM. organic-world.net—The World of Organic Agriculture 2022—Download PDF. 2022. Available from: https://www.organic-world.net/yearbook/yearbook-2022/yearbook-2022-download-pdf.html

[21] Uysal A, Demir S, Sayilgan E, Eraslan F, Kucukyumuk Z. Optimization of struvite fertilizer formation from baker's yeast wastewater: Growth and nutrition of maize and tomato plants. Environmental Science and Pollution Research. 2014;**21**:3264-3274. Available from: http://link.springer.com/10.1007/s11356-013-2285-6

[22] Chew KW, Chia SR, Yen H-W, Nomanbhay S, Ho Y-C, Show PL. Transformation of biomass waste into sustainable organic fertilizers. Sustainability. 2019;**11**:2266. Available from: https://www.mdpi.com/2071-1050/11/8/2266

[23] Roubík H, Mazancová J, Banout J. Current approach to manure management for small-scale southeast Asian farmers-using Vietnamese biogas and non-biogas farms as an example. Renewable Energy. 2018;**115**:362-370. Available from: https://www.sciencedirect.com/science/article/pii/S0960148117308261

[24] Sharara MA, Sadaka SS. Opportunities and barriers to bioenergy conversion techniques and their potential implementation on swine manure. Energies. 2018;**11**:957. Available from: https://www.mdpi.com/1996-1073/11/4/957

[25] Wolka K, Melaku B. Exploring selected plant nutrient in compost prepared from food waste and cattle manure and its effect on soil properties and maize yield at Wondo Genet, Ethiopia. Environmental Systems Research. 2015;**4**:9. Available from: http://www.environmentalsystemsresearch.com/content/4/1/9

[26] Yadav A, Garg VK. Biotransformation of bakery industry sludge into valuable product using vermicomposting. Bioresource Technology. 2019;**274**:512-517. Available from: https://www.sciencedirect.com/science/article/pii/S0960852418316675

[27] Mahaly M, Senthilkumar AK, Arumugam S, Kaliyaperumal C, Karupannan N. Vermicomposting of distillery sludge waste with tea leaf residues. Sustainable Environment Research. 2018;**28**:223-227. Available from: https://www.sciencedirect.com/science/article/pii/S2468203917302030

[28] Soobhany N. Remediation potential of metalliferous soil by using extracts of composts and vermicomposts from municipal solid waste. Process Safety and Environmental Protection. 2018;**118**:285-295. Available from: https://www.sciencedirect.com/science/article/pii/S0957582018302830

[29] Ayyobi H, Olfati J-A, Peyvast G-A. The effects of cow manure vermicompost and municipal solid waste compost on peppermint (*Mentha piperita L.*) in Torbat-e-Jam and Rasht regions of Iran. International Journal of Recycling of Organic Waste in Agriculture. 2014;**3**:147-153. Available from: https://link.springer.com/article/10.1007/s40093-014-0077-8

[30] Karimi H, Mokhtari M, Salehi F, Sojoudi S, Ebrahimi A. Changes in microbial pathogen dynamics during vermicomposting mixture of cow manure–organic solid waste and cow manure–sewage sludge. The International Journal of Recycling of Organic Waste in Agriculture. 2017;**6**:57-61. Available from: http://link.springer.com/10.1007/s40093-016-0152-4

[31] Awasthi SK, Kumar M, Sarsaiya S, Ahluwalia V, Chen H, Kaur G, et al. Multi-criteria research lines on livestock manure biorefinery development towards a circular economy: From the perspective of a life cycle assessment and business models strategies. Journal of Cleaner Production. 2022;**341**:130862. Available from: https://www.sciencedirect.com/science/article/pii/S0959652622005005

[32] Ugwu SN, Harding K, Enweremadu CC. Comparative life cycle assessment of enhanced anaerobic digestion of agro-industrial waste for biogas production. Journal of Cleaner Production. 2022;**345**:131178. Available from: https://www.sciencedirect.com/science/article/pii/S0959652622008095

[33] Pichlak M, Szromek AR. Linking eco-innovation and circular economy—A conceptual approach. Journal of Open Innovation: Technology, Market, and Complexity. 2022;**8**:121. Available from: https://www.sciencedirect.com/science/article/pii/S2199853122007223

[34] Willer H, Trávníček J, Meier C, Schlatter B. The World of Organic Agriculture 2021-Statistics and Emerging Trends. 2021. Available from: https://orgprints.org/40014/

Chapter 4

Evaluating Land Management Options to Optimize Soil Organic Matter Retention in Oxisols: Case Studies Involving Brazil and Hawaii

Michael Aide

Abstract

Oxisols are extremely weathered soils developed largely across equatorial regions. Emerging agricultural land management techniques address soil degradation caused by conventional tillage, nutrient neglect, and other agronomic issues. In Brazil, no-till systems support the maintenance of an epipedon residue protective layer that improves soil organic matter contents, strengthens soil structures, fosters nutrient provision, and the development of a robust soil biology community. Growing awareness by Brazilian producers focuses on soil enrichment to enhance crop productivity but also to support ecosystem services and other facets associated with soil health. In Hawaii, the nearly total decline of sugarcane and pineapple production has left large acreages of land that have suffered long-term soil erosion and nutrient impoverishment. The Hawaiian agricultural community is attempting to support the state's food security, primarily by creating small-acreage farms that produce an amazing diversity of foods and center soil sustainability around utilizing locally proven soil health techniques. Both entities have community-based extension services, compelling research components, and producer-oriented digital tool offerings.

Keywords: soil carbon sequestration, ferralsols, oxic horizons, no-till, soil structure

1. Introduction

1.1 The oxisol soil profile and its diagnostic soil horizons

The complex interaction of a warm and moist climate, living organisms, earth materials, landscape position, and protracted time intervals produces intensively weathered soils, where only the most recalcitrant minerals persist and soil fertility is perceived as low. The "Keys of Soil Taxonomy" is the official soil taxonomy classification for the United States Department of Agriculture, and oxisols are the soil order representing intensively weathered soils [1]. Similarly, the Reference Soil Group, supported by the Food and Agriculture Organization of the United Nations, provides ferralsols as deeply weathered, red or yellow soils of the humid tropics as a comparable classification for oxisols. However, some ferralsols have kandic horizons that are not associated with oxisols. Most oxisols are in Africa (Democratic Republic of the Congo, Central African Republic, Angola, Guinea, Madagascar, and adjacent nations) and South America (Brazil, especially in the Cerrado). Most oxisols are located between the Tropic of Cancer and the Tropic of Capricorn, with climate providing isothermic and isohyperthermic soil temperature regimes and udic to ustic soil moisture regimes. However, oxisols are also in portions of Central America, China, Puerto Rico, Hawaiian Islands, Indonesia, Colombia, multiple nations across Southeast Asia, and other locations.

The typical oxisol soil profile reflects the location, climate, ancestral and current vegetation, topographic position, and other factors; however, generalized statements describing a typical soil profile are appropriate. Oxisols are soils having a sandy loam or finer texture with less than 10% weatherable minerals in the 50–200 μm range and a low apparent cation exchange capacity [1, 2]. Oxisols typically feature an oxic horizon within 0.150 m of the mineral soil surface and do not have a kandic horizon within that depth range. Oxic horizons (Bo) are subsurface horizons (unless exposed because of recent erosion) with chemical and mineralogical properties reflective of advanced weathering [1]. The following additional characteristics include: (i) sandy loam or a finer texture, (ii) thickness (30 cm or more), (ii) less than 10% weatherable minerals in the 0.05–0.2 mm fraction, (iii) typically the volume rock structure is less than 5%, (iv) a limited clay content increase at the upper boundary transition, and (v) an apparent cation exchange capacity of 16 cmol(+) or less kg-clay^{-1} [1]. In the near surface horizons, oxisols typically have a fine or very fine granular (spherical) structure and exhibit subangular to prismatic structures in deeper soil horizons. Macropores are common, supporting a rapid saturated hydraulic conductivity and a reasonably low soil bulk density (1.0–1.3 g cm^{-3}). Available water contents typically range from 0.05 to 0.15 cm^{-1} [2].

1.2 Oxisol genesis and pertinent soil profile forming processes

Pedogenic processes critical to the genesis of oxisols include: (i) mineral desilication to concentrate Al- and Fe-oxyhydroxides and phyllosilicates (kaolinite), (ii) melanization and humification, (iii) bioturbation, principally Anthropoda, and (iv) in some locations with fluctuating water tables, the formation of plinthite [1–3]. The silicon loss rate is largely dependent on the water residence time, the mineralogy of the silicate minerals, and the exposed surface areas. Plinthite is an iron-rich, humus-poor mixture of clay with quartz, and other minerals, and it commonly occurs as discrete dark red redox-derived minerals [2, 3]. Thus, regions with diverse parent materials (earth materials) may extensively weather to provide a clay mineralogy that is strikingly similar.

Desilication reactions are examples of incongruent weathering where the chemical reaction results in the synthesis of secondary minerals. Examples of desilication reactions are as follows:

i. Potassium feldspar ($KAlSi_3O_8$) plus water (H_2O) altering to K^+, montmorillonite ($(Na,Ca)_{0.3}(Al,Mg)_2Si_4O_{10}(OH)_2{\bullet}n(H_2O)$), and silicic acid ($H_4SiO_4$),

ii. Montmorillonite and water altering to kaolinite ($Al_2Si_2O_5(OH)_4$) and silicic acid,

iii. Kaolinite and water altering to gibbsite (Al_2O_3) and silicic acid.

Le Châtelier's principle states that if a dynamic equilibrium is disturbed by changing the conditions, the position of equilibrium shifts to counteract the change to reestablish an equilibrium that is, adding water as a reactant and the removal of silicic acid by leaching shifts the equilibrium to produce weathered minerals. These environmental conditions are pervasive across subtropical and tropical ecosystems.

Melanization is soil darkening because of the accumulation of soil organic matter, and humification is the synthesis of humus because of the abiotic and biotic reactions acting on decomposing particulate plant and animal materials. There is an emerging paradigm in soil science that soil structure formation and clay/oxyhydroxide organic material sorption provide a protective barrier that partially isolates microbial activity and inhibits microbial soil organic matter decomposition. The classical viewpoint is that soil organic matter accumulation occurs because of an increase of recalcitrant organic materials. Perhaps both pedogenic pathways are operative. Bioturbations are soil horizon disturbances caused by living organisms that promote isotropy, causing the various parts of the soil to become more similar (haploidization). In oxisols, bioturbation is commonly associated with soil inhabiting insects [2].

Key pedogenic soil pathways important to improving soil carbon abundance include: (i) immobilization and mineralization, (ii) carbon adsorption onto phyllosilicates, and (iii) soil structure formation and its maintenance. The intensity of these pathways is influenced by land management, soil hydrology and available water capacity, pH, oxidation-reduction status, soil biology, including microbial activity, and other soil and environmental factors.

1.3 Purpose

This manuscript is a review of oxisols and land management options to effectively maintain or increase soil organic carbon. This manuscript places an emphasis on two regions: Brazil and Hawaii. Brazil is an immense nation with an expanding agricultural base, a dedicated agriculture research infrastructure, an agriculture extension and education outreach program, and a history of research focusing on tillage and residue management changes to support soil carbon accumulation. Conversely, Hawaii is a collection of seamounts in the Pacific Ocean that have an agricultural history of extensive pineapple (*Ananas comosus*) and sugarcane (*Saccharum officinarum*) plantations that has given way to an assortment of small-acreage landholdings that provide food for the local economy. Hawaii, like Brazil, has dedicated agriculture research infrastructure, an agriculture extension, and education outreach program, that support soil health attainment, including soil carbon sequestration.

2. Forest net primary productivity and soil carbon responses

To assess oxisol genesis and delineate the importance of key pedogenic soil-forming processes, consideration of the net primary productivity of an ecosystem is integral to the discussion. The establishment of greater soil carbon concentrations may be estimated as:

Soil carbon content at time t = Soil carbon content at initial time t = 0 + Carbon entering the system—carbon exiting the system, where time t is the duration interval.

Estimates for carbon entering the soil system includes consideration of root and residue/litter incorporation rates. Estimation of the soil carbon loss rates frequently include soil respiration measurements and analysis of selected soil indicators to indicate the intensity of soil pedogenic processes involving the carbon cycle. Soil management that fosters soil structure formation or at least optimizes the preservation of soil structure limits soil carbon loss rates [4–11]. Similarly, soil organic carbon adsorption involving Al- and Fe-oxyhydroxides and phyllosilicates limits soil carbon loss rates [8–11]. Concentrating on Rhodic Haplustox and Anionic Acrustox soils, Pessoa and Libardi [12] demonstrated that iron and aluminum oxides, kaolinite, and the biological activity collectively acted to improve microaggregate formation and subsequently the soil's hydraulic conductivity.

Schrumpf et al. [4] noted that soil organic matter stability was a complex function of plant residue abundance, soil aggregate carbon occlusion, organo-mineral complexes with Fe- and Al-oxyhydroxides, and soil profile landscape position. Schmidt et al. [5] noted that soil organic matter accumulation and persistence is a function of (i) root type and root biomass, (ii) physical separation from microbial activity, (iii) very long turnover rates involving deep soil profile carbon, (iv) microbial activities, and (v) stabilized multi-tiered trophic levels. Kumar et al. [6] discussed the capability of plant roots and residues to develop micro- and macroaggregates, which subsequently sequester soil organic carbon. Witzgall et al. [7] conducted an incubation study involving litter amendments to coarse-textured and fine-textured soils. Microbial activity and fungal growth were supported to a greater extent in the coarse-textured soils, and soil organic matter persistence was attributed to particulate soil organic matter occlusion in soil aggregates, and by mineral surface adsorption. Evaluating Brazilian oxisols in a series of microcosm experiments, Souza et al. [10] documented that litter-carbon retention increased with soil depth and litter-nitrogen retention decreased with increasing soil depth. Noncrystalline and crystalline Al- and Fe-oxyhydroxides strongly correlated with the retention of isotopically labeled carbon and nitrogen in litter amendments, thus influencing the neoformation of mineral-organic associations.

Schweizer et al. [13] reported that finer soil textures correlated with particulate soil organic matter preservation. Lal [14] noted that carbon emissions are predicated on microbial decomposition of forest litter and soil mineralization rates. Net primary productivity strategies to increase soil carbon stocks include (i) support forest regrowth activities, (ii) replace conventional tillage options with conservation tillage options, including no-till options, (iii) include cover crops as a portion of the rotation, (iv) implement emerging nutrient management programs, including manure and sludge application, (v) improve grazing programs, (vi) encourage water conservation and (v) integrate efficient and sustainable irrigation practices. In China, Liu et al. [15] investigated soil organic matter accumulation in mixed and Larch (Larix) forests. Soil organic carbon in the organic near surface horizons was preferentially preserved in macroaggregates, whereas soil organic matter in deeper soil horizons was preferentially preserved in microaggregates. Subsequently, Liu et al. [16] observed mixed forests and Larch plantation forests and documented that for both forest types, aggregate formation and mineral association contributed to soil organic carbon storage and considerable carbon storage, was associated deep within the soil profile. Deng et al. [17] reviewed the influence of land use on soil carbon stocks from numerous forest types and many nations and documented that soil organic carbon stocks were generally decreased by conversion from forest to cropland; however, some positional carbon sequestration was reported for conversions from farmland to grassland and forest to grassland.

In a review of agroforestry, Shi et al. [18] reported that soil carbon stocks after agroforestry establishment were greater than associated croplands or pastures.

Utilizing the Nation Forest Inventory in China, Wang et al. [19] estimated forest soil carbon from 1999 to 2008. The results showed that carbon sequestration was 78 Tg carbon $yr.^{-1}$ and the typical soil carbon sequestration values were 49.9 g C m^{-2} $yr.^{-1}$; however, large spatial differences occurred across China. Kleber et al. [20] described soil processes involving soil organic matter transformations. Soil structures do create distinctive microsites that support diverse microbial communities; however, this complexity involving soil organic carbon adsorption by mineral surfaces may create a multitude of soil reactions related to soil organic matter microbial oxidation, each with a different carbon management outcome.

3. Potential of soil carbon sequestration in oxisols

For a specific soil order, the soil carbon saturation deficit is equal to an estimate of the maximum soil organic carbon content, also termed the saturated soil organic matter capacity, minus the actual soil organic matter content. Estimates for the maximum soil carbon content are not quantitively available because of the complexity of soil dynamics. Soil organic carbon contents are influenced by (i) the soil physical and chemical properties (soil drainage, soil permeability, pH, oxidation-reduction episodes, the soil's thermal and moisture regime, texture, soil structure, soil classification, clay mineralogy, and other features), (ii) the soil biology and established vegetation, (iii) the influence of climate, and (iv) human land management.

Cerri et al. [21] reported literature citations involving soil carbon accumulation rates for various Brazilian ecosystems, ranging from 0.6 to 5.9 Mg C ha^{-1} $yr.^{-1}$. In Brazil, Gomes et al. [22] estimated that Luvisols (Alfisols) had the least soil organic carbon density (6.45 kg m^{-2}), whereas histosols had the greatest soil organic carbon density (14.87 kg m^{-2}). Deep and nutrient poor ferralsols and acrisols contain approximately 57% of the total soil organic carbon. Across Brazil, these soils store approximately 71.3 PgC within the top 1 m, where the upper 0.3 m of soil contains approximately 36 PgC. Cook-Patton et al. [23] inferred that soil carbon accumulation has significant potential, whereas Harris et al. [24] concluded global forests were a carbon sink. Zhao et al. [25] supported the premise that increased soil temperatures encourage soil organic matter decomposition; however, the improved net primary productivity would not offset the carbon loss. In their review, Cerri et al. [21] discussed the CO_2 fertilization effect, noting that increased atmospheric CO_2 may support soil carbon accumulation; however, carbon input to oxisols in tropical rainforests is a complex function of the floristic composition, plant biomass, and soil fertility and predictions of carbon sequestration are not yet fully understood.

Schaefer et al. [26] employed soil and vegetation databases from Western Amazonia to indicate that substantial whole soil carbon stocks were associated with high nutrient status, high clay activity soils, and deeply weathered latosols (oxisols), and (ii) reduced whole soil carbon stocks were associated with coarse-textured soils. Cerri et al. [21] reported literature concerning the Brazilian Amazon estimating the potential for carbon sequestration. These authors separated the literature into the following land management classifications: (i) primary forest at 344 million ha, (ii) primary forest converted to well-managed pasture at 5 million ha, (iii) degraded to well-managed pasture at 13 million ha, (iv) degraded pasture to secondary forest at 13 million ha, (v) existing secondary forest at 18 million ha, and (vi) degraded pasture to agroforestry at 13 million ha. The mean rates of carbon sequestration are (i) primary forest at 1 Mg C ha^{-1} $yr.^{-1}$, (ii) primary forest converted to well-managed pasture at 0.27 Mg C ha^{-1} $yr.^{-1}$, (iii) degraded to well-managed pasture at 0.27 Mg C ha^{-1} $yr.^{-1}$, (iv) degraded pasture to secondary forest at 4.0 Mg C ha^{-1} $yr.^{-1}$, (v) existing secondary forest at 4.0 Mg C ha^{-1} $yr.^{-1}$, and (vi) degraded pasture to agroforestry at 2.7 Mg C ha^{-1} $yr.^{-1}$.

Lehmann et al. [27] strongly supported the premise that soil organic carbon management has the potential to aid climate change mitigation through drawdown of atmospheric carbon dioxide. Soil carbon enrichment management must justify for the soil processes influencing carbon sequestration and emission with changing soil conditions. These authors proposed that the microbial community capabilities for soil organic matter decomposition alters in response to soil conditions that in turn, if appropriately understood, may predict sequestration verses emission.

4. Oxisols in Hawaii

In Hawaii, oxisols occupy mountainous slopes where mafic and ultramafic rocks have undergone intensive and prolonged weathering, especially on the geologically older islands. The local geology consists of volcanic domes formed by bedded flows of pahoehoe and Aa lava. Most of the mafic material is basaltic with some inclusions of andesitic rocks and ash/cinder cones. The climate is dependent on the island's topographic influences. Open ocean precipitation measures 0.76–1.02 m $yr.^{-1}$; however, seasonal orographic rainfall patterns may depart dramatically from the open marine rainfall, creating semi-arid to lush tropical vegetation zones within an island [28]. For the majority of the Hawaiian oxisol sites, the vegetation was irrigated sugarcane and dryland pineapple; however, most of the field sites have reverted to local vegetation, including: guava (*Psidium guajava*), koa haole (*Leucaena glauca*), lantana (*Lantana camara*); joee (*Stachytarpheta jamaicensis*), Japanese tea (*Cassia leschenaultiana*), bermudagrass (*cynodon dactylon*), honohono (*Commelina diffusa*), kiawe (*Prosopis pallida*), pitted beardgrass (*Bothriochloa barbinodis perforatus*), feather fingergrass (*Chloris virgata*), ilima (*Sida cordifolia*), and buffelgrass (*Cenchrus ciliaris*), cactus (*Opuntia megacantha*), and fingergrass (*Chloris sp.*) [29]. Unlike Brazil, Hawaii is a much smaller land mass and the traditional agriculture of sugarcane and pineapple plantations have relocated elsewhere. Thus, a considerable portion of this case study involves small family-operated landholdings developed to support the local food economy.

As examples, three Hawaiian oxisol soils are presented, (i) the Lahaina soil series, (ii) the Molokai soil series, and (iii) the Wahiawa soil series [29, 30]. The Lahaina series (very-fine, kaolinitic, isohyperthermic Rhodic Eutrustox) consists of very deep, well-drained soils that formed in basaltic residuum on low elevation, mountain slopes of 0–40%. The Lahaina soil profile has an Ap – Bo – BC horizon sequence featuring an ochric epipedon (Ap1, Ap2) and an oxic horizon (Bo1, Bo2, Bo3) on the Hawaiian Islands of Lanai, Maui, Molokai, and Oahu. The typical pH is moderately acid in the near surface and slightly acid in the deeper soil horizons. The clay content is approximately 50% from 0 to 0.8 m and 35% in the deeper horizons [30]. The soil organic matter content ranges from near 1.4% in the upper 0.2 m to less than 0.3% in the deeper soil horizons. The available water content ranges from 0.11 to 0.12 cm^{-1} throughout the soil profile. The cation exchange capacity is 12 cmol $kg\text{-}soil^{-1}$ for 0–0.2 m and 10 cmol $kg\text{-}soil^{-1}$ for 0.2–1.4 m. The soil structure transitions from weak fine granular structure in the Ap horizons to moderate, medium, and course subangular block structures in the deeper soil horizons. The saturated hydraulic conductivity

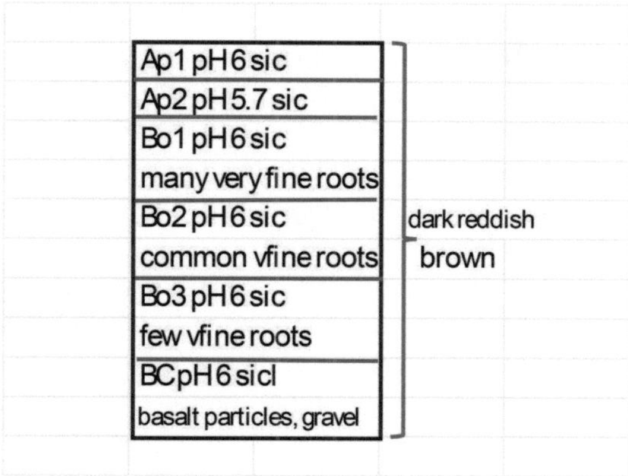

Figure 1.
Soil profile sketch of the Lahaina soil series. Vfine is very fine, sic is silty clay, sicl is silty clay loam.

is approximately 90 mm $hr.^{-1}$ for the 0–0.4 m depth, whereas the saturated hydraulic conductivity is approximately 15 mm $hr.^{-1}$ for the 0.4 to 1.5 m depth [30]. A soil profile illustration is in **Figures 1** and **2**.

The Molokai series (very-fine, kaolinitic, isohyperthermic Typic Eutrotorrox) consists of very deep, well-drained soils that formed in basaltic residuum on low elevation, mountain slopes of 0–25%. The Molokai soil profile has an Ap – Bto horizon sequence featuring an umbric epipedon (Ap1, Ap2) and an oxic horizon (Bto1, Bto2, Bto3) having patchy clay films on the Hawaiian Islands of Molokai, Lanai, Maui, and Oahu. The pH is very strongly acid in the Ap horizon, transitioning to slightly acid and neutral in the Bto horizons. The epipedon's weak, very fine to fine granular structure changes too weak to strong, medium to coarse, and subangular structures in the deeper soil profile. Mean annual rainfall is about 0.635 meter, and mean annual temperature is about 23°C [29, 30].

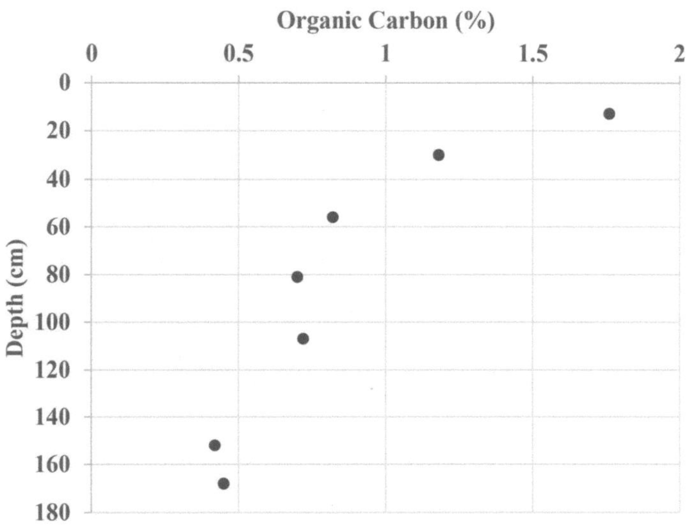

Figure 2.
Organic carbon (%) in a pedon of the Lahaina soil series. Source: Ref. [29].

The soil organic matter content ranges from near 2.5% in the upper 0.45 m to less than 0.2% in the deeper soil horizons. The available water content is approximately 0.1 cm^{-1} from 0 to 0.4 m and approximately 0.12 cm^{-1} from 0.4 to 1.8 m. The clay content is near 38% and the cation exchange capacity is 10–12 cmol kg-$soil^{-1}$. The saturated hydraulic conductivity is approximately 32 mm $hr.^{-1}$ for the 0–0.4 m depth, whereas the saturated hydraulic conductivity is approximately less than 8 mm $hr.^{-1}$ for the 0.4–1.8 m depth [30].

The Wahiawa series (very-fine, kaolinitic, isohyperthermic Rhodic Haplustox) consists of very deep, well-drained soils that formed in basaltic residuum and alluvium. Wahiawa soils are on uplands and have slopes of 0–25%. The Wahiawa soil profile has an Ap-Bo horizon sequence featuring an mollic epipedon (Ap1, Ap2) and an oxic horizon (Bto1, Bto2, Bto3). The pH is moderately acidic in the epipedon, which gradually changes to neutral in the deeper soil horizons. The moderately very fine and fine granular structures in the Ap horizons convert to moderate, medium to coarse subangular blocky structures in the deeper Bo horizons. This soil series is located on Oahu, Hawaii. The available water content is rather uniform throughout the soil profile at 0.12 cm^{-1}. The saturated hydraulic conductivity is approximately 33 mm $hr.^{-1}$ from 0 to 0.3 m and dramatically decreases to 25 mm $hr.^{-1}$ from 0.3 to 1.5 m. The soil organic matter content is approximately 3.5% in the upper 0.3 m, whereas at deeper soil horizons the soil organic matter content is less than 1%. The cation exchange capacity is 12 cmol kg-$soil^{-1}$ from 0 to 0.3 m and is 8 cmol kg-$soil^{-1}$ for deeper soil horizons. The pH is approximately 5.7 in the Ap horizons and is approximately 6.5 in the deeper soil profile. Clay content is approximately 60% in the upper 0.2 m, whereas deeper soil horizons have clay contents from 80 to 90% [29, 30].

For reference, 15 Hawaiian oxisol soils are presented along with their taxonomic classification (**Table 1**) [31]. To provide a measure of distinctiveness, the taxonomic

Soil series	Taxonomic classification
Halii	Fine, ferruginous, isothermic Anionic Acroperox
Helemano	Very-fine, kaolinitic, isohyperthermic Rhodic Eutrustox
Holomua	Fine, kaolinitic, isohyperthermic Typic Eutrotorrox
Kapaa	Very-fine, sesquic, isohyperthermic Anionic Acrudox
Kunuweia	Clayey-skeletal, ferritic, isothermic Typic Hapludox
Lahaina	Very-fine, kaolinitic, isohyperthermic Rhodic Eutrustox
Lihue	Very-fine, ferruginous, isohyperthermic Rhodic Eutrustox
Makapili	Very-fine, sesquic, isohyperthermic Anionic Acrudox
Molokai	Very-fine, kaolinitic, isohyperthermic Typic Eutrotorrox
Niu	Very-fine, kaolinitic, isohyperthermic Rhodic Eutrustox
Pooku	Very-fine, ferritic, isohyperthermic Anionic Acrudox
Puhi	Very-fine, ferruginous, isohyperthermic Humic Kandiustox
Uwala	Very-fine, kaolinitic, isohyperthermic Typic Haplotorrox
Wahiawa	Very-fine, kaolinitic, isohyperthermic Rhodic Haplustox
Waikapu	Fine, kaolinitic, isohyperthermic Typic Haplotorrox

Table 1.
Hawaiian oxisols and their classification.

classification may require a brief introduction. At the end of the classification, the suffix "ox" refers to the oxisol order. The suborder and great group indicators are: (i) Hapl- implies minimum horizonation, (ii) Perox implies an oxisol with a sombric horizon (low base saturation and accumulation of humus), (iii) Acro- implies extreme weathering, (iv) Torr- implies a torric moisture regime, (v) Ust- implies a ustic moisture regime, (vi) ud- implies a udic moisture regime, (vii) Kandic- implies an abundance of 1:1 layer phyllosilicates (kaolinite), (viii) Eutr-implies a high base saturation [2].

In Maui et al. [32] observed the storage of soil organic carbon and inferred soil carbon dynamics for soils having a 100-year history of sugarcane production. Increased negative surface charge, water-stable aggregation, organo-Fe complexation, and root inputs were instrumental factors associated with soil organic carbon protection. Across the Hawaiian Islands, Selmants et al. [33] provided carbon storage and flux baseline estimates and provided how carbon cycling and storage may respond to projected climate, land use, land cover, and disturbance changes. The Big Island (Hawaii Island), which is lacking oxisols, is projected to continue being an atmospheric carbon sink, largely because of the robust fertility of relatively young volcanic parent material.

Crow et al. [34] documented that investment is required to create data resources sufficient to employ decision support tools for Hawai'i's diverse soil expression and discrete communities to accurately assess mitigation, adaption, and resilience to climate change. Crow et al. [34] further noted that climate smart practices improve soil health, soil carbon sequestration and selected ecosystem services; however, site-specific knowledge needs to be cataloged for modeling tools integral to decision support and policy. The authors cataloged estimates of selected island's area (km^2) and soil organic carbon contents (Tg carbon): (i) Hawaii at 10,453 km^2 and 146.1 Tg carbon, (ii) Kauai at 1428 km^2 and 33.5 Tg carbon, (iii) Lanai at 365 km^2 and 2.7 Tg carbon, (iv) Maui at 1898 km^2 and 32.5 Tg carbon, (v) Molokai at 672 km^2 and 10.8 Tg carbon, (vi) Oahu at 1539 km^2 and 31.5 Tg carbon, and (vii) total at 16,470 km^2 and 257.9 Tg carbon. The authors discussed the newly established leading soil health indicators: (i) biological [24-hour CO_2 burst, ß-glucosidase, ß-glucosaminidase, potentially mineralizable nitrogen], (ii) chemical [pH, % organic carbon, DOC:DON ratio, hot water extractable carbon], and (iii) physical [water holding capacity, water stable mega-aggregates, bulk density].

In Hawaii, Bashkin and Binkley [35] measured soil carbon following afforestation of sugarcane fields with fast-growing Eucalyptus (*Eucalyptus globulus*). Compared to an abandoned sugarcane cane site (untreated check), the total soil carbon 10–13 years after afforestation was strikingly similar (113–114 Mg ha^{-1}). Eucalyptus establishment increased total soil C in the 0–0.10 m layer by 11.5 Mg ha^{-1}; however, the 0.10–0.6 m layer lost 10.1 Mg ha^{-1}. The net effect on soil carbon because of afforestation depended on both newly gained carbon in the epipedon and lost carbon in the deeper soil profile.

5. Policy and producer network options in Hawaii to support soil health and carbon sequestration

Soil carbon databases are becoming increasing utilized for planning. The global community may participate in the International Soil Carbon Network (ISCN). The International Soil Carbon Network is a "science-based network that facilitates data sharing, assembles databases, identifies gaps in data coverage, and enables spatially explicit assessments of soil carbon in context of landscape, climate, land use, and biotic variables. The network provides (i) scientific and logistical infrastructure for sharing knowledge, information, and data, (ii) shared data and recognition, and (iii) opportunities for synergistic interactions, (iv) products beneficial to stakeholders and scientists, (v) a framework for common scientific protocols and collaborative decision support tools" [36]. The International Soil Carbon Network was initially created by the United States Department of Agriculture; however, the organizing board currently reflects an international posture.

In Hawaii, the Crow Soil Ecology & Biogeochemistry Laboratory of the University of Hawaii Manoa "studies carbon dynamics and sequestration, greenhouse gas flux, nutrient, and water management in natural and managed (forested or agriculture) ecosystems and how these relate to global change pressures—such as land-use, climate change, and invasive species—sustainable agriculture, and renewable energy production." Research scientists quantify and model carbon tradeoffs to create resilience against climate change, energy reduction, and water accessibility [37]. This outreach program by the University of Hawaii Manoa supports soil testing, producer education and research dedicated to soil health.

Soil quality has been defined by the Soil Science Society of America as "the capacity of a specific kind of soil to function, within natural or managed ecosystem boundaries, to sustain plant and animal productivity, maintain or enhance water and air quality, and support human health and habitation." Currently, various definitions of soil health exist, including: (i) "the continued capacity of soil to function as a vital living ecosystem that sustains plants, animals, and humans" [37], and (ii) "the continued capacity of soil to function as a vital living system, within ecosystem and land-use boundaries, to sustain biological productivity, maintain the quality of air and water environments, and promote plant, animal, and human health" [38, 39].

Soil quality focuses on the physical and chemical properties of a soil pedon that establish the soil's potential in terms of soil organic matter, nutrient provision, soil structure and other attributes. Soil health expands soil quality to include the biological integrity of the soil community, referencing the significance and importance of diverse and well-populated trophic levels to stabilize ecosystem services.

To further the aims of Hawaiian producers in establishing soil health programs, the Hawai'i soil health tool was established [40]. The tool, when backed by soil data by the Hawaiian soil health laboratory, provides ratings for a soil health rating protocol. The soil indicators are those previously described in this manuscript by Crow et al. [34].

6. Oxisols in Brazil

The Amazonian tropical rainforest is the largest, continuous tropical rainforest, spanning from the crystalline mass of Guyana, the savannas of Colombia and Venezuela, the plateau of Mato Grosso (Brazilian Highlands), to the Atlantic Ocean. In Brazil, oxisols mostly reside on Mid- to Late-Tertiary geomorphic surfaces, where earth materials have been repeatedly weathered, eroded, transported, and subsequently redeposited to form fluvial terraces and pediments [41–43]. In the Brazilian Amazon ecosystem, oxisols have great ecological significance because comprise more than 40% of its total soil base [32].

The relationships between soils attributes, soil carbon stocks and vegetation carbon stocks are poorly known in Amazonia [26]. However, in an important and detailed review, Cerri et al. [21] provided data on the organic carbon content of the Brazilian Amazon, that is 47 Pg C in the upper 2 m of soil and 21 Pg C in the upper

0.2 m of soil. Considering the influence of climate change in the Southwestern Brazilian Amazon, Neill et al. [44] examined changes to soil bulk density and carbon-nitrogen stocks in multiple chronosequences, with paired representatives of different aged intact forest and pastures. These authors documented that some pastures exhibited reduced carbon stocks, whereas other sites exhibited an increase in the carbon stocks (loss of 0.72 kg C m^{-2} to an increase of 1.77 kg m^{-2}). In Kenya, Verchot et al. [45] investigated long-term soil organic matter storage in an intensively managed agroforestry system and proposed that the formation of microaggregates is critically important to long-term carbon storage.

In Brazil, Denef et al. [46] researched macroaggregate-occluded microaggregates under native forest vegetation, conventional tillage, and no-tillage systems for long-term carbon stabilization. The authors concluded that in these oxisols, (i) the carbon content in macroaggregate-occluded microaggregates was more responsive to soil management changes, (ii) the carbon accumulation was greater for forest vegetation and no-till agricultural systems, and (iii) macroaggregate-occluded microaggregates are an important indicator parameter for predicting land-use induced total SOC changes. In Rio Grande do Sul State (Brazil), de Oliveira Ferreira et al. [47] investigated no-till croplands and native vegetation and noted that intensive soil fertility management and with a high soil input carbon crop rotation supported soil carbon recovery.

In soil studies involving alfisols, ultisols, and oxisols in Rio Grande Sul and Santa Catarina, Amado et al. [48] assessed the potential of C accumulation in soybean and maize crop lands having no-till cropping systems. These authors reported that physical protection of soil organic matter within aggregates was important for carbon accumulation in some no-till soils and corresponded with higher particulate organic carbon to total organic carbon ratios. Pigeon pea (*Cajanus cajan*) and velvet bean (*Stizolobium cinereum*) in no-till maize cropping systems had the highest soil carbon accumulation rates, suggesting that legume cover crops are import to supporting soil carbon sequestration. In Brazil, de Oliveira Ferreira et al. [49] assessed the soil macroaggregate distribution and the soil organic carbon and particulate organic carbon stocks in no-till soils compared to adjacent native vegetation. The crop rotation systems involved soybean (*Glycine max*), maize (*Zea mays*) beans (*Phaseolus vulgaris*) black oat (*Avena strigosa*), white oat (*Avena sativa*), vetch (*Vicia sativa*), and wheat (*Triticum aestivum*). The extra-large macroaggregate classes in sandy clay and clay loam oxisols exhibited the greatest soil organic carbon and particulate organic carbon contents.

The effects of conventional-tillage and no-tillage on a Typic Haplorthox and three crop rotations was studied to assess aggregate stability and aggregate carbon contents [50]. The no-tillage system had the better aggregation indices for the 0–0.20 m layer because of increased organic carbon retention, especially in the 2 mm aggregate size class. The crop rotations studied did not affect the aggregate stability indices. In oxisols cultivated under no-till systems in southern Brazil, Nunes et al. [51] investigated soil relationships involving chemical management and structural degradation. Selecting no-till cultivated oxisols with adjacent noncultivated areas, the authors demonstrated that dispersed clay in the cultivated soil and the noncultivated soil were similar; however, in the cultivated soil the subsurface layer had higher dispersed clay levels and water percolation was slower.

In Colombia's eastern high plains in the Orinoquia region involving oxisols, Moreno-Conn et al. [52] evaluated land use management's influence on selected ecosystem indicators. The experimental design involved native savanna converted to (i) annual crops rotation with sugarcane, (ii) maize-soybean, (iii) sugarcane, (iv) rubber with cover crops, (v) pasture, and (vi) native savanna (untreated check). Indicators of ecosystem services included soil fertility, water regulation, macroinvertebrate diversity, climate control related to greenhouse gas emissions, and carbon stock. Soil fertility increased most-notably in the pasture production system, whereas the maize-soybean system showed better water regulation indicators. The results suggest that different cropping systems provide different ecosystem indicator responses. The Century Model simulations underscore the importance of Brazilian oxisol-based cropping systems needing to yield greater biomass productions to increase soil organic carbon contents [53].

Assessing a Cerrado located oxisol, Ngolo et al. [54] focused on microstructural changes because of (i) disc plowing, (ii) no-tillage, and a (iii) disc harrow with a subsoiler relative to a native Cerrado area as an untreated check. Micromorphology revealed some changes in the pedological features of soil aggregates, but the granular structure showed good resistance even after two decades of use and management. It also indicated a decrease in larger pores and an increase in the surface soil layer micropores for the disc plowing and no-tillage treatments and granular soil aggregates displayed resilience even after long-term management. In Brazil, de Freitas et al. [55] established soil quality indicators for oxisols cultivated to sugarcane and referenced to native forest. The soil quality indicators showed that conservation management practices were critical to maintain key soil attributes, including soil organic matter.

In a ferralsol in Paraná State (Brazil), Rosset et al. [56] quantified the organic carbon, organic nitrogen, and C:N ratio from pastures, native forests, and no-till fields and no-till fields with some production of Congo grass (*Brachiaria ruziziensis*). The carbon and nitrogen soil attributes from these pastures, native forests, and no-till fields were variable, illustrating the importance of the influence of land management on soil health. The introduction of Brachiaria, during the production cycle, effectively increased the carbon stocks in those cultivated soils. Using six Rhodic Hapludox soils with multiple land uses, Marinho et al. [57] documented soil depth distributions of soil organic carbon and investigated soil organic carbon variability under these different land uses. Soil carbon storage and stratification ratios for the surface horizons were greater long term under sugarcane cultivation than native forest.

Reflecting on the premise that tropical and subtropical land-use changes because of deforestation are a major contributor to greenhouse gas emission, Don et al. [58] investigated that afforestation programs may off-set CO_2 emissions from land-use change. In a global meta-analysis, the authors noted that the greatest soil organic losses were attributed to the conversion of primary forest into cropland (−25%), perennial crops (−30%), and grasslands (−12%). Soil organic carbon SOC losses were partially reversible with afforestation (+29%). Selecting mollisols and oxisols in South America, Berhongaray and Alvarez [59] supported the premise that silvopastoral programs support soil C sequestration.

7. Policy and producer network options in Brazil to support soil health

The evolution of the concept of soil health is the result of a research-oriented international dialog, of which Brazil has been a robust and energetic partner. During the last 30 years, the concept of soil fertility was extended to include soil quality, which was further transformed to include soil health. The scientific community recognizes that soil health incorporates biological components which mediate an array of soil processes that permit the soil to maintain full multifunctionality; that is soils must sustainably function beyond the needs of crop production. These collected functionalities include ecosystem services: (i) provision (food, feed, biomass, others), (ii) regulation (erosion control, climate change, others), (iii) biodiversity (habitat, nutrient cycling, others), and (iv) cultural functions (esthetics, cultural heritage, education/research, and others) [41].

Modern Brazilian agriculture initiated around the 1960's, with the establishment of (i) subsidized credit, (ii) investments in science and technology, and (iii) public and private rural extension. In the 1970s Brazil because a major food exporter, contributing to the national economy. Conventional agriculture with monocultures caused accelerated soil degradation. Further, rapid agricultural expansion in the Cerrado region was occurring. Research programs focused on soil conservation, including minimum and no-till systems. No-till soil preparation evolved to form no-till systems. No-till systems may be defined as "soil is managed by an ordered complex of interrelated and interdependent agricultural practices that must occur together" [41]. These interrelated and interdependent agricultural practices include crop rotation, cover crops/forage grasses, to form a dense lasting crop residue cover. These systems increased residual biomass, soil organic matter, and provided more robust and diverse soil trophic levels. Currently, advances in soil testing include utilizing the soil enzymes arylsulfatase and β-glucosidase. The crop-livestock-forestry systems and successional agroforestry systems are further expanding and gaining producer acceptance.

8. Climate impact analysis

Strengthening climate resilience with adoption of no-tillage systems occurs because of the resultant soil organic matter accumulation, which increases maintenance and formation of the soil structure. Improved soil structure fosters (i) improved water infiltration, (ii) greater available water content and elevated field capacity, (iii) expanded root proliferation, (iv) greater aeration, and (v) a more robust microbial activity that provides reduced pathogenicity of soil-borne plant diseases. Reduced tillage limits tractor usage thus reduces fuel consumption. The maintenance of residue covers limits soil erosion, especially when combined with cover crops.

Recently, Mondal et al. [60] performed a global meta review, documenting that no-till systems (i) increased soil organic carbon in the 0–s5 cm layer, (ii) soil organic stock increased 5.4 Mg C ha^{-1} over 0–30 cm soil profiles, (iii) deeper subsoils had a net soil organic carbon accumulation, (iv) microbial biomass carbon and soil aggregate-associated carbon increased; however, the carbon sequestration may not offset emissions from global fossil fuel consumption.

9. Producer education and community engagement

No-till will support (i) water infiltration, (ii) reduced fuel consumption, tractor usage and associated labor, and minimize soil erosion. However, as a rule, conventional tillage will (i) reduce weed pressure and reduce the reliance on herbicides, (ii) remove re-existing soil compaction, (iii) create a quality seed bed, (iv) reduce incidence of selected insects and diseases, (v) incorporate soil amendments and fertilizers, and (vi) permit selected farming activities such as furrow irrigation and raised beds. Thus, producer information for understanding the advantages and disadvantages is critical. Incorporating newer practices, such as cover crops, also requires producer education. Thus, extension activities must appropriately respond to new production options.

Both Brazil and Hawaii have extension activities dedicated to supporting proposed agronomic activities and guaranteeing their benefits. However, the agronomic protocols must be adopted to local conditions, Many Brazilian farms are large-scale operations concentrating on beef or row-crops, whereas in Hawaii many operations are small-acreage enterprises concentrating on products intended for local consumption. In Brazil, Fluentes-Llanillo et al. [61] noted that no-till expanded to 33.0 million ha or 61% of total crop land by 2017; however, no-till without additional complementary practices may not be completely sustainable. Policy initiatives are desired to incorporate soil mulch cover and crop diversification into the no-till systems. Production and soil monitoring systems are required to showcase soil improvement to encourage producer adoption.

10. Concluding statements

Oxisols are intensely weathered soils whose conventional agriculture practices have substantially degraded the provision of ecosystem services. Two case studies from two very different regions highlight emerging and innovative agricultural development programs that support sustainable food security, soil health, carbon sequestration, and other desirable features. In Brazil, agriculture production has grown considerable, permitting Brazil to become a major food exporting nation. To address soil degradation, systems involving no-till and other practices such as cover crops to create carbon-enriched epipedons that support soil fertility and offer other ecosystem services.

To address soil degradation and improve agricultural productivity, both Brazil and Hawaii are implementing innovative soil management techniques. In Brazil, the adoption of no-till has been integrated with cover crops and other practices to enhance soil organic matter accumulation and foster a robust soil biology. Similarly, Hawaii has focused on family-managed, small-acreage farms to provide food security using locally developed techniques. Both regions maintain extension services, producer education programs, and tools for soil conservation and soil health.

Critical needs remain, including agriculture friendly policy initiatives, continued research, and market availability. Research programs for understanding carbon sequestration and the underlying soil mechanisms are needed to continue developing land management options.

Author details

Michael Aide
Southeast Missouri State University, Cape Girardeau, MO, United States America

*Address all correspondence to: mtaide@semo.edu

References

[1] Soil Survey Staff. Keys to Soil Taxonomy. 13th ed. USDA-Natural Resources Conservation Service. Washington DC: US Gov't Printing Office; 2022

[2] Buol SW, Southard RJ, Graham RC, McDaniel PA. Soil Genesis and Classification. Iowa State Press, Ames IA: Blackwell Publication Co; 2003

[3] Aide MT, Pavich Z, Lilly ME, Thorton R, Kingery W. Plinthite formation in the coastal plain region of Mississippi. Soil Science. 2004;**169**:613-624. DOI: 10.1097/01.ss.0000142636.77482.e0

[4] Schrumpf M, Kaiser K, Guggenberger G, Persson T, Kogel-Knabner I, Schulze ED. Storage and stability of organic carbon in soils as related to depth, occlusion within aggregates, and attachment to minerals. Biogeosciences. 2013;**10**:1675-1691. DOI: 10.5194/bg-10-1675-2013

[5] Schmidt MW, Torn MS, Abiven S, Dittmar S, Guggenberger G, Janssen IA, et al. Persistence of soil organic matter as an ecosystem property. Nature. 2011;**478**(7367):49-56. DOI: 10.1038/nature10386

[6] Kumar R, Rawat KS, Singh J, Singh A, Rai A. Soil aggregation dynamics and carbon sequestration. Journal of Applied and Natural Science. 2013;**5**(1):250-267. DOI: 10.31018/jans.v5i1.314

[7] Witzgall K, Vidal A, Schubert DI, Hoschen C, Schweizer SA, Buegger F, et al. Particulate organic matter as a functional soil component for persistent soil organic carbon. Nature Communications. 2021;**12**:4115. DOI: 10.1038/s41467-021-24192-8

[8] Ye C, Bai T, Yang Y, Zhang H, Guo H, Li Z, et al. Physical access for residue-mineral interactions controls organic carbon retention in an oxisol soil. Scientific Reports. 2017;**7**:6317. DOI: 10.1038/s41598-017-06654-6

[9] Kramer MG, Lajtha K, Aufdenkampe AK. Depth trends of soil organic matter C:N and 15N natural abundance controlled by association with minerals. Biogeochemistry. 2017;**136**:237-248. DOI: 10.1007/s10533-017-0378-x

[10] Souza IF, Almeida LFJ, Jesus GL, Pett-Ridge J, Nico PS, Kleber M, et al. Carbon sink strength of subsurface horizons in Brazilian oxisols. Soil Science Society of America Journal. 2018;**82**:76-86. DOI: 10.2136/sssaj2017.05.0143

[11] Tan C, Wang C, Zhou T, Pang X, Zhao K, Zhou Z. Mineral protection mediates soil carbon temperature sensitivity of nine old-growth temperate forests across the latitude transect. Science of the Total Environment. 2024;**15**:171754. DOI: 10.1016/j.scitotenv.2024.171754

[12] Pessoa TN, Libardi PP. Physical-hydric properties of oxisols as influenced by soil structure and clay mineralogy. Catena. 2022;**211**:106009. DOI: 10.1016/j.catena.2021.106009

[13] Schweizer S, Mueller C, Höschen C, Ivanov P, Kögel-Knabner I. The role of clay content and mineral surface area for soil organic carbon storage in an arable toposequence. Biogeochemistry. 2021;**156**:401-420. DOI: 10.1007/s10533-021-00850-3

[14] Lal R. Soil carbon sequestration impacts on global climate change and food security. Science.

2004;**304**(5677):1623-1627. DOI: 10.1126/science.1097396

[15] Liu Q, Wang P, Xue Z, Zhou Z, Liu J, An S. Is the change of soil carbon capacity persistence rising or remain stable with maturity of vegetation restoration? Frontiers in Soil Science. 2021;**1**:663910. DOI: 10.3389/fsoil.2021.663910

[16] Liu D, Li S, Zhu W, Wang Y, Zhang S, Fang Y. Storage and stability of soil organic carbon in two temperate forests in Northeastern China. Land. 2023;**12**(5):1019. DOI: 10.3390/land12051019

[17] Deng L, Zhu G, Tang Z, Shangguan Z. Global patterns of the effects of land-use changes on soil carbon stocks. Global Ecological Conservation. 2016;**5**:127-138. DOI: 10.1016/j.gecco.2015.12.004

[18] Shi L, Feng W, Xu J, Kuzyakov Y. Agroforestry systems: Meta-analysis of soil carbon stocks, sequestration processes, and future potentials. Land Degradation and Development. 2018;**29**:3886-3897. DOI: 10.1002/ldr.3136

[19] Wang B, Liu M, Zhou Z. Preliminary estimation of soil carbon sequestration of China's forests during 1999-2008. Journal of Resources and Ecology. 2022;**13**(1):17-26. DOI: 10.5814/j.issn.1674-764x.2022.01.002

[20] Kleber M, Bourg IC, Coward EK, Hansel CM, Myneni SB, Nunan N. Dynamic interactions at the mineral–organic matter interface. Nature Reviews Earth & Environment. 2021;**2**:402-421. DOI: 10.1038/s43017-021-00162-y

[21] Cerri CEP, Cerri CC, Bernoux M, Volkoff B, Rondon MA. Potential of soil carbon sequestration in the Amazonian tropical rainforests. In: Lai R, Cerri BM, Etchevers J, Cerri E, editors. Carbon Sequestration in Soils of Latin America. New York: Haworth Press; 2006. pp. 245-266

[22] Gomes LC, Faria RM, Eliana S, Veloso GV, Schaefer CEGR, Filho EIF. Modelling and mapping soil organic carbon stocks in Brazil. Geoderma. 2019;**340**:337-350. DOI: 10.1016/j.geoderma.2019.01.007

[23] Cook-Patton SC, Leavitt SM, Gibbs D, Harris NL, Lister K, Anderson-Teixeira KJ, et al. Mapping carbon accumulation potential from global natural forest regrowth. Nature. 2020;**585**(7826):545-550. DOI: 10.1038/s41586-020-2686-x

[24] Harris NL, Gibbs DA, Baccini A, et al. Global maps of twenty-first century forest carbon fluxes. Nature Climate Change. 2021;**11**:234-240. DOI: 10.1038/s41558-020-00976-6

[25] Zhao F, Wu Y, Hui J, Sivakumar B, Meng X, Liu S. Projected soil organic carbon loss in response to climate warming and soil water content in a loess watershed. Carbon Balance and Management. 2021;**16**:24. DOI: 10.1186/s13021-021-00187-2

[26] Schaefer C, Amaral E, Mendonça B, Oliveira H, Lani J, Costa L, et al. Soil and vegetation carbon stocks in Brazilian Western Amazonia: Relationships and ecological implications for natural landscapes. Environmental Monitoring and Assessment. 2008;**140**:279-289. DOI: 10.1007/s10661-007-9866-0

[27] Lehmann J, Hansel CM, Kaiser C, Kleber M, Maher K, Manzoni S, et al. Persistence of soil organic carbon caused by functional complexity. Nature Geoscience. 2020;**13**:529-534. DOI: 10.1038/s41561-020-0612-3

[28] Giambelluca TW, Chen Q, Frazier AG, Price JP, Chen Y-L, Chu P-S, et al. Online rainfall atlas of Hawai'i. Bulletin of the American Meteorological Society. 2013;**94**:313-316. DOI: 10.1175/BAMS-D-11-00228.1

[29] Foote DE, Hill EL, Nakamura S, Stephens F. Soil Survey Kayai, Oahu, Maui, Molokai, and Lanai, State Hawaii. USDA-Soil Conservation Service in cooperation with University Hawaii Agriculture Experiment Station. Washington DC. US Govt. Printing Office; 1972. Available from: https://archive.org/details/usda-islandsHI1972

[30] National Cooperative Soil Survey. National Cooperative Soil Characterization Database. USDA-NRCS. Washington DC: US Gov't Printing Office; 2013. Available from: http://ncsslabdatamart.sc.egov.usda.gov [Accessed: May 10, 2024]

[31] Soil Survey Staff. Official Soil Series Descriptions. USDA-NRCS. Available from: http://soils.usda.gov/technical/classification/osd/index.html [Accessed: May 10, 2024]

[32] Wells JM, Crow SE, Sierra CA, et al. Edaphic controls of soil organic carbon in tropical agricultural landscapes. Scientific Reports. 2022;**12**:21574. DOI: 10.1038/s41598-022-24655-y

[33] Selmants PC, Giardina CP, Jacobi JD, Zhu Z. Baseline and Projected Future Carbon Storage and Carbon Fluxes in Ecosystems of Hawai'i. U.S. Geological Survey Professional Paper 1834. Reston, VA: U.S. Department of the Interior, U.S. Geological Survey; 2017. 134 p

[34] Crow SE, Rivera-Zayas J, Tallamy-Glazer C, Vizka E, Silva J. Hawai'i Natural and Working Lands Baseline and Benchmarks, 2021 (Final Report). State of Hawai'i Office of Planning and Sustainable Development. Soil Ecology and Biogeochemistry Laboratory: Department of Natural Resources and Environmental Management, University of Hawai'i Mānoa; 2021. Available from: https://planning.hawaii.gov/wp-content/uploads/UH-CTAHR-Baselines-and-Benchmarks-Final-Report.pdf

[35] Bashkin MA, Binkley D. Changes in soil carbon following afforestation in Hawaii. Ecology. 1998;**79**:828-833. DOI: 10.2307/176582

[36] Nave L, Johnson K, van Ingen C, Agarwal D, Humphrey M, Beekwilder N. International Soil Carbon Network (ISCN) Database v3-1. 2016. Web. DOI: 10.17040/ISCN/1305039

[37] Crow Soil Ecology & Biogeochemistry Lab. Department of Natural Resources and Environmental Management, University of Hawaii Manoa. Available from: https://soilandcarbon.com/

[38] Doran JW, Sarrantonio M, Liebig M. Soil health and sustainability. In: Sparks DL, editor. Advances in Agronomy. Vol. 56. San Diego: Academic Press; 1996. pp. 1-54. DOI: 10.1016/S0065-2113(08)60178-9

[39] Doran JW, Zeiss MR. Soil health and sustainability: Managing the biotic component of soil quality. Applied Soil Ecology. 2000;**15**:3-11. DOI: 10.1016/S0929-1393(00)00067-6

[40] Hawai'i Soil Health Resource Guide. O'ahu Resource Conservation and Development Council. Available from: Oahurcd.org/soilhealthplans/

[41] Vezzani FM, Anghinoni I, Cherubin MR, Mendes IC. Soil health and modern Brazilian agriculture. In: Karlen DL, editor. Soil Health and

Sustainable Agriculture in Brazil. Soil Science Soc Am. Madison, WI: John Wiley and Sons (Hoboken, NJ); 2024. pp. 1-18

[42] De Moraes Sa JC, Amado TJC, Oliveira FA, Lai R. Soil organic carbon restoration as the key driver to promote soil health in no-till systems of the tropics. In: Karlen DL, editor. Soil Health and Sustainable Agriculture in Brazil. Soil Science Soc Am. Madison, WI: John Wiley and Sons (Hoboken, NJ); 2024. pp. 62-102

[43] Bernoux M, Arrouays D, Cerri CC, Bourennane H. Modelling vertical distribution of carbon in oxisols of western Brazilian Amazon (Rondonia). Soil Science. 1998;**163**(12):941-951

[44] Neill C, Melillo JM, Steudler PA, Cerri CC, de Moraes JFL, Piccolo MC, et al. Soil carbon and nitrogen stocks following forest clearing for pasture in the Southwestern Brazilian Amazon. Ecological Applications. 1997;7(4):1216-1225. DOI: 10.2307/2641209

[45] Verchot LV, Dutaur L, Shepherd KD, Albrecht A. Organic matter stabilization in soil aggregates: Understanding the biogeochemical mechanisms that determine the fate of carbon inputs in soils. Geoderma. 2011;**161**:182-193. DOI: 10.1016/j.geoderma.2010.12.017

[46] Denef K, Zotarelli L, Boddey RM, Six J. Microaggregate-associated carbon as a diagnostic fraction for management-induced changes in soil organic carbon in two oxisols. Soil Biology Biochemistry. 2007;**39**:1165-1172. DOI: 10.1016/j.soilbio.2006.12.024

[47] de Oliveira FA, Amado TJC, Rice CW, Ruiz Diaz DA, Briedis C, Inagaki TM, et al. Driving factors of soil carbon accumulation in oxisols in long-term no-till systems of South Brazil. Science of The Total Environment. 2018;**622-623**:735-742. DOI: 10.1016/j.scitotenv.2017.12.019

[48] Amado TJ, Bayer C, Conceição PC, Spagnollo E, de Campos BH, da Veiga M. Potential of carbon accumulation in no-till soils with intensive use and cover crops in southern Brazil. Journal of Environmental Quality. 2006;**35**(4):1599-1607. DOI: 10.2134/jeq2005.0233

[49] de Oliveira FA, de Moraes Sá JC, Lal R, Tivet F, Briedis C, Inagaki TM, et al. Macroaggregation and soil organic carbon restoration in a highly weathered Brazilian oxisol after two decades under no-till. Science of The Total Environment. 2018;**621**:1559-1567. DOI: 10.1016/j.scitotenv.2017.10.072

[50] Filho CD, Lourenço AC, Guimarães MD, Fonseca IC. Aggregate stability under different soil management systems in a red latosol in the state of Parana, Brazil. Soil & Tillage Research. 2002;**65**:45-51. DOI: 10.1016/S0167-1987(01)00275-6

[51] Nunes MR, da Silva Alvaro P, Denardin JE, Giarola NFB, Vaz Carlos MP, van Es Harold M, et al. Soil chemical management drives structural degradation of oxisols under a no-till cropping system. Soil Research. 2017;**55**:819-831. DOI: 10.1071/SR17063

[52] Moreno-Conn LM, Rodríguez-Hernandez NS, Arguello JO, Gallo Gordillo O, Bernal-Riobo JH, Arango M, et al. Land use change and its effect on ecosystem services in an oxisol of the eastern High Plains of meta department in Colombia. Frontiers in Environmental Science. 2022;**10**:687804. DOI: 10.3389/fenvs.2022.687804

[53] Tornquist CG, Mielniczuk JO, Cerri CEP. Modeling soil organic carbon dynamics in oxisols of Ibirubá (Brazil)

with the century model. Soil & Tillage Research. 2009;**105**:33-43

[54] Ngolo AO, Oliveira FS, Oliveira MF, Fernandes RBA. Microstructural changes in oxisols under long-term different management systems. Revista Brasileira de Ciência do Solo. 2023;**47**:e0230051. DOI: 10.36783/18069657rbcs20230051

[55] de Freitas L, Filho MVM, Casagrande JC, de Oliveira IA, Silva DG. Soil quality indicator of oxisols grown with sugarcane and native forest in northeastern São Paulo state, Brazil. Environmental Earth Sciences. 2018;**77**:642-642. DOI: 10.1007/s12665-018-7830-7

[56] Rosset JS, Lana MC, Schiavo JA, Piccolo MC, Pinto LASR, Ziviani MM, et al. Organic matter and isotopic composition of soils under different management systems in western Paraná State, Brazil. Environmental Earth Sciences. 2022;**81**:136. DOI: 10.1007/s12665-022-10261-8

[57] Marinho MA, Pereira MWM, Vázquez EV, Lado M, Paz GA. Depth distribution of soil organic carbon in an oxisol under different land uses: Stratification indices and multifractal analysis. Geoderma. 2016;**2016**. DOI: 10.1016/j.geoderma.2016.09.021

[58] Don A, Schumacher J, Freibauer A. Impact of tropical land-use change on soil organic carbon stocks – A meta-analysis. Global Change Biology. 2010;**17**:1658-1670. DOI: 10.1111/j.1365-2486.2010.02336.x

[59] Berhongaray G, Álvarez R. Soil carbon sequestration of Mollisols and oxisols under grassland and tree plantations in South America - A review. Geoderma Regional. 2019;**18**. DOI: 10.1016/j.geodrs.2019.e00226

[60] Mondal S, Chakraborty D, Paul RK, Mondal A, Ladha JK. No-till is more of sustaining the soil than a climate change mitigation option. Agriculture, Ecosystems & Environment. 2023;**352**:108498. DOI: 10.1016/j.agee.2023.108498

[61] Fuentes-Llanillo R, Telles TS, Soares D, Rodrigues de Melo T, Friedrich T, Kassam AH. Expansion of no-tillage practice in conservation agriculture in Brazil. Soil Tillage Research. 2021;**208**:104877. Available from: https://api.semanticscholar.org/CorpusID:230585651

Chapter 5

Cover Crops for More Sustainable Soil Fertility Management: Examples from Brazilian Cerrado

Leonardo Santos Collier and Karin Ferretto Santos Collier

Abstract

The growth of agriculture in Central Brazil brought opportunities as well as treats when we think about natural resources. Soil's lack of organic matter and plant nutrients became challenging to many agricultural practices adopted so far. Cover crops and agroforestry systems are able to improve soil fertility management in Brazilian Cerrado lands, strategically to face recent adverse weather conditions. As we have water limiting nitrogen and phosphorus nutrient efficiency, research was done to understand how several leguminous and grass species can interfere with soil quality for plant production. However, many crops in successional systems were not enough to guarantee better quality soil for a long time, so the intercropping and consortiated systems started to be adopted, as an option for some agricultural systems. *Crotalaria* and *Urochloa* as a cover crop can improve nitrogen fertilizers effect in production by 50%. We intend to present a review of how these both groups of agricultural practices linked to cover crops can influence nutrient and water use, in agriculture of this part of Brazil.

Keywords: cash crops, cover crops, organic matter, phosphorus cycling, nitrogen cycling, agroforestry

1. Introduction

Brazil's agricultural occupation began on the coast and remained that way until the beginning of the twentieth century. The climate and soil conditions in the Central region of the country have always been challenging due to lack of infrastructure, chemically poor soil, and absence of rain for up to 6 months [1–3]. Government management encouraged the occupation of this region, which initially occurred with extensive cattle farming. The change of the country's capital and the technological development of agriculture made the cultivation of grains and fibers the main axis of the economy in this region.

The striking physiognomy of the vegetation that named the Biome is the Cerrado. It is possible to observe smaller trees that have graminoid species covering the soil and many variations depending on the level of water and nutrients [1, 3, 4].

The viability of agriculture led to the production of important commodities for the region, such as soybeans, corn, sugar cane, cotton, and coffee, with space for cattle farming and pig and poultry farming.

This movement was only possible through a set of land use technologies with chemical correction (liming and gypsum) and the use of fertilizers, in addition to the genetic breeding of the crops [5]. This change occurred over a period of two decades, bringing a large suppression of existing vegetation and an expansion [6] of crops towards even more environmentally fragile biomes (Amazon and Caatinga).

Researchers and farmers realized the importance of improving soil coverage with plant residues, and thus, the use of the no-tillage planting system [7–9], which was already adopted in the south of the country, became an almost mandatory practice to preserve the levels of organic matter in the soil.

The study of cover crops used in cultivation systems without soil disturbance has gained more prominence. In the last two decades, water crises resulting from climate change and Brazil's dependence on synthetic fertilizers [10] have intensified the use of cover crops and the understanding of the factors that affect their management [9, 11–23].

Despite being productive, monocultures generate many biotic and abiotic imbalances, and with this, consortia and crop rotations also began to present themselves as necessary alternatives [24–28], both due to environmental pressure and due to cost. Even in these systems, soil cover with annual or perennial plants is absolutely necessary.

This text is a review of studies on cover crops in grain cultivation systems in the Brazilian Cerrado, and in integrated systems such as Integration-Crop-Livestock-Forest [29] (where species such as *Eucalyptus* are widely used) and Agroforestry Systems (systems that use native trees and greater species diversity [30]). The focus of this study was the contribution to soil fertility management, aiming for more efficient use of fertilizers.

2. How is the Brazilian Cerrado

The environment known in Brazil as "Cerrado" is a territorial extension located between the Amazon Rainforest and the Brazilian coast, with more than 2 million km^2 [1], (approximately 23% of the Brazilian territory). The native vegetation formation is of the savannah type, but there are particularities due to the variation in soils and rainfall totals. In the areas further east, close to the Caatinga Biome, rainfall totals reach 1100 mm per year and many soils can be derived from sandstones and other quartzose materials, which defines more graminoid formations. In the northwest direction towards the Amazon basin, annual rainfall can exceed 1800 mm, the soils can be more clayed, and the vegetation becomes denser and with taller tree size [2]. Throughout its extension, the Cerrado has a dry period during the southern hemisphere winter that can reach 4–6 months.

The phytophysionomy of the Cerrado is composed of "cerrado in the strict sense," "cerrado park" (**Figure 1**), palm grove, and "vereda". Floristic compositions are related to water resources and soil type. In the cerrado "stricto sensu," the grasses and herbaceous plants are sparsely distributed. In the "cerrado park," the trees are concentrated on small elevations known as "murundus", surrounded and interspersed by areas of grassy fields. In the "cerrado park," the trees are concentrated in small elevations known as "murundus," surrounded and interspersed by areas of grassy

Figure 1.
Two areas with contrasting difference on soil cover. South of Tocantins State. Above: area after sugar cane harvesting, October 2008 (high temperatures on soil). Below: area soil with cover of residues of maize and grasses before soybean sowing, November 2008. Both on Ferralsols.

fields. The palm grove is distinguished by the predominance of species of the Araceae family in areas with good drainage or with flooding. Finally, the "vereda" is characterized by the presence of the Buriti palm tree (*Mauritia flexuosa*) surrounded by aristos and small trees [4, 7].

The soils formed according to the WRB classification [31] show the predominance of Ferralsols. These soils are the most occupied by agriculture and livestock in the biome, followed by Arenosols and Plinthosols, which also stand out for the anthropization of their areas in the last 20 years. These soils are generally found on flat relief, have a thickness greater than 100 cm, with variable texture, and a structure of the clods from granular to subangular blocks, which greatly facilitates the use of machinery for planting grains and pastures. The formation environment favors weathering, and therefore, they are acidic, poor in basic cations [1], and when more clayey, more exchangeable Al, factors that do not prevent agricultural use through soil correction practices with liming, accessible to most farmers and ranchers. The formation environment facilitates microbial activity under the surface layer of the soil, resulting in low levels of organic matter. Some soil management practices can further reduce the levels of organic matter in these soils, resulting in a series of chemical, physical, and biological limitations in the soil, strongly affecting the viability of the expanding agricultural production sector in the Cerrado.

3. Agricultural expansion towards Brazilian Cerrado and no-tillage cultivation system

The agricultural occupation of the Cerrado occurred in stages. In the 1930s, the government promoted the "March to the West" in the last century, but land use was restricted to beef cattle farming due to the still deficient infrastructure. With the construction of the new federal capital in the 1950s and 1960s, access to the Central region of the country improved, but it was only after the creation of EMBRAPA in 1973 that agricultural growth intensified. From a "chemical fertility" point of view, Cerrado soils were seen as a major challenge [32]. The costs of correcting acidity and phosphate and nitrogen fertilization could make the expansion of agricultural crops unfeasible.

Since the 1990s onwards, with the country's economic stability, the Cerrado experienced an intense growth in its occupation, with a deforestation of 1/3 of its native vegetation cover. The country's grain production axis shifted from the South to the Center-West of the country until the early 2000s. Recently, this region has held 60% of the national grain and fiber production [3], which represented an 87% growth in the agricultural area between 2000 and 2015. According to Zalles et al. [33], this expansion initially occurred in Mato Grosso (2000–2005), with more recent expansion under the states of Maranhão, Tocantins, Piauí, and Bahia, represented by the acronym MATOPIBA (**Figure 2**). In both cases, although there is a significant loss of existing vegetation cover (Cerrado), the occupation of grain agriculture occurred in many pasture areas with some level of degradation.

All this growth was accompanied by technical challenges in the use of natural resources. In the 2000s, producers realized that cultivation with soil disturbance resulted in the loss of already low levels of organic matter, with an effect on the water balance of crops and low efficiency in the use of fertilizers. The use of the no-till system grew dramatically and the cultivation of soybeans during the harvest period and corn at the end of the rainy season now occurs mainly without soil disturbance with harrows or plows. Brazil grew from 20 million hectares in 2003 to 36 million hectares under no-till in 2022 [33, 34], of which approximately 17 million are in the Cerrado. Meanwhile, keeping soil coverage is very difficult in these tropical areas (**Figure 3**).

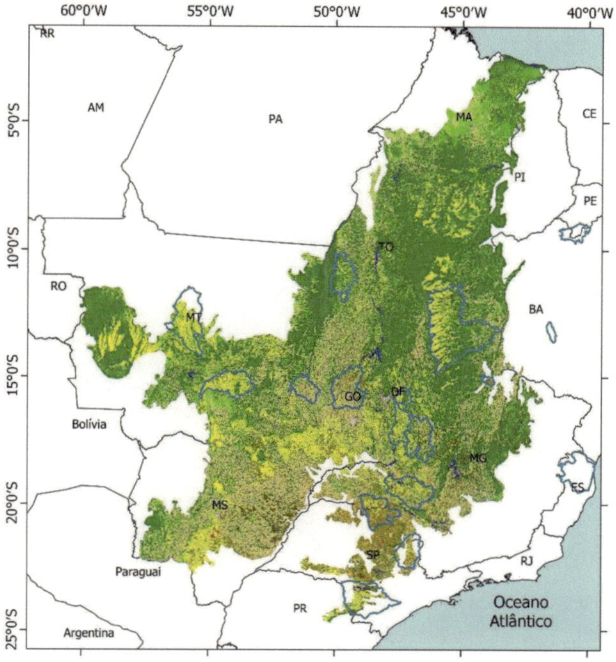

Figure 2.
Area of Cerrado Biome vegetation in the brazilian territory. The more green color corresponds to more preserved vegetation area. Yellow areas has more agriculture. Intermediate colors represents pastures. ANA—Agência Nacional de Águas. Atlas da irrigação: *uso da água na agricultura. 2ª. ed. Brasília: ANA. 2021. 130 p. INPE.* TerraClass: *organização, acesso e transparência—Bioma Cerrado. [ca. 2018]. Disponível em: https://www.terraclass.gov.br/geoportal-cerrado/ [Accessed: November 16, 2022].*

Figure 3.
Overview of soybean monoculture in Garganta region, border of Bahia and Tocantins states, January 2008. Source—the author.

Figure 4.
Two situations where no-tillage presents different viability. On brachiaria straw, the top photo shows the layer of soil cut for soybean planting (December 2007). Highlight the dark color of the soil. The photo below shows the cotton crop that has a planting season in Brazilian Cerrado that does not allow to introduce a cover crop before or after its cultivation. It leaves very little residue to protect the soil, what turns no-tillage not feasible under these conditions. Source, the authors.

Given the climate behavior of recent decades, the challenges for agriculture in the Cerrado have not ceased. The no-till system, as it is closely linked to soybean monoculture (**Figure 4**), involves intensive use of herbicides and mechanization that does not prevent soil compaction, which harms crop performance. The search for systems that are less sensitive to phytosanitary problems, the even greater concern for water resources, and the search for biological quality in agricultural systems have encouraged integrated production systems, where the principle of intercropping between species becomes possible and desirable. Integrated systems called Crop-Livestock-Forest Integration and Agroforestry Systems have been gaining prominence in the last 15 years [8].

4. Cover crops in Cerrado's agriculture

With the consolidation of the no-till system for grain production, the use of cover crops as green manure and mulch is the main practice to increase soil organic matter (SOM) levels in the Cerrado [34].

The use of plants to form mulch in production systems is based on knowledge of their use as green manure.

Green manure is based on the incorporation of plant residues into the soil, with the aim of preserving and/or restoring the productivity of arable land [11, 12].

The use of green manure practices aims to enhance some functions performed by the soil. To achieve this, the crops used for this purpose must have high biomass production, low susceptibility to pests and diseases, uniform seeds, rapid growth, tolerance to acidity, and low nutritional requirements [9, 13, 33].

Cover crops and/or green manures can influence some aspects related to the physical properties of the soil, such as protection against erosion, increased water infiltration rate, increased water availability in the soil, and increased soil aggregation, in addition to conditioning the rupture of dense and/or compacted layers [12, 29].

However, this practice may affect the chemical aspects of the soil, such as increasing the MOS content, promoting the mobilization and recycling of nutrients, increasing the efficiency of the use of mineral fertilizers, and reducing the leaching rate of nutrients, especially N and K [14].

Regarding the biological aspects of the soil, this practice may promote the supply of N through biological fixation carried out by legumes, or through the mineralization of plant residues, in addition to creating environmental conditions favorable to the biological activity of the soil [15, 16].

In general, the use of these plants can suppress a portion of invasive plants due to their rapid and aggressive growth (allelopathic and/or suppressive effect, as exemplified by crotalarias, reducing the incidence of nematodes) and can be used in animal feed [clover (*Trifolium repens* L.), pigeon pea (*Cajanus cajan*), cowpea (*Vigna unguiculata* L.), labe-labe (*Lablab purpureus*)], recovery of degraded areas, production of nectar and pollen for bees [lupine (*Lupinus albus*), clover (*Trifolium repens* L.), sunflower (*Helianthus annuus*), crotalaria (*Crotalaria* sp.), pigeon pea (*Cajanus cajan*)], in addition to producing a large amount of biomass, which will be used as mulch [11, 13–16, 22, 35].

Based on the above, it is possible to arrive at a misinterpretation that only legumes are sources of residues for the contribution of MOS from the biomass produced. However, some grasses can also be used as cover crops and/or green manure in agricultural crops.

With the growth of areas with no-tillage, there is an increasing demand for species capable of maintaining straw on the soil during the off-season. Thus, species such as millet (*Pennisetum glaucum*), sorghum (*Sorghum bicolor*), goosefoot grass (*Eleusine indica*), oats (*Avena sativa*) and, more recently, B*rachiaria* sp. have stood out in terms of providing soil cover (**Figure 5**), in addition to being a source of animal feed, mainly in integrated production systems, especially in crop-livestock integration (iCL) and crop, livestock, and forestry (iLPF) systems [14, 17, 18, 36].

Figure 5.
Landscape of "cerrado park" with grassy cover, and on deep of the photo the "Serra Geral plateau" formed by the weathering of Urucuia sandstones Geological group, typical in the "Matopiba" region. January, 2008, Jalapão State Park, Tocantins State, Brazil.

In systems of pasture crops alternated with soybean crops (*Glicine max*), in a no-tillage system, higher soil C stocks occurred, when compared to monoculture systems [29].

The use of cover crops during 3 cycles in the Cerrado-Amazon transition indicates that the greater the biomass of the plants, the more significant the effect on microporosity in subsurface layers. The contribution of the use of cover crops to the availability of nutrients to crops is expected, but it is worth noting that in sandy tropical soil, although there are effects on soil microbial biomass in the first year of use of cover crops, the contributions to soil chemical attributes only manifested themselves from the third crop onwards [11, 12].

In smallholders farming systems where the objective of using cover crops is to release nutrients, their incorporation into the soil can bring short-term benefits [19], where the use of *Crotalaria ochroleuca* and *Mucuna pruriens* stands out in relation to other species. The search for systems capable of promoting an increase in soil organic matter content has gained even greater importance than the economy of synthetic fertilizers [9] report that the lower the lignin content and the lower the lignin:N ratio and the higher the soluble fractions such as hemicellulose in cover crop residues, the greater the presence of labile fractions of organic matter and soluble humic substances. *Urochloa ruziziensis* and *Canavalia brasiliensis* (**Figure 6**) stood out in this aspect as options for improving the quality of soil organic matter.

Plant residues rate of decomposition has strong influence on soil nutrients dynamics. Torres et al. [37] highlight the high rate of decomposition of millet and sun hemp residues, associated with high biomass and nitrogen content.

There is no policy tool that obliges farmers or ranchers to use cover crops. There are still situations where the use of cover crops was not always a technical choice seeking to improve soil quality. Often the seed most found on the market is adopted. Millet was a widely used species, but it did not always provide good soil coverage, and its low biomass production did not have satisfactory effects on organic matter levels

Figure 6.
Nitrogen management via tropical leguminous species. Legumes as cover crops. Its N_2 fixation capacity makes its waste natural organic fertilizers. In addition to the importance of N provided by a more accelerated decomposition process, it can contribute to the bioavailability of P, Ca, Mg and micronutrients. Many tropical species have dry biomass production that can reach more than 12 tons per hectare. Top photo shows the growth of Canavalia ensiformis *plants 30 days into the rainy season. Lower photo shows the nodulation in the root system, showing the colonization of diazotrophic bacteria. Garganta region Tocantins/Bahia State border, 2008. Source – the author.*

and nutrient cycling. The climatic events of the last decade have practically forced farmers to look for more efficient management options, and the use of more than one species of cover crop is quite common.

4.1 Soil acidity changes with plant residues

Although liming is a well-established practice for neutralizing acidity and toxic Al in tropical and subtropical soils, some studies highlight the beneficial effects of soil organic matter in this regard. After conducting a study in no-till areas, Miyazawa et al. [20] state that, despite presenting potential acidity, these soils do not require liming as do areas without the use of organic waste. Meanwhile, Ziglio et al. [38] demonstrate the formation of organometallic compounds through organic acids, which release OH-, immobilizing H+ and Al^{+3} and mobilizing basic cations (Ca^{+2}, Mg^{+2} and K^{+}). This occurs due to the formation of organic acids and humic substances in the process of waste decomposition.

Lange et al. [39] observed the movement of Ca in the soil profile when limestone was applied on the surface, which is explained by the decomposition of the straw that generates OH^- and HCO_3^-. On the other hand, nitrogen fertilization in these areas promotes soil acidification, both on the surface and in the subsurface, which justifies the application of limestone in these areas, even under a no-till system on the straw. However, this effect is restricted to the surface layer of the soil where the residues are applied and there was no evidence of the movement of organic acids in the soil profile [40, 41].

Veronese et al. [14] highlight that the use of millet (*Pennisetum glaucum*) and Brachiaria (*Brachiaria* sp.) straw in areas cultivated with soybean (*Glycine max*) in no-tillage systems facilitates the formation of an alkalinization front as limestone doses increase, with an impact on crop productivity. Similar results were obtained with the use of millet (*Pennisetum glaucum*) and Brachiaria (*Brachiaria* sp.) straw in the Cerrado of Maranhão [17].

The application of soil acidity correctives also influences the decomposition of cover crop residues. Ferrari Neto et al. [42] observed an increase in microbial biomass carbon and mineralizable carbon after the use of limestone when *Urochloa* spp. residues were intercropped with the main crops.

4.2 Leguminous as cover crops: Nitrogen supply effects

The use of cover crops is one of the principles of the no-till system in straw, resulting in several transformations of the soil organic matter and, in turn, in the dynamics of N and several other nutrients. The use of straw for soil cover promotes changes in the transformations of preexisting organic matter, as well as soil protection, nutrient recycling from the return of plant biomass to the soil, leading to a stimulus of biological activity, which makes residue management important in the dynamics of soil nutrients.

Most studies with cover crops and/or green manures that evaluate their effect on the management of nitrogen fertilization are associated with greater release of N from their decomposition, due to their ability to perform BNF, when legumes. However, Torres et al. [21] and Pacheco et al. [43] report that straw from grasses such as millet (*Pennisetum glaucum*) and Brachiaria (*Urochloa* sp.) can also present high N contribution, as well as rapid mineralization. These authors emphasize that these species can absorb considerable amounts of N and that, and its decomposition was favored under edaphoclimatic conditions of Cerrado.

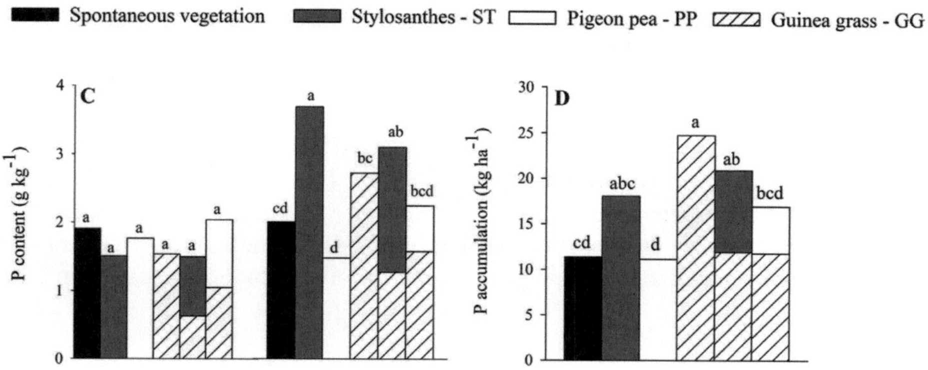

Figure 7.
Phosphorus (P) contents (gkg^{-1}) (C) and accumulation (kgha^{-1}) (D) on aerial part of cover crops in single systems: pigeon pea (PP), stylosanthes (ST) and guinea grass (GG), and in intercropping systems: (PP + GG, and ST + GG), and spontaneous vegetation. Treatments were compared by the Tukey-test ($p < 0.05$), results represented by lowercase letters. Adapted from [44].

Collier et al. [22], when conducting a study in the south of the state of Tocantins, Brazil, observed a very expressive response of crotalaria straw (*Crotalaria* sp.) in relation to jack bean straw (*Canavalia ensiformis*). The former favored the corn crop (*Zea mays*) to reach maximum productivity with the use of 100 kg ha^{-1} of N, while the latter required 250 kg ha^{-1} of N (**Figure 7**). According to Torres and Pereira [23], the benefits promoted using crotalaria straw (*Crotalaria juncea*) in the productivity of corn are evident, associated with the assimilation of atmospheric N.

When evaluating the use of jack bean (*Canavalia ensiformis*) as a cover crop, Collier et al. [15] found a significant effect of mulch and cropping when intercropped with corn (*Zea mays*), allowing the use of half the N dose to achieve higher productivity (**Figure 7**). Corroborating these results, Heinrichs et al. [16], when evaluating corn plants (*Z. mays*) intercropped with jack beans (*C. ensiformis*), had favorable production results in the second year of cultivation. The biological fixation capacity conditions the accumulation of N in the plant tissue, as well as the high biomass production, which justifies the increase in N and, consequently, the increase in corn production.

Among the agroecosystems that employ practices that favor nutrient cycling are AFSs, in which tree species, when absorbing nutrients from deeper layers, deposit the residues in the form of litter, leading to nutrient cycling by its decomposition. This may be associated with the use of species capable of performing BNF, making the system even less dependent on synthetic nitrogen fertilizers.

According to Daudin and Sierra [45], the AFS that uses *Gliricidia sepium* in the French Antilles assimilates 56 kg ha^{-1} of N by the surrounding grasses, due to the tree species' ability to fix atmospheric N and cycle nutrients. As already mentioned in other references, climate and soil conditions will regulate N absorption by grasses.

In general, in traditional crops using legumes in succession, there is an expectation of improvement in the N balance for the subsequent crop. According to Peoples et al. [46], the use of legumes may not mean a significant contribution of N, due to the possibility of export by grains [soybean (*Glicine max*), beans (*Phaeseolus vulgaris* L.), peanuts (*Arachis hypogaea* L.)] or by the rapid decomposition of residues, followed by N mineralization, common in tropical and subtropical conditions. The benefit of legumes may also be associated with the release of organic substances linked to the H+ dynamics during the BNF process, and not specifically to the amount of N that remains in the subsequent crop system.

Sometimes the use of cover crops can immobilize N during the decomposition of residues, having a negative effect on cash-crops. Echer et al. [47] highlights the effect of *Brachiaria ruziziensis* delaying the availability of N to cotton in the Center-South of Brazil.

Integration systems between grains and forage have been gaining prominence in many areas of Central Brazil. The use of palisade grass (*Urochloa brizantha*) or guinea grass (*Megathyrsus maximum*) intercropped with corn could suggest competition for N affecting corn. However, work by Mateus et al. [48] demonstrated that single and intercropped crops had the same productivity for nitrogen fertilizer used in top dressing.

4.3 Cover crops and phosphorus dynamics in agricultural systems

Soil management practices affect the dynamics of soil organic matter, as well as the dynamics of P. The dynamics of this macronutrient in tropical and subtropical soils are quite complex, due to the ease of becoming unavailable due to adsorption processes, considering that P fertilization is based on the use of mineral sources [49]. However, the organic fraction of the soil has a major influence on the availability of P.

The complexity of P management in soils in tropical and subtropical regions is due to the occurrence of highly weathered soils, with a predominance of 1:1 clay mineral and iron and aluminum oxides, resulting in an increase in its adsorption capacity together with the mineral fraction.

Studies involving minimum cultivation, based on the maintenance of straw on the soil, confirm the reduction in the levels of phosphate fertilization for crops. This effect is more associated with organic colloids in P adsorption, than the actual supply of nutrients by the residues. This effect is highlighted by Pavinato and Rosolem [50] in their review on the subject.

In this sense, Guareschi et al. [51] highlighted that the C levels in the no-tillage system resulted in an increase in organic forms of P and, consequently, are responsible for the increase in less crystalline forms of iron (Fe), described as iron oxalate (Feo). With the increase in C levels, the reduction in P adsorption by the mineral fraction is evident. Although interference in P adsorption is the main argument for the effect of straw, there are also records of P release from the decomposition of residue tissues.

According to Pacheco et al. [18], straws of *Brachiaria ruziziensis*, *B. brizantha*, millet (*Pennisetum glaucum*), and pigeon pea (*Cajanus cajan*) intercropped with Brachiarias present a rapid release of P from plant tissues, even without the influence of the legume. In turn, Teixeira et al. [13] obtained contrasting results, in which the straw from the consortium of millet (*Pennisetum glaucum*) with crotalaria (*Crotalaria* sp.) presented greater P availability, better than the single millet (*Pennisetum glaucum*), reaching 80% of the total absorbed by the plants in 24 days.

Chagas et al. [52] observed that the release of P from bean crop residues was consistent with their decomposition, varying with respect to the decomposition of leaves, pods, and stems of the plants. Crusciol et al. [53] observed that the P released in black oat tissues occurred at a constant rate, reaching 58% of the total accumulated by the plant after 53 days.

Arruda et al. [44] observed significant accumulation of phosphorus in the biomass of guinea grass alone and intercropped with stylosants, reaching 25 $kg.ha^{-1}$ (**Figure 8**).

Regarding aspects related to the use of residues of animal origin, there is also the possibility of supplying P, but to a lesser extent when compared to N [54]. However,

Figure 8.
*Agroforestry system with cerrado native trees and fruit species. In the foreground image leguminous cover crops (*Pueraria phaseoloides*). Picture was taken in the beginning of dry season, May, 2010. Federal University of Tocantins, Gurupi, Brazil.*

adequate management of organic residues can enhance the efficient use of phosphorus in soils in tropical and subtropical regions.

In general, the use of organic residues influences the sources and methods of application of P. In the late 1990s, areas of the Cerrado where corn (*Zea mays*) and/or soybeans (*Glicine max*) were grown under straw in the no-tillage system began to introduce broadcast application of phosphate fertilizers, some of which had low solubility. However, this practice was common in sandy soils, where the adsorption exerted by the mineral fraction is lower. However, the efficiency of use of P from natural phosphates has been increased, due to the influence of straw decomposition, making broadcast application of phosphates before crops a viable alternative.

Collier et al. [55], in studies with acidic and dystrophic soils of medium texture in the Cerrado region, observed significant results in corn (*Zea mays*) productivity, with the use of natural phosphate under straw of jack beans (*Canavalia ensiformis*). On this occasion, the activity of soil microorganisms, enhanced by the addition of residues, results in greater solubilization of phosphate, making it more available to plants. Within similar subject, Leite et al. [56] found that the cultivation of sunnhemp, lablab, and velvet bean in corn pre-cropping increases the moderately labile organic P (P extracted with NaOH). Phosphorus inorganic and organic forms were redistributed between labile and moderately labile compartments, which is important for plant availability.

Rodrigues et al. [57], after studying soils that had been under soybean and cotton cultivation for 42 years, obtained better recovery of labile P, when this crop rotation used brachiaria as a cover crop. The soil cultivated with brachiaria recovered 5–20% of labile forms of P, and 10–25% of organic forms over the study period.

Ramos et al. [36] observed greater use of natural phosphate applied in pots and under straw of grasses [brachiaria (*Brachiaria* sp.), millet (*Pennisetum glaucum*), and sorghum (*Sorghum bicolor*)], with no difference between them and using beans (*Phaseolus vulgaris* L.) as the main crop. These same authors observed different responses regarding the efficiency of use of natural phosphate in soybean (*Glicine max*) crops with the presence of straw.

Due to climatic restrictions that avoid autumn-winter harvest, leguminous cover crops can be intercropped with the main crop. Benicio and Collier [58] observed that even in competition with corn, crotalaria and pigeon pea, when each one was intercropped with corn, there was a better use of P by plants even with less soluble fertilizer sources.

4.4 Cover crops and potassium dynamics in agricultural systems

The dynamics of K in tropical and subtropical soils are associated with fewer loss processes when compared to N and P. The greatest concern is related to the speed of its release from the solid mineral phase, which can lead to losses due to the leaching process [59].

In the management of K fertilization, there is a concern regarding the most favorable balance, due to the amount of this nutrient required by the plant. In addition, another major aggravating factor is the cost of acquiring this mineral, due to Brazil's need to import >90% of potassium sources [10]. Given this scenario, there is a discussion about how management practices involving organic waste could contribute to the more efficient use of this nutrient.

In general, SOM can supply a portion of the K demand of plants; however, its contribution is also related to the mobilization of K *via* the generation of new electrical charges in the soil.

A study on 9-year-old cotton cultivation systems found that the introduction of *Brachiaria ruziziensis* into the rotation promoted an increase in organic matter levels with a direct impact on improving the retention of potassium fertilizers [60].

The release of K from the decomposition of plant tissues is predictable, due to its ability to not establish stable bonds with organic complexes in plants. Crusciol et al. [61] observed a release of almost all K from the tissues of forage radish (*Raphanus sativus* L.) in up to 53 days. Pacheco et al. [18] found significant accumulation and release of K in *Brachiaria ruziziensis* straw, alone or intercropped with pigeon pea (*Cajanus cajan*) in the Cerrado region. However, Torres and Pereira [62] obtained a greater release of K in grass straw, especially millet (*Pennisetum glaucum*), when evaluating the total K that the plant could accumulate and the speed of release of this nutrient with the decomposition of the straw in the soil.

Although the amount of K released by plant tissues does not immediately meet crop demand, there are situations in which the management of these residues can demonstrate the opposite. Rosolem et al. [63] observed that successive cuts of *B. ruziziensis* can supply K to crops, ensuring the residual effect of potassium fertilization for soybean (*Glycine max*) crops. This result is important, especially for integrated cropping systems, since forage can represent savings in the supply of K to the crop. The same authors emphasize the ability to mobilize K at low levels in the soil and the existence of "luxury consumption" during fertilization. Thus, straw management can supply exchangeable K, since there is a lower chance of mobilizing K from non-exchangeable forms linked to the solid mineral phase of the soil.

Firmiano et al. [64], evaluating the effect of millet and brachiaria predecessor straw on potassium fertilization of soybeans, realized that where there were brachiaria residues, they provided higher productivity for soybeans, but without the fertilizer savings obtained with millet, which achieved 16% lower productivity.

Thus, Almeida et al. [65] obtained an important result when they realized that the productivity of sugarcane (*Saccharum officinarum*) did not differ between fertilizations of 65 and 195 kg ha^{-1}. According to the authors, in addition to the release of K in the straw, there is the effect of organic matter on the soil CEC from the release of organic acids that interfere with the retention of K in the soil, remaining in the exchangeable fraction. The increase in the levels of organic matter in the surface layer suggests the same—this resulted in better use of K by the sugarcane crop.

5. Integrated agricultural systems and soil chemical parameters

The various results presented in this chapter mostly involve practices that interfere with the dynamics of MOS in systems with single or monoculture agricultural crops. However, it is important to highlight the participation of agroecosystems, where crops of agronomic interest coexist simultaneously or in succession with tree species (**Figure 9**).

Crop systems based on these principles bring to the same area a set of practices that tend to interfere in a very complex way in the dynamics of organic matter. Thus, there is still much to be elucidated regarding the aspects related to these agroecosystems. Most of the reports and scientific works in Brazil are concentrated in the Amazon and the Atlantic Forest. Particularly in the Cerrado region, some studies have been directed towards enhancing the use of pasture areas that were previously poorly managed.

Thus, it is believed that in the Cerrado region, the integrated crop-livestock (iLP) and integrated crop-livestock-forest (iLPF) systems have great potential to help increase organic matter content [24], as well as nutrient cycling [46]. These advantages can be provided by the greater input of organic residues due to the straw of annual crops, forage species, biomass of forest species, and the deposition of animal waste involved in the integration.

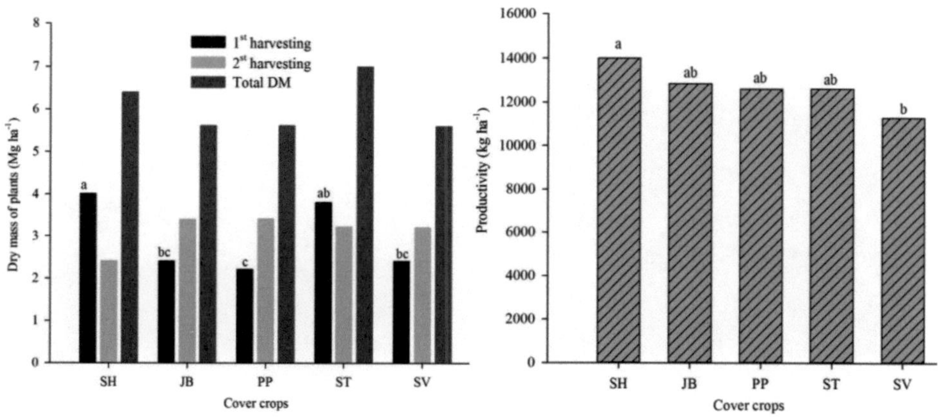

Figure 9.
*Left—Dry mass production (Mgha^{-1}) of cover crops in an agroforestry system. Sunn hemp (SH), jack beans (JB), pigeon peas (PP), stylosanthes (ST), and spontaneous vegetation (SV). 1st harvesting DM (F = 6.08**; CV = 26.4), 2nd harvesting DM (F = 1.47n.s.; CV = 24.5), total DM (F = 1.32n.s.; CV = 20.5). Right - Productivity of green maize cobs in intercropping with baru trees (*Dypterix alata*), grown on leguminous cover crops residues: sunn hemp (SH); jack beans (JB); pigeon peas (PP); stylosanthes (ST) e spontaneous vegetation (SV). F = 4.88**; CV = 7.74%. Adapted from [30].*

Salton et al. [29], in studies of the iLP system, observed that the use of pastures in rotation with crops increased the carbon content in the particulate fraction and the lability of SOM, while the lowest accumulation rates and C stocks occurred in monoculture systems. Mendonça et al. [35] obtained good results in the consortium of corn with other forage grasses, achieving an accumulation of up to 150 $kgha^{-1}$ of potassium, the release of which occurs up to 100% in the 90 days after sowing the soybean.

The forestry component, when used in the integration system, can reduce the effect of soil compaction, conditioned by the high stocking of animals, due to their aggressive root system, in addition to contributing to the supply of nutrients from cycling. In this sense, Gatto et al. [66] report the nutrient balance after eucalyptus harvesting, considering that the removal of the trunk (bark + wood) from the system is positive, allowing us to state that the cultivation of eucalyptus (*Eucalyptus* sp.), when managed correctly, can bring benefits to the system. However, the same authors, when considering a second forest cycle, noted that the contents of K, Ca, and Mg would be adequate when presenting nutrient stocks in the system, while for P and S, there would be a deficit of 10.66 and 23.16 kg ha^{-1}, respectively, requiring replacement through fertilization.

Freitas et al. [26], comparing monoculture of brachiaria, agroforestry systems (eucalyptus+brachiaria+maize), and agroforestry systems (eucalyptus+acacia+brachi aria+maize), observed increases in N, S, K, Ca, and Mg in the integrated systems, but with reductions in P levels. The same authors emphasize that the greater contribution of litter and nutrients, specifically N, in the agroforestry system with eucalyptus and acacia, has the potential for recovery and maintenance of pastures. In the integrated systems, the adoption of leguminous species with the potential for biological fixation of N should be considered.

Another production system, no less important, in which plants coexist harmoniously, forming an agroecosystem that brings benefits in terms of production and soil quality,

Figure 10.
*Left: Maize cultivated over crotalaria straw. Right: Maize intercropped with jack beans (*Canavalia brasiliensis*). Both photos from Federal University of Tocantins. Rainy season 2008. Source: the author.*

concerns the agroforestry system (AFS). This system has been more targeted at smallholders' agricultural systems but has potential for use in larger areas of land (**Figure 10**).

With a purpose more focused on less technically literate farmers in the Cerrado of Piauí, the work of Iwata et al. [27] demonstrates that the diversification of species in the agroforestry system and the time of introduction of the system bring improvements in soil fertility attributes, with increases in exchangeable bases, P, reduction of active acidity, exchangeable acidity, and potential acidity of the soil.

Collier et al. [30], used 3 leguminous and spontaneous vegetation in an agroforestry system in the Cerrado of the state of Goiás, intercropped in a 5-year-old trees. Dry mass production and decomposition rates of these species did not show any different standard from open areas. Crotalaria (Sunn hemp) residues showed an increase in P levels in the soil and promoted a 24% increase in the production of corn cobs in relation to spontaneous vegetation (**Figure 10**).

Agroforestry systems with Cerrado native tree species are generally associated with low fertility soils. The use of cover crops to increase soil fertility was studied by Arruda et al. [44] with satisfactory biomass production from guinea grass followed by the same intercrop with stylosants (**Figure 11**).

Menezes et al. [28] showed that intercropping of native and exotic species from tropical rainforests, with the presence of crops such as corn (*Zea mays*), beans (*Phaeseolus vulgaris* L.), rice (*Oryza sativa*), and cassava (*Manihot esculenta*) at the beginning of planting and without correction and fertilization, after 5 years, promoted a reduction in acidity levels and increased the levels of exchangeable Ca and Mg in relation to areas maintained with native forest vegetation. These authors emphasize that even though there is an effect of residues from the felling of species, the result is promising, considering the demands of the crops.

If we could assume that integrated systems value diversity practices between plant species, the table presented below (**Table 1**) aims to summarize the main effects of integrated systems on soil fertility. However, what we present may not occur in a standard way for many systems that do not have satisfactory soil management.

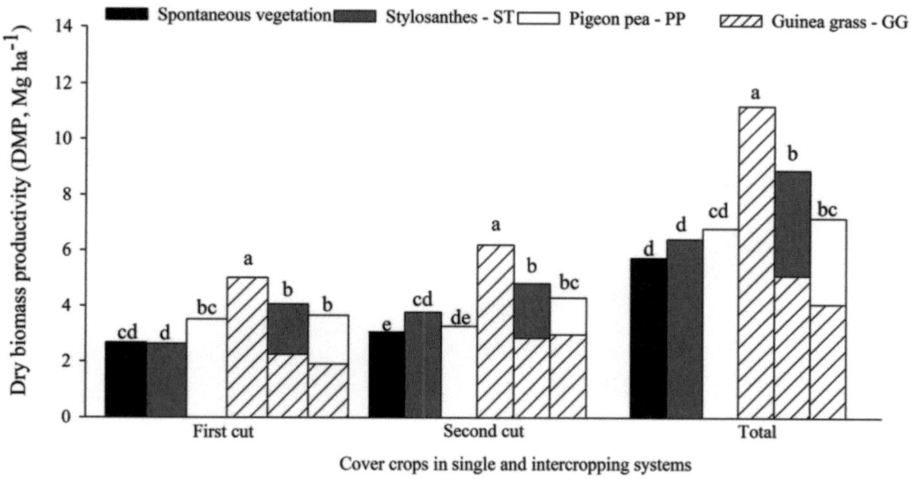

Figure 11.
Dry biomass productivity (DBP, Mgha^{-1}) of cover crops in single system in single systems: pigeon pea (PP), stylosanthes (ST) and guinea grass (GG), and in intercropping systems:(PP + GG, and ST + GG), and spontaneous vegetation. Treatments were compared by the Tukey-test ($p < 0.05$)*, results represented by lowercase letters in each cut. Adapted from [44].*

Soil chemical attributes	Main effects in Cerrado tropical conditions
Soil pH—soil acidity	Buffering soil acidification reactions. Smaller pH fluctuations.
N transformations in the soil-plant atmosphere system	Reduction in losses due to NH_3 volatilization and NO_3^- leaching processes, mainly. Increase in the period of microbial immobilization.
Phosphorus dynamics in the soil	Less impact of P fixation processes by Fe and Al oxides from tropical soils. Microbial action through phosphatases improving the efficiency of P absorption by plants grown in these systems.
Potassium dynamics in the soil	Reduction of leaching losses. Attenuation of imbalance in the presence of other basic cations. Less dependence on the use of mineral fertilizers.
Sulfur cycling in soil	Favorable environment for sulfate oxidation in organic residues in the soil, reducing dependence on mineral fertilizers.
Micronutrients in the soil	The imbalance between some micronutrients, such as metals, can be reduced. The efficiency of using some micronutrients such as boron can be improved.
Use of synthetic organic agrochemicals	A worsening of the desired effect may occur, such as the inefficiency of a dosage due to the chance of interaction with chemical compounds existing in the agroecosystem.

Table 1.
Summary of the main effects obtained in integrated cultivation systems in the Cerrado on chemical aspects of the soil related to plant nutrition.

6. Conclusion

The use of cover crops in agricultural systems in the Brazilian Cerrado is a reality.

The diversity of species is great, but among the grasses, brachiaria stands out and among the legumes, crotalarias.

Tackling the water problem by conserving soil cover is vital. More diverse systems such as agroforestry, despite being more complex, facilitate the adoption of cover crops and other technologies for soil management. The climate events of the last decade have been driving this work.

The use of cover crops brings efficiency in the use of fertilizers along with less use of fossil fuels and expands the search for possible biological interactions to maintain the viability of food production in this region.

The adoption of cover crops also involves a change in the way in which species that are great commodities are cultivated. The increase in biodiversity through crop rotation and integrated systems increases soil management in search of quality and resilience.

Author details

Leonardo Santos Collier[1*] and Karin Ferretto Santos Collier[2]

1 Federal University of Goiás State, Goiânia, Brazil

2 Plant Production, Goiás State, Goiânia, Brazil

*Address all correspondence to: leonardo@ufg.br

References

[1] Salgado AAR, Assis WL, Magalhães Junior AP, de Sordi MV, Oliveira FV. Semi-humid: Landscape of Central Brazil. In: Salgado AAR et al., editors. The Physical Geography of Brazil, Geography of the Physical Environment. 2019. DOI: 10.1007/978-3-030-04333-9_6

[2] Ruggiero PGC, Batalha MA, Pivello VR, Meirelles ST. Soil-vegetation relationship in Cerrado (Brazilian savanna) and semideciduous forest, Southeastern Brazil. Plant Ecology. 2002;**160**:1-16. DOI: 10.1023/A:1015819219386

[3] Reis T, Russo G, Ribeiro V, Moutinho P, Guimarães SM, Alencar A, et al. Climate Challenges and Opportunities in the Brazilian Cerrado: What is the Cerrado and Why is it Important? Police Brief. Brasilia: IPAM; 2017. DOI: 10.13140/RG.2.2.36452.48004

[4] Borghetti F, Barbosa E, Ribeiro L, Ribeiro JF, Maciel E, Walter BMT. Fitogeografia das savanas sul-americanas. Heringeriana. 2023;**17**(1):e918014. DOI: 10.17648/heringeriana v17i1.918014

[5] Lopes AS, Guilherme LRG. Fertilidade do solo e produtividade agrícola. In: Novais RF, Alvarez VH, Barros NF, Fontes RLF, Cantarutti RB, Neves JCL, editors. Fertilidade do solo. Viçosa, MG: SBCS; 2007. pp. 1-64

[6] Map Biomas. Bioma já perdeu quase metade de sua vegetação nativa com isso está ficando mais quente e seco. Available from: http://brasil.mapbiomas.org/2022/09/11/soja-ocupa-10-do-cerrado [Accessed: July 21, 2024]

[7] Greschuk LT, Demattê JAM, Silvero NEQ. A soil productivity system reveals most Brazilian agricultural lands are below their maximum potential. Scientific Reports. 2023;**13**:14103. DOI: 10.1038/s41598-023-39981-y

[8] Ofstehage A, Nehring R. No-till agriculture and the deception of sustainability in Brazil. International Journal of Agricultural Sustainability. 2021;**19**(3-4):335-348. DOI: 10.1080/14735903.2021.1910419

[9] Carvalho AM, Ribeiro RLP, Marchão RL, Oliveira AD, Pulrolnik K, Figueiredo CC. Chemical composition of cover crops and soil organic matter pools in no-tillage systems in the Cerrado. Soil Use and Management. 2022;**38**:940-952. DOI: 10.1111/sum.12746

[10] Alves EA. Produção de potássio. Panorama no Brasil e no Mundo. 2008. Available from: http://www.cetem.gov.br/agrominerais/livros/producaopotassio.pdf [Accessed: December 30, 2015]

[11] Cordeiro CFS, Echer FR, Araujo FF. Cover crops impact crops yields by improving microbiological activity and fertility in sandy soil. Journal of Soil Science and Plant Nutrition. 2021;**21**:1968-1977. DOI: 10.1007/s42729-021-00494-0

[12] Soares MB, Lima JP, Tavanti RFR, Freddi OS, Rigotti AR, Petter FA. Use of cover crops in the southern Amazon region: What is the impact on soil physical quality. Geoderma. 2021;**384**:114796. DOI: /10.1016/j.geoderma.2020.114796

[13] Teixeira CM, Carvalho GJ, Andrade MJB, Silva CAS, Pereira JM. Decomposição e liberação de nutrientes das palhadas de milheto e milheto+crotalária no plantio direto do

feijoeiro. Acta Scientiarum Agronomy. 2009;**31**(4):647-653. DOI: 10.4025/actasciagron.v31i4.1356

[14] Veronese M, Francisco EAB, Zancanaro L, Rosolem CA. Plantas de cobertura e calagem na implantação do sistema plantio direto. Pesquisa Agropecuária Brasileira. 2012;**47**(8):1158-1165. DOI: 10.1590/S0100-204X2012000800017

[15] Collier LS, Correia MAR, Ramos LN, Prado RM, Flores RA, Nunes TV. Adubação fosfatada no sulco e em faixas sob palhada de leguminosa e produtividade de milho em plantio direto no Tocantins. Revista Ceres. 2008;**55**(2):109-116. ISSN 0034-737X

[16] Heinrichs R, Vitti GC, Moreira A, Figueiredo PAM, Fancelli AL, Corazza EJ. Características químicas de solo e rendimento de fitomassa de adubos verdes e de grãos de milho, decorrente do cultivo consorciado. Revista Brasileira de Ciência do Solo. 2005;**29**(1):71-79. DOI: 10.1590/S0100-06832005000100008

[17] Bressan SB, Nóbrega JCA, Nóbrega RSA, Barbosa RS, Sousa LB. Plantas de cobertura e qualidade química de Latossolo Amarelo sob plantio direto no Cerrado maranhense. Revista Brasileira de Engenharia Agrícola e Ambiental. 2013;**17**(4):371-378. DOI: 10.1590/S1415-43662013000400003

[18] Pacheco LP, Barbosa JMB, Leandro WM, Machado PLOA, Assis RL, Madari BE, et al. Produção e ciclagem de nutrientes por plantas de cobertura nas culturas de arroz de terras altas e de soja. Revista Brasileira de Ciência do Solo. 2011;**35**(5):1787-1799. DOI: 10.1590/S0100-06832011000500033

[19] Souza GAVS, Santos D, Rios ES, Souza GJL. Agronomic evaluation of legume cover crops for sustainable agriculture. Russian Agricultural Sciences. 2018;**44**:31-38. DOI: 10.3103/S1068367418010093

[20] Miyazawa M, Pavan MA, Franchini JC. Neutralização da acidez do perfil do solo por resíduos vegetais. Piracicaba, SP: POTAFÓS; 2000. 8 p (Encarte Técnico, n. 92)

[21] Torres JLR, Pereira MG, Andrioli I, Polidoro JC, Fabian AJ. Decomposição e liberação de nitrogênio de resíduos culturais de plantas de cobertura em um solo de Cerrado. Revista Brasileira de Ciência do Solo. 2005;**29**(4):609-618. DOI: 10.1590/S0100-06832005000400013

[22] Collier LS, Castro DV, Dias Neto JJ, Brito DR, Ribeiro PAA. Manejo da adubação nitrogenada para o milho sob palhada de leguminosas em plantio direto em Gurupi, TO. Ciência Rural. 2006;**36**(4):1100-1105. DOI: 10.1590/S0103-84782006000400009

[23] Torres JLR, Pereira MG. Produção e decomposição de resíduos culturais antecedendo milho e soja num Latossolo no Cerrado mineiro. Comunicata Scientiae. 2014;**5**(4):419-426. DOI: 10.14295/cs.v5i4.508

[24] Rodrigues RdAR, Ferreira IGM, da Silveira JG, da Silva JJN, Santos FM, da Conceição MCG. Crop-livestock-Forest integration systems as a sustainable production strategy in Brazil. In: Søndergaard N, de Sá CD, Barros-Platiau AF, editors. Sustainability Challenges of Brazilian Agriculture. Environment & Policy. Vol. 64. Cham: Springer; 2023. DOI: 10.1007/978-3-031-29853-0_9

[25] Ferreira EVO, Anghinoni I, Andrighetti MH, Martins AP, Carvalho PCF. Ciclagem e balanço

de potássio e produtividade de soja na integração lavoura-pecuária sob semeadura direta. Revista Brasileira de Ciência do Solo. 2011;**35**(1):161-169. DOI: 10.1590/S0100-06832011000100015

[26] Freitas ECS, Oliveira Neto SN, Fonseca DM, Santos MV, Leite HG, Machado VD. Deposição de serapilheira e de nutrientes no solo em sistema agrossilvipastoril com eucalipto e acácia. Revista Árvore. 2013;**37**(3):409-417. DOI: 10.1590/S0100-67622013000300004

[27] Iwata BF, Leite LFC, Araújo ASF, Nunes LAPL, Gehring C, Campos LP. Sistemas agroflorestais e seus efeitos sobre os atributos químicos em Argissolo Vermelho-Amarelo do Cerrado piauiense. Revista Brasileira de Engenharia Agrícola e Ambiental. 2012;**16**(7):730-738. DOI: 10.1590/S1415-43662012000700005

[28] Menezes JMT, van Leeuwen J, Valeri SV, Cruz MCP, Leandro RC. Comparação entre solos sob uso agroflorestal e em florestas remanescentes adjacentes, no norte de Rondônia. Revista Brasileira de Ciência do Solo. 2008;**32**(4):893-898

[29] Salton JC, Mielniczuk J, Bayer C, Fabrício AC, Macedo MCM, Broch DL. Teor e dinâmica do carbono no solo em sistemas de integração lavoura-pecuária. Pesquisa Agropecuária Brasileira. 2011;**46**(10):1349-1356. DOI: 10.1590/S0100-204X2011001000031

[30] Collier LS, Arruda EM, Campos LFC, Nunes JNV. Soil chemical attributes and corn productivity grown on legume stubble in agroforestry systems. Revista Caatinga. 2018;**31**(2):279-289. DOI: 10.1590/1983-21252018v31n203rc

[31] WRB. Soils of the World According to the World Reference Base for Soil Resources. 2015. Available from: https://www.isric.org/sites/default/files/WRBSoilMap.pdf [Accessed: August 06, 2024]

[32] Buol SW. Soils and agriculture in central west and North Brazil. Scientia Agricola. 2009;**66**(5):697-707. DOI: 10.1590/S0103-90162009000500016

[33] Zalles V, Hansen MC, Potapov PV, Stehman SV, Tyukavina A, Pickens A, et al. Near doubling of Brazil's intensive row crop area since 2000. PNAS. 2019;**116**(2):428-435. DOI: 10.1073/pnas.1810301115

[34] Landers JN. Plantio Direto no Brasil Central e a Associação de Plantio Direto no Cerrado. In: Encontro nacional de Plantio Direto na palha and Encontro mundial do Sistema Plantio Direto. 2022. 271 p. FEBRAPDP, Foz do Iguaçu. Anais[...], Londrina: Madrepérola; 2022. ISBN: 978-65-87269-95-5

[35] Mendonça VZ, Mello LMM, Andreotti M, Pariz CM, Yano EH, Pereira FCBL. Liberação de nutrientes da palhada de forrageiras consorciadas com milho e sucessão com a soja. Revista Brasileira de Ciência do Solo. 2015;**39**:183-193. DOI: 10.1590/01000683rbcs20150666

[36] Ramos SJ, Faquin V, Rodrigues CR, Silva CA, Ávila FW, Sampaio RA. Utilização de fósforo e produção do feijoeiro: Influência de gramíneas forrageiras e fontes de fósforo. Revista Brasileira de Ciência do Solo. 2010;**34**(1):89-96. DOI: 10.1590/S0100-06832010000100009

[37] Torres JLR, Pereira MG, Fabian AJ. Produção de fitomassa por plantas de cobertura e mineralização de seus resíduos em plantio direto. Pesquisa agropecuária brasileira.

2008;**43**(3):421-428. DOI: 10.1590/S0100-204X2008000300018

[38] Ziglio CM, Miyazawa M, Pavan MA. Formas orgânicas e inorgânicas de mobilização do cálcio no solo. Brazilian Archives of Biology and Technology. 1999;**42**(2):257-262. DOI: 10.1590/S1516-89131999000200016

[39] Lange A, Carvalho JLN, Damin V, Cruz JC, Marques JJ. Alterações em atributos do solo decorrentes da aplicação de nitrogênio e palha em sistema semeadura direta na cultura do milho. Ciência Rural. 2006;**36**(2):460-467. DOI: 10.1590/S0103-84782006000200016

[40] Amaral AS, Anghinoni I, Heinrichs R, Bertol I. Movimentação de partículas de calcário no perfil de um cambissolo em plantio direto. Revista Brasileira de Ciência do Solo. 2004;**28**(2):359-367. DOI: 10.1590/S0100-06832004000200014

[41] Moraes MF, Cantarella H, Quaggio JA, Coscione AR. Mobilidade de íons em solo ácido com aplicação de calcário, ácido orgânico e material vegetal em superfície. Revista Brasileira de Ciência do Solo. 2007;**31**(4):673-684. DOI: 10.1590/S0100-06832007000400008

[42] Ferrari Neto J, Franzluebbers AJ, Crusciol CAC, Rigon JPG, Calonego JC, Rosolem CA, et al. Soil carbon and nitrogen fractions and physical attributes affected by soil acidity amendments under no-till on Oxisol in Brazil. Geoderma Regional. 2021;**24**:e00347. DOI: 10.1016/j.geodrs.2020.e00347

[43] Pacheco LP, Monteiro MMS, Silva RF, Soares LS, Fonseca WL, Nóbrega JCA, et al. Produção de fitomassa e acúmulo de nutrientes por plantas de cobertura no Cerrado piauiense. Bragantia. 2013;**72**(3):237-246. DOI: 10.1590/brag.2013.041

[44] Arruda EM, Collier LS, Oliveira KR, Flores RA, Barros LR, Ferraz-Almeida R, et al. Cover plants can contribute on macronutrient accumulation in agroforestry systems during off-season. Agroforestry Systems. 2023;**97**:1087-1096. DOI: 10.1007/s10457-023-00849-9

[45] Daudin D, Sierra J. Spatial and temporal variation of below-ground N transfer from a leguminous tree to an associated grass in an agroforestry system. Agriculture, Ecosystems and Environment. 2008;**126**(3-4):275-280. DOI: 10.1016/j.agee.2008.02.00

[46] Peoples MB, Brockwell J, Herridge DF, Rochester IJ, Alves BJR, Urquiaga S, et al. The contributions of nitrogen-fixing crop legumes to the productivity of agricultural systems. Symbiosis. 2009;**48**(1):1-17. DOI: 10.1007/BF03179980

[47] Echer FR, Castro GSA, Bogiani JC, Rosolem CA. Crescimento inicial e absorção de nutrientes pelo algodoeiro cultivado sobre a palhada de *Brachiaria ruziziensis*. Planta Daninha, Viçosa-MG. 2012;**30**(4):783-790. DOI: 10.1590/S0100-83582012000400012

[48] Mateus GP, Crusciol CAC, Pariz CM, Costa NR, Borghi E, Costa C, et al. Corn intercropped with tropical perennial grasses as affected by sidedress nitrogen application rates. Nutrient Cycling and Agroecosystems. 2020;**116**:223-244. DOI: 10.1007/s10705-019-10040-1

[49] Novais RF, Smyth TJ, Nunes FN. Fósforo. In: Novais RF, Alvarez VH, Barros NF, Fontes RLF, Cantarutti RB, Neves JCL, editors. Fertilidade do solo. Viçosa: SBCS; 2007. pp. 471-550

[50] Pavinato PS, Rosolem CA. Disponibilidade de nutrientes no solo—decomposição e liberação de compostos orgânicos de resíduos vegetais. Revista Brasileira de Ciência do Solo. 2008;**32**(3):911-920. DOI: 10.1590/S0100-06832008000300001

[51] Guareschi RF, Pereira MG, Perin A. Adsorption of P and forms of iron in notillage areas in the 'Cerrado' biome. Acta Scientiarum. Agronomy. 2015;**37**(1):109-116. DOI: 10.4025/actasciagron.v37i1.17686

[52] Chagas E, Araújo AP, Teixeira MG, Guerra JG. Decomposição e liberação de resíduos da cultura do feijoeiro. Revista Brasileira de Ciência do solo. 2007;**31**(4):723-729. DOI: 10.1590/S0100-06832007000400013

[53] Crusciol CAC, Moro E, Lima EV, Andreotti M. Taxas de decomposição e de liberação de macronutrientes da palhada de aveia preta em plantio direto. Bragantia. 2008;**67**(2):481-489. DOI: 10.1590/S0006-87052008000200024

[54] Santos GA, Silva LS, Canellas LP, Camargo FAO. Fundamentos da Matéria Orgânica do Solo: Ecossistemas Tropicais & Subtropicais. 2nd ed. Porto Alegre: Metrópole; 2008. 645 p

[55] Collier LS, Kikuchi FY, Benício LPF, Sousa SA. Consórcio e sucessão de milho e feijão-de-porco como alternativa de cultivo sob plantio direto. Pesquisa Agropecuária Tropical. 2011;**41**(3):306-313. DOI: 10.5216/pat.v41i3.8706

[56] Leite JNF, Cruz MCP, Ferreira ME, Andrioli I, Braos LB. Frações orgânicas e inorgânicas do fósforo no solo influenciadas por plantas de cobertura e adubação nitrogenada. Pesquisa Agropecuária Brasileira. 2016;**51**(11):1880-1889. DOI: 10.1590/S0100-204X2016001100010

[57] Rodrigues M, Withers PJA, Soltangheisi A, Vargas V, Holzschuh M, Pavinato PS. Tillage systems and cover crops affecting soil phosphorus bioavailability in Brazilian Cerrado Oxisols. Soil and Tillage Research. 2021;**205**:104770. DOI: 10.1016/j.still.2020.104770

[58] Benício LPF, Collier LS. Mixtures of phosphorus sources of different solubilities and uptake efficiency in corn intercropped with cover crops. Agri-Environmental Sciences. 2022;**8**:e022009. DOI: 10.36725/agries.v8i2.7913

[59] Werle R, Garcia RA, Rosolem CA. Lixiviação de potássio em função da textura e da disponibilidade do nutriente no solo. Revista Brasileira de Ciência do Solo. 2008;**32**:2297-2305. DOI: 10.1590/S0100-06832008000600009

[60] Ferreira ACB, Borin ALDC, Lamas FM, Ferreira GB, Resende AV. Exchangeable potassium reserve in a Brazilian savanna Oxisol after nine years under different cotton production systems. Scientia Agricola. 2022;**79**(4):e20200339. DOI: 10.1590/1678-992X-2020-0339

[61] Crusciol CAC, Cottica RL, Lima EV, Andreotti M, Moro E, Marcon E. Persistência de palhada e liberação de nutrientes do nabo forrageiro no plantio direto. Pesquisa Agropecuária Brasileira. 2005;**40**(2):161-168. DOI: 10.1590/S0100-204X2005000200009

[62] Torres JLR, Pereira MG. Dinâmica do potássio nos resíduos vegetais de plantas de cobertura no Cerrado. Revista Brasileira de Ciência do Solo. 2008;**32**(6):1609-1618. DOI: 10.1590/S0100-06832008000400025

[63] Rosolem CA, JPTMM V, Steiner F. Suprimento de potássio em função da

adubação potássica residual em um Latossolo vermelho do Cerrado. Revista Brasileira de Ciência do Solo. 2012;**36**(5):1507-1515. DOI: 10.1590/S0100-06832012000500015

[64] Firmiano RS, Mesquita GM, Rios AD, Buso WHD. Yield traits of soybean cultivated under brachiaria and millet straw and potassium doses. Revista de Agricultura Neotropical. 2022;**9**(1):e6547. DOI: 10.32404/rean.v9i1.6547

[65] Almeida HJ, Cruz FJR, Pancelli MA, Flores RA, Vasconcelos RL, Prado RM. Decreased potassium fertilization in sugarcane ratoons grown under straw in different soils. Australian Journal of Crop Science. 2015;**9**(7):596-604. ISSN:1835-2707

[66] Gatto A, Bussinguer AP, Ribeiro FC, Azevedo GB, Bueno MC, Monteiro MM, et al. Ciclagem e balanço de nutrientes no sistema solo-planta em um plantio de *Eucalyptus* sp., no Distrito federal. Revista Brasileira de Ciência do Solo. 2014;**38**(3):879-887. DOI: 10.1590/S0100-06832014000300019

Section 2

Precision Agriculture and Agroecological Practices

Chapter 6

Advancements in Precision Nitrogen Management for Sustainable Agriculture

Sandeep Kumar, Tanu Sharma, Sanjay Singh Rathore and Vinod Kumar Singh

Abstract

The chapter on precision nutrient management addresses critical issues stemming from imbalanced fertilization practices and poor nutrient management, which compromise crop productivity and environmental health. It explores cutting-edge techniques like Green Seeker, Leaf Color Chart (LCC), and Soil-Plant Analysis Development (SPAD) meters, coupled with site-specific nutrient management (SSNM), to optimize nitrogen applications. Strategies such as slow-release N fertilizers, integrated nutrient management (INM), and nitrogen splitting, This SPAD-based N management saved 33.3% N in rice and 18.8% N in wheat while improving agronomic N use efficiency to 58.5% and ANR to 32.2% in rice and 15.1% in wheat, compared to conventional fixed time N management. LCC saved an average of 51, 20, and 29 kg N ha^{-1} in rice, maize, and wheat, respectively. Innovations like nano urea and urease inhibitors, enhance nitrogen use efficiency (NUE). Research showed that using these inhibitors is beneficial for increasing both crop yields and NUE, with an average increase of 7.5% in yields and 12.9% in NUE. Application methods such as foliar spray and fertigation minimize environmental losses by targeted nutrient delivery. Decision support systems (DSS) and software platforms like Nutrient Experts and Crop Manager enable tailored nutrient applications based on soil tests and crop response, promoting sustainable agriculture practices and reducing reliance on imported N fertilizers.

Keywords: green seeker, LCC, nano urea, nitrogen use efficiency, nutrient experts, SPAD, SSNM

1. Introduction

The global use of fertilizers significantly enhances food production, contributing to 30–50% of crop yield [1]. However, concerns about nutrient use efficiency (NUE), environmental pollution, and global warming arise from current practices. Fertilizer nutrient recovery efficiency is low: 30–50% for N, 15–20% for P, and 40–50% for K, with secondary and micronutrients even lower at 5–12% [2, 3]. Major issues include nutrient depletion due to imbalanced fertilizer use, reduced use of organic sources like farm yard manure (FYM) and compost, lack of green manure integration, and poor irrigation management, leading to soil degradation. Imbalanced fertilization and poor nutrient management have led to stagnation or a decrease in crop productivity [4]. Nitrogen use efficiency (NUE) is poor, ranging from 30 to 50%. As a result, more than 50% of the nitrogen not assimilated by plants becomes a potential source of environmental pollution. This pollution includes groundwater contamination, eutrophication, acid rain, ammonia redeposition, global warming, and stratospheric ozone depletion [2]. Furthermore, the high expenditure on nitrogenous fertilizer imports imposes a financial burden on the central government [5].

The Green Revolution of the 1960s brought economic growth and food self-sufficiency, alleviating hunger through inorganic fertilizer use. Despite its negative impacts on soil and environmental health [6], the revolution boosted farming capacity and the economy. Recent advancements in nutrient application control, management, and prediction are particularly noteworthy. Effective nutrient management is vital to meet the growing global food demand and ensure sustainable production. Proper nutrient supply is crucial for plant growth and development, making it a key focus area in agricultural science.

In India, where agriculture is the backbone of the economy [7], precision nitrogen management is crucial. Nutrient management and recommendation processes in India still rely on response data averaged over extensive geographic regions. Traditional nitrogen application methods, often characterized by blanket recommendations and indiscriminate use of fertilizers, have led to several challenges, including nutrient imbalances, soil degradation, and environmental pollution [2, 8]. Indian agriculture is characterized by highly fragmented holdings, with each farmer managing small fields individually. This farming pattern results in significant variability between fields due to differences in farmer expertise, fertilization history, crop sequences, farm management practices, and resource availability. Generalized nutrient recommendations across such diverse, small-scale farming areas can lead to over- or under-application of nutrients, posing economic and environmental challenges. Addressing these concerns is crucial for the sustainability of agriculture. Precision nitrogen management offers a solution to these challenges by promoting site-specific nutrient management (SSNM) practices [9]. Individualized nutrient management is necessary for these farms to achieve optimal yields. Adapting the precision nutrient management concept to fit India's unique farming systems is anticipated to help reverse trends in productivity and soil fertility.

Precision nutrient management involves the precise application of nutrients, taking into account variations in soil composition and micro-climatic conditions within a field. Recommendations for nutrient management can vary from broad regional scales down to individual farms, single fields, or even specific parcels within a larger field. Precision nitrogen management is vital in modern agriculture. It aims to optimize nitrogen use efficiency while minimizing environmental impacts [9]. Geo-statistical analysis and GIS-based mapping offer opportunities to assess variability in native nutrient distribution and other yield-limiting or enhancing soil parameters across large areas. These tools can support the development of tailored nutrient management strategies, leading to improved yields and environmental stewardship. Research indicates a direct correlation between variability in production conditions and improvements in yield and profitability through effective variability management [10].

One of the key advancements in precision nitrogen management in India is the use of remote sensing technologies. Satellite imagery and drones equipped with multispectral and hyperspectral sensors enable farmers to monitor crop health and nutrient status in real time [11]. These technologies provide valuable data on plant nitrogen levels, allowing for precise and timely interventions. For instance, the use of the Normalized Difference Vegetation Index (NDVI) and other vegetation indices helps in assessing the nitrogen status of crops and guiding the application of nitrogen fertilizers based on actual crop needs [12, 13].

Another significant development is the adoption of decision support systems (DSS) and mobile applications that assist farmers in making informed nitrogen management decisions. These tools integrate weather data, soil information, and crop growth models to provide site-specific fertilizer recommendations. For example, Nutrient Expert [14] and Crop Nutrient Manager [15] offer customized nitrogen management advice to farmers, enhancing NUE and reducing wastage [16].

Soil health management plays a crucial role in precision nitrogen management. Soil testing and the use of soil health cards have gained prominence in India, enabling farmers to understand the nutrient status of their fields and apply fertilizers accordingly. The government's Soil Health Card Scheme has been instrumental in promoting balanced fertilization practices, thereby improving soil fertility and reducing the overuse of nitrogen fertilizers. By incorporating soil test results into precision nitrogen management plans, farmers can optimize nitrogen application rates, ensuring that crops receive the right amount of nutrients at the right time.

The development and commercialization of new fertilizer molecules such as nano urea, urease inhibitors, nitrification inhibitors, slow-release nitrogen fertilizers, urea super granules, sulfur-coated fertilizers, mud ball techniques, and customized fertilizers can enhance nitrogen use efficiency and minimize environmental degradation.

The integration of precision irrigation techniques with nitrogen management is another noteworthy advancement. Technologies such as drip irrigation and fertigation allow for the precise application of water and nutrients directly to the root zone [17–19], minimizing losses due to leaching and volatilization. This approach not only enhances NUE but also conserves water, a critical resource in many parts of India.

The development and promotion of nitrogen-efficient crop varieties also contribute to sustainable nitrogen management [20]. Research institutions in India are focusing on breeding crops that can utilize nitrogen more efficiently, reducing the need for excessive fertilizer inputs.

Integrating advanced technologies and innovative practices in nitrogen management can revolutionize Indian agriculture, ensuring sustainability and productivity.

2. Objectives of precision nutrient management

The need to follow the 4R principles (right dose, right source, right time, and right method) in nutrient management is crucial for several reasons:

1. *Optimized nutrient use*: by applying the right dose, we ensure that crops receive the exact amount of nutrients they need, avoiding both deficiency and excess. This optimizes nutrient use efficiency.

2. *Enhanced crop yields*: using the right source ensures that plants receive nutrients in the most effective form, which can significantly enhance crop yields and quality.

3. *Environmental protection*: applying nutrients at the right time prevents losses due to leaching, runoff, or volatilization, thus protecting water quality and reducing environmental pollution.

4. *Cost-effectiveness*: using the right method of application reduces wastage and ensures that nutrients are available to crops when they need them, making nutrient management more cost-effective for farmers.

5. *Sustainable agriculture*: the 4R principles support sustainable agricultural practices by maintaining soil health and fertility, thus ensuring long-term agricultural productivity.

3. Principles of precision nutrient management

Precision nutrient management (PNM) aims to align nutrient supply with crop demand more effectively. This approach involves evaluating soil fertility variations and recommending nutrient prescriptions based on the 4R principles (right dose, right source, right time, and right method). Research conducted at both research stations and farmers' fields has highlighted the importance of various PNM practices, such as soil test-based recommendations, integrated plant nutrient supply (IPNS), site-specific nutrient management (SSNM), soil test crop response (STCR), in-season real-time nitrogen supply, and decision support tools.

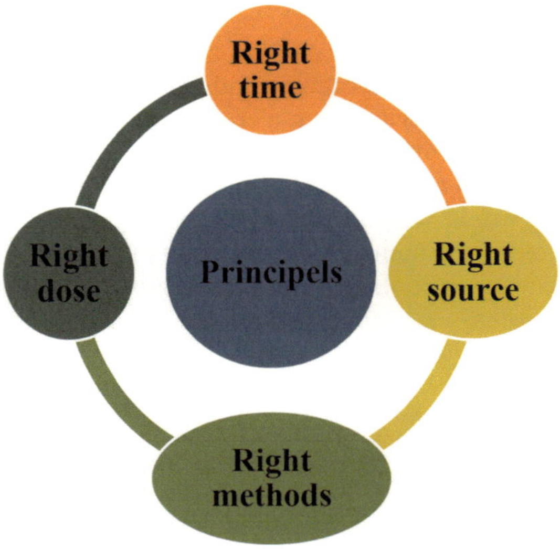

3.1 Right dose

The right dose of nitrogen application is crucial to optimize crop yields, enhance plant health, and prevent environmental pollution. Over-application can lead to run-off, water contamination, and greenhouse gas emissions, while under-application can limit crop growth and reduce agricultural productivity. Various techniques for calculating nitrogen have been developed so far.

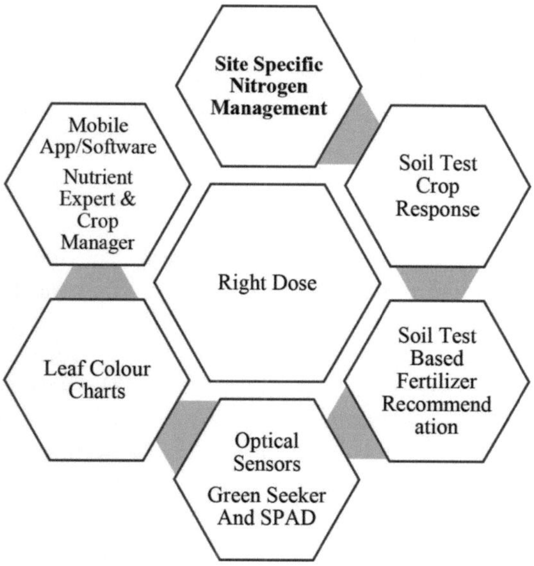

3.1.1 Site-specific nutrient management (SSNM)

Precision nitrogen management through site-specific nutrient management is essential to bridge the gap between the nutrient needs of high-yielding crops and the indigenous nutrient supply. By establishing a realistic yield target, estimating actual yield responses to fertilizer N application, and calculating agronomic nitrogen efficiency, farmers can determine the optimal nitrogen dose. This will be done through following steps:

Steps of SSNM:

1. *Establish a yield target*: the yield target (attainable in an average season) is the average yield of the past 3–5 years where nutrients are not limiting and good crop management is practiced and should not be more than 70–80% of potential yield.

2. *Estimate actual yield responses to fertilizer N application and associated agronomic efficiencies*: determine nitrogen-limited yield with nitrogen omission plots, which is the yield for a crop not fertilized with the N but fertilized with other nutrients. Calculate the N fertilized yield (NPK) and N-limited yield (PK) + No N.

3. *Estimate the agronomic N efficiency (Kg seed yield increased/kg of N applied)*

$$= \frac{\text{Seed yield in N fertilized plot (kg/ha)} - \text{Seed yield in N limited plot(kg/ha)}}{\text{Amount of N applied in N fertilized plot (kg/ha)}}$$

4. *Calculation of dose (kg/ha)*:

$$= \frac{\text{Seed yield in N fertilized plot (kg/ha)} - \text{Seed yield in N limited plot(kg/ha)}}{\text{Agronomic N efficiency (Kg seed yield increased/kg of N applied)}}.$$

This approach not only enhances crop yield and resource use efficiency but also minimizes environmental impact and ensures sustainable agricultural practices. By applying these steps, farmers can achieve higher productivity and profitability while safeguarding the environment.

3.1.2 Soil test crop response (STCR)

The STCR approach aims to establish a foundation for precisely adjusting fertilizer doses based on varying soil test values and targeted crop production levels. This method, as described by Ramamoorthy and Velayutham [21], provides a relationship between soil test values and crop yield. To calculate the nitrogen (N) dose using STCR, users need information on the average yield of the variety, average nutrient uptake (kg/ton), apparent nutrient recovery (ANR) of N fertilizer (kg/ha) for the crop, and the available N in the soil (kg/ha). After gathering the necessary information, the N dose (kg/ha) is calculated using the following formula:

$$\text{N dose (kg/ha)} = \text{Average yield (kg/ha)} \times \text{N uptake} - \text{Available N insoil (kg/ha)} \times \frac{100}{\text{ANR (\%)}} \tag{1}$$

The STCR approach is a robust and precise method for adjusting fertilizer doses based on soil test values and targeted crop yields. By leveraging the relationship between soil nutrient levels and crop yield, STCR enables farmers to optimize fertilizer application, ensuring that crops receive the necessary nutrients without overapplication. This method enhances crop productivity, improves nutrient use efficiency, and reduces environmental impacts, contributing to more sustainable and cost-effective agricultural practices.

3.1.3 Soil test-based recommendation/nutrient indexing

In these methods, available nitrogen needs to be analyzed using suitable techniques. The categories for low, medium, and high nitrogen levels are as follows: low: less than 280 kg/ha; medium: 280–560 kg/ha; high: greater than 560 kg/ha. Then, nutrient indexing will be performed using the following formula:

$$NI = \frac{\left(\text{No of samples in high category} \times 3\right) + \left(\text{no of samples in medium cateogory} \times 2\right) + \left(\text{no of samples in low cateogory}\right)}{\text{Total numbers of samples tested}} \tag{2}$$

Based on the nutrient index category (low, medium, or high), nutrient recommendations will be made:

- *Low category:* if the soil fertility status is in the low category, 25% more fertilizer will be recommended.
- *Medium category:* for soils in the medium category, the normal fertilizer recommendation will be followed.

- *High category:* if the soil fertility status is in the high category, 25% less fertilizer will be recommended.

This approach aims to achieve optimum productivity and profitability while ensuring sustainability.

3.1.4 Green Seeker™ technology

Green Seeker™ is a tool that measures the NDVI value of crop vegetation. NDVI serves as a valuable tool for assessing nitrogen content indirectly by measuring vegetation health and vigor, which are closely linked to nitrogen availability and utilization in crops. High NDVI values are a positive indicator of vegetation health, photosynthetic activity, and overall crop productivity potential. The procedure for calculating nitrogen using a Green Seeker™ involves several steps: First, the instrument is calibrated for the specific crop and growth stage by measuring crop reflectance at different nitrogen levels or stages. Once calibrated, the Green Seeker™ measures the reflectance of the crop canopy in near-infrared and red wavelengths, which correlate with nitrogen levels due to their sensitivity to chlorophyll content. These reflectance values are then analyzed to derive an index such as NDVI (**Figure 1**).

$$\text{NDVI} = \frac{\text{Near infrared} - \text{Red}}{\text{Near infrared} + \text{Red}} \tag{3}$$

Using a calibration curve or specific algorithm, these index values are converted into estimated nitrogen levels in the crop. A simple, farmer-friendly android mobile application was created to utilize NDVI values from Green Seeker™ for accurately calculating the required amount of urea for topdressing, aimed at enhancing nitrogen use efficiency. Based on these calculations, fertilizer application rates are adjusted to meet crop requirements while optimizing nitrogen use efficiency. Throughout the growing season, nitrogen levels are monitored using the Green Seeker™ at key growth

Figure 1.
Green Seeker. Green Seeker™ tool might help reduce your nitrogen costs - Corn (msu.edu).

Treatments	Grain yield (t/ha)	Cost of cultivation (₹/ha)	Net returns (₹/ha)	Net B:C	Agronomic efficiency (kg grain increased/kg N added)	Apparent N recovery (%)
N_o	2.73	32,066	21,100	0.72	—	—
N_1	6.3	34,996	87,286	2.55	24.2	35.4
N_2	6.5	34,508	91,616	2.71	28.8	42.6
N_3	6.9	34,570	97,927	2.88	26.9	46.3
SEm±	0.08	—	1497	0.05	0.43	0.62
LSD ($P \leq 0.05$)	2.73	—	4293	0.13	1.23	1.79

No is no-N applied, N_1 is 33% N as basal, 33% N at knee high stage and 34% N at silking stage, N_2 is 50% was basal + rest N as guided by Green Seeker (GS), N_3 is 75% as basal + rest N as guided by GS (source: [4, 8, 22]).

Table 1.
Effect of Green Seeker™ based nitrogen management on maize productivity, profitability and nitrogen use efficiency.

stages to fine-tune application rates as needed. This process ensures precise nitrogen management, aiming to maximize crop yield and quality while minimizing environmental impacts and input costs (**Table 1**).

3.1.5 Soil-plant analysis development (SPAD)

The SPAD meter measures the relative chlorophyll content in leaves, which can correlate with the nitrogen level in plants. Using a SPAD meter to estimate nitrogen levels involves several steps. First, calibrate the meter as per the manufacturer's instructions to suit the specific crop. Next, select healthy, fully expanded leaves that represent the overall nitrogen status of the crop. Place the SPAD meter probe on the leaf surface, ensuring it covers the sensor area completely, and record the SPAD value displayed on the meter. Repeat measurements on several leaves across different parts of the field for a representative sample. Use these SPAD values to estimate the crop's nitrogen status—higher values typically indicate sufficient nitrogen, while lower values may suggest nitrogen deficiency. Adjust nitrogen application rates based on these readings to optimize efficiency and ensure proper crop growth (**Figure 2**).

Ghosh et al. [23] reported that the top dressing with 25 kg N ha^{-1} at medium SPAD levels (S_{36} for rice and S_{40} for wheat) resulted in higher soil N availability, leaf N content, and grain yields, with rice yielding 5.21 t ha^{-1} and wheat 4.48 t ha^{-1}. This approach saved 33.3% N in rice and 18.8% N in wheat, while improving agronomic N use efficiency to 58.5% and ANR to 32.2% in rice and 15.1% in wheat, compared to conventional fixed time N management. The SPAD-based management strategy demonstrated significant potential for efficient N fertilizer management, with the optimal SPAD thresholds estimated at 37.5 for rice and 41.8 for wheat. However, research suggests that maintaining a leaf SPAD reading of around 37–50 [24, 25] is generally beneficial for maize crops.

3.1.6 Leaf Color Chart (LCC)

The Leaf Color Chart is utilized for assessing nitrogen fertilizer requirements in rice crops. Featuring four green strips spanning from yellow-green to dark green, the

DOI: http://dx.doi.org/10.5772/intechopen.1006470

Figure 2.
SPAD. Chlorophyll fluorometer—chlorophyll meter latest price, manufacturers & suppliers (indiamart.com).

Figure 3.
Leaf Color Chart. http://www.knowledgebank.irri.org/.

LCC evaluates the leaf's greenness to gauge its nitrogen content. How to use the Leaf Color Chart (**Figure 3**)

1. *Select plants for testing*

    - Randomly select at least 10 disease-free rice plants or hills in a field, where plant population is uniform.

2. *Match the leaf to the chart*

    - Select the topmost, youngest, fully expanded leaf from each hill or plant. This part best reflects the N status of the plants.

    - Place the middle part of the leaf on the LCC and compare its color with the color panels. Do not detach or destroy the leaf.

3. *Measure the leaf color*

- Measure the leaf color under the shade of your body. Direct sunlight affects leaf color readings.
- If possible, the same person should read the LCC at the same time of the day, every time.
- If the color of a rice leaf is in between two shades, take the average of the two values as the reading. For example, if the color is in between 3 and 4, the reading should be 3.5.

4. *Determine the average LCC*

- Take the reading of the 10 leaves, and determine the average. If the color is more or less than 3, N fertilizer top dressing is needed.

Apply the Leaf Color Chart every 7–10 days, beginning from the onset of tillering (14 days after transplanting), and maintain this schedule until 5–10 days after panicle initiation.

A long-term (2000–2011) field experiments were conducted by Singh et al. [26] to evaluate the effectiveness of using threshold leaf color greenness to guide in-season, need-based fertilizer N top-dressings for various rice, wheat, and maize varieties across diverse agro-climatic conditions in Punjab. The evaluation of this need-based fertilizer N management strategy, using a Leaf Color Chart, involved 461 on-farm experiments in rice, 96 in wheat, and 23 in maize. Results indicated that blanket recommendations failed to adequately synchronize N supply with the actual crop N demand. In contrast, LCC-based N management effectively addressed field-to-field and temporal variations in N supply to plants, showing promise for improving fertilizer N use efficiency in rice, wheat, and maize. LCC-based N management consistently produced grain yields equal to or greater than those achieved with blanket N recommendations while saving an average of 51, 20, and 29 kg N ha^{-1} in rice, maize, and wheat, respectively. The LCC, being a simple, economical, and farmer-friendly tool, can be used even by illiterate farmers to make immediate decisions on N supply based on crop needs (**Figure 4**).

3.1.7 IPNI Nutrient Expert® for android tablets and smartphones

The International Plant Nutrition Institute (IPNI) has developed android-based versions of the Nutrient Expert® app for various crops and regions, including maize, rice, and wheat for China and South Asia; soybean for China; and maize for the Philippines. This application aims to promote the dissemination of 4R-based nutrient management recommendations—applying the right source of nutrients at the right rate, at the right time, and in the right place—which is a major priority for IPNI. Available in English and local languages such as Chinese, Hindi, and Tagalog, the app can be downloaded for free from http://software.ipni.net. Once installed on a mobile device, it provides nutrient recommendations even without an internet connection. A summary recommendation can be sent to farmers via SMS, while a full report can be emailed. The Nutrient Expert® app allows crop advisers and agricultural extension workers to obtain fertilizer recommendations tailored to specific field conditions,

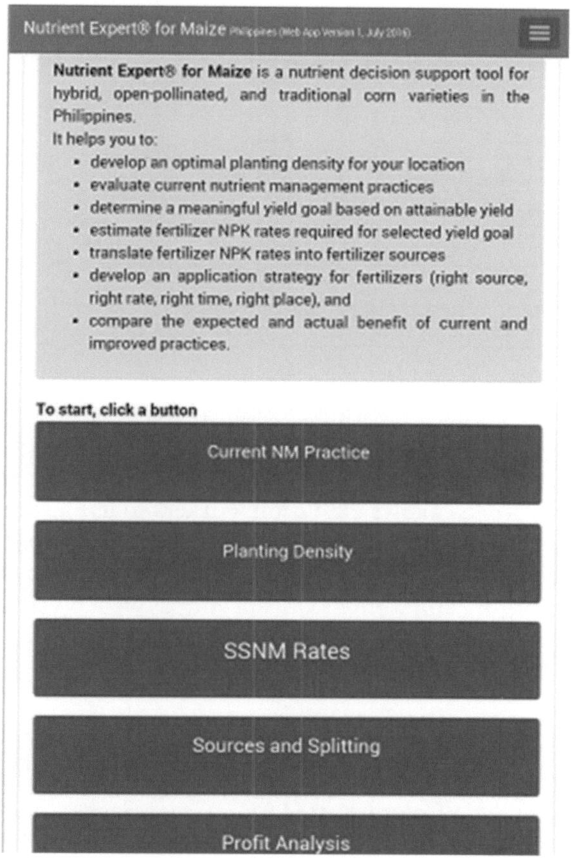

Figure 4.
Nutrient expert. http://software.ipni.net.

helping farmers increase yield, profit, and fertilizer use efficiency. Additionally, the app includes a 'Profit Analysis' module that compares the projected profit of current farming practices with those recommended by Nutrient Expert®. The Nutrient Expert® for Maize (Philippines) also features a 'Net Profit Analysis' calculator and a 'Cost-Profit Scenarios' table to assist farmers in making informed fertilizer investment decisions. All Nutrient Expert® apps are based on field-validated models, with underlying concepts and principles widely published in scientific journals. Upcoming releases include Nutrient Expert® apps for Sub-Saharan Africa and North Africa, which are currently undergoing field validation and are expected to be available in 1–2 years (**Table 2**).

3.1.8 Crop manager

A web-based platform for field-specific information on crop and nutrient management to increase yields and income of rice farmers (https://www.irri.org/crop-manager). The International Rice Research Institute (IRRI) and its partners have developed site-specific nutrient management (SSNM) principles to determine the optimal nutrient management practices for individual rice fields. Key components of these principles include the timing, quantities, and types of fertilizers used. Recommendations generated by the Rice Crop Manager (RCM) are provided to farmers via a

Treatments	Grain yield (t/ha)	Harvest index (%)	Grain N content (%)
Soil test-based	4.75	38.92	1.58
Nutrient Expert® based	4.86	39.08	1.58
Nutrient Expert based + Green Seeker™ based	5.05	39.36	1.60
SEm±	0.02	0.14	0.006
LSD ($P \leq 0.05$)	0.06	NS	0.019

Source: [27].

Table 2.
Effect of precision nitrogen management on maize productivity, harvest index and nitrogen content.

one-page printout and short messaging service (SMS) to enhance and guide crop management. This approach allows extension workers to use a computer or smartphone to offer farmers crop management advice tailored to their specific field conditions. IRRI has developed crop managers for three states *viz.,* Uttar Pradesh (http://webapps.irri.org/in/up/rwcm/), Odisha (http://webapps.irri.org/in/od/rcm/; [16]) and Bihar (http://webapps.irri.org/in/br/cmrs/) (**Figure 5**).

3.1.9 Integration of precision nitrogen management tools

To optimize nutrient management effectively, integrate the Nutrient Expert® Decision Support Tool, Leaf Color Chart (LCC), Green Seeker, and SPAD meter. Begin with a thorough soil test to assess essential nutrient levels. Utilize the Nutrient Expert® tool to calculate crop-specific nutrient requirements and provide initial recommendations.

To ensure precise nitrogen management throughout the growing season, adopt the following approach: Regularly monitor critical growth stages using standardized LCC, Green Seeker™, and SPAD meters. Adjust nitrogen application rates based on meter readings to optimize crop health and maximize yield potential. This integrated approach enhances efficiency, reduces input costs, and promotes sustainable agricultural practices (**Table 3**).

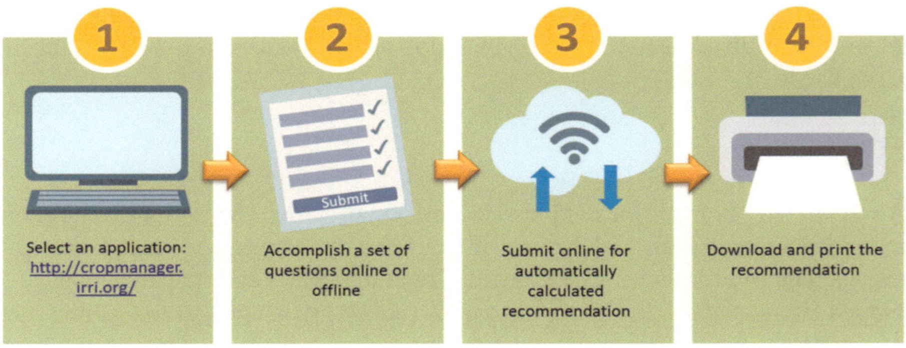

Figure 5.
Crop manager. https://www.irri.org/crop-manager.

Treatments	Grain yield (kg/ha)	Net returns (₹/ha)	Net B:C	Harvest index	Test weight (g)
T_1	5415.2	59,916	2.86	34.6	276.6
T_2	5741.3	67,643	3.26	35.6	283.6
T_3	6217.7	75,194	3.47	37.2	299.6
T_4	5538.3	64,486	3.17	34.8	277.7
T_5	5960.9	71,284	3.37	36.9	287.9
T_6	5152.9	55,616	2.74	34.7	272.2
SEm±	159.9	2718	0.09	0.3	5.2
LSD ($P \leq 0.05$)	479.4	8150	0.28	1.0	15.5

Where: T_1 consists of state recommendation with 150-80-60 kg/ha. T_2 and T_3 consist of Nutrient Expert recommendations for 6 and 7 t/ha respectively. T_4 and T_5 include Nutrient Expert recommendation with Green Seeker™ reading for second split for 6 and 7 t/ha respectively. T_6 State recommendation with Green Seeker for second split comprised (source: [28]).

Table 3.
Effect of Green Seeker™ based nitrogen management on maize productivity and profitability.

3.1.10 Nutrient management based on nutrient management zones

Nutrient management based on nutrient management zones (NMZs) is a precision agriculture approach that optimizes the application of fertilizers and soil amendments by dividing a field into distinct zones with similar nutrient needs (**Figure 6**).

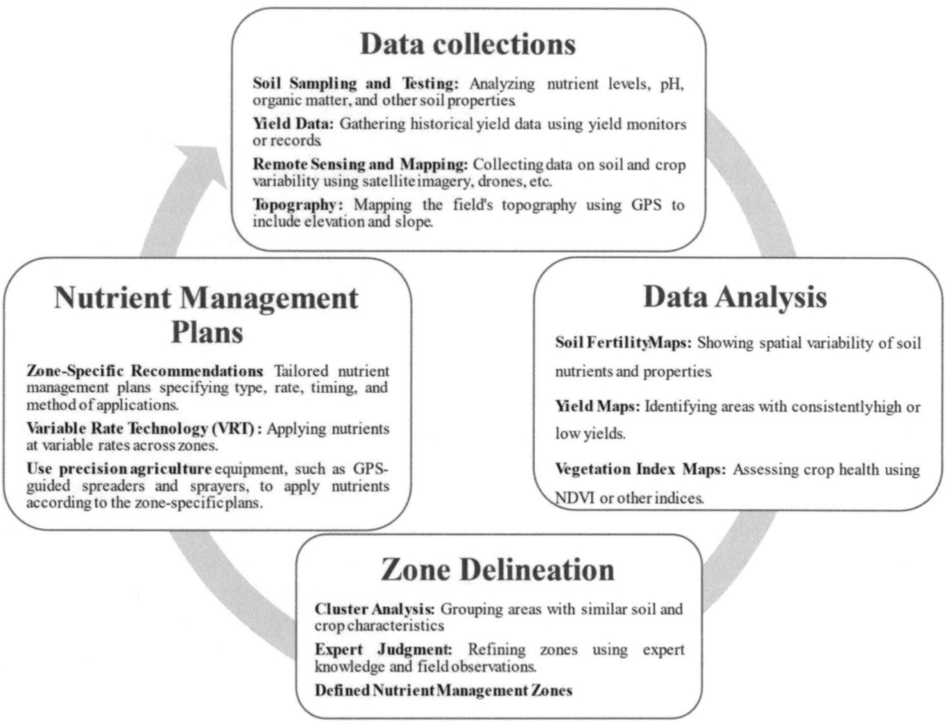

Figure 6.
Steps to develop and implement nutrient management zones.

Nutrient management based on nutrient management zones optimizes fertilizer application by tailoring it to specific soil fertility variations across fields. This precision approach, guided by soil testing and crop nutrient requirements, enhances agricultural productivity while minimizing environmental impact. By strategically allocating nutrients where they are most needed, farmers can improve crop yield, quality, and profitability. Moreover, it supports sustainable practices by reducing over-application of fertilizers, which can lead to nutrient runoff and environmental degradation. Ultimately, nutrient management zones enable efficient resource use, contributing to resilient and environmentally conscious agricultural systems.

3.2 Right source

Using the right source of nitrogen is essential because it ensures that the nutrient is efficiently available to plants, enhancing their growth and yield. The correct source minimizes environmental impact, reduces nitrogen loss through leaching or volatilization, and matches the specific needs of different crops and soil types, thereby promoting sustainable agriculture.

3.2.1 Coated urea

Coated urea is a type of fertilizer where the urea granules are coated with various materials to control the release of nitrogen into the soil, enhancing its efficiency and reducing nitrogen losses. Some examples of coated urea: *Neem-coated urea*: urea granules are coated with neem oil or neem extracts. Neem acts as a natural urease inhibitor, reducing nitrogen loss through volatilization. Enhances nitrogen use efficiency and reduces the need for frequent fertilization. The neem coating also has pesticidal properties, helping to control soil-borne pests and diseases. *Sulfur-coated urea:* urea granules are coated with sulfur to slow the release of nitrogen). *Polymer-coated urea*: urea granules are coated with a polymer material. Provides a more controlled and predictable release of nitrogen. *Polymer-sulfur-coated urea*: a combination of sulfur and polymer coatings. *Urease and nitrification inhibitor-coated urea*: dicyandiamide (DCD) and 3,4-dimethylepyrazole phosphate (DMPP) and the urease inhibitor N-(n-butyl) thiophosphoric triamide (NBPT). *Biodegradable polymer-coated urea*: urea granules coated with biodegradable polymers.

3.2.2 Urease and nitrification inhibitors

Nitrogen and urease inhibitors are chemicals that slow down the conversion of nitrogen fertilizers into forms that can be lost to the environment. These inhibitors help retain nitrogen in the soil for a longer period, enhancing its availability to plants and improving NUE. Nitrification and urease inhibitors are suggested as methods to reduce nitrogen losses and enhance crop nitrogen use efficiency. However, their impact on crop yield varies. A meta-analysis was conducted by Abalos et al. [29] to assess their effectiveness in improving NUE and crop productivity. The study focused on commonly used nitrification inhibitors—DCD and DMPP and the urease inhibitor NBPT, which are considered among the best available options. The results indicated that using these inhibitors is beneficial for increasing both crop yields and NUE, with an average increase of 7.5% in yields and 12.9% in NUE. However, their effectiveness varied depending on environmental and management factors in the evaluated studies. The most significant responses were observed in coarse-textured soils, irrigated systems, and crops receiving high nitrogen fertilizer rates. In particular, the urease inhibitor NBPT had the largest effect size in alkaline soils (pH $\geq$ 8).

3.2.3 Slow-release nitrogen fertilizers

Slow-release nitrogen fertilizers are designed to release nutrients gradually over time, matching the nutrient uptake patterns of crops. This reduces the risk of nitrogen leaching and volatilization, leading to improved NUE and better crop yields.

3.2.4 Nano urea

Nano urea is a nanotechnology based revolutionary Agri-input which provides nitrogen to plants (https://nanourea.in/en/nano-urea). Nano urea is a sustainable option for farmers toward smart agriculture and combat climate change. These fulfill the plant nutrient requirement as a fertilizer since Nano urea is bio available to plants because of its desirable particle size of about 20–50 nm and more surface area (10,000 times over 1 mm urea prill) and number of particles (55,000 nitrogen particles over 1 mm urea prill). Hence, nano urea increases its availability to crops by more than 80% resulting in higher nutrient use efficiency. In addition to this, nano urea helps in minimizing the environmental footprint by reducing the loss of nutrients from agriculture fields in the form of leaching and gaseous emissions which are used to cause environmental pollution and climate change [17–19].

Upadhyay et al. [30] reported that using 75% of the recommended nitrogen with conventional fertilizer combined with a nano-urea spray (N_{75}PK + nano-urea) reduced energy requirements by approximately 8–11% and increased energy use efficiency by about 6–9% compared to the standard 100% nitrogen application with prilled urea fertilizer. Additionally, the N_{75}PK + nano-urea treatment resulted in ~14% higher economic yields across all crops compared to the N_{50}PK + nano-urea treatment. This approach also maintained comparable soil nitrogen and dehydrogenase activities to conventional fertilization (N_{100}PK), indicating that foliar spraying nano urea with 75% N is supportive of soil health. Moreover, two foliar sprays of nano urea reduced nitrogen usage by 25% without compromising yield and significantly lowered greenhouse gas emissions from 164.2 to 416.5 kg CO_2-eq ha^{-1} across different crops.

3.2.5 Urea super granules

Urea super granules (USG) are large, discrete particles of ordinary urea ($(NH_2)_2CO$) [31]. Each USG particle typically weighs between 1 and 2 grams. They come in spherical shapes produced through melt granulation or pillow shapes with broken edges formed by briquetting. USG offers several benefits: It enhances nitrogen use efficiency. The use of nitrification inhibitors reduces nitrogen leaching. USG dissolves completely in water, ensuring 100% solubility. It provides broader spread patterns, allowing for more efficient coverage of larger areas compared to other nitrogen fertilizers like UAN and anhydrous ammonia.

3.2.6 Customized fertilizers

Customized fertilizers are specifically formulated to meet the unique nutrient requirements of particular crops, soil types, and climatic conditions. These fertilizers are tailored to optimize plant growth, enhance yield, and improve nutrient use efficiency. Customized fertilizers contain a specific blend of macro and micronutrients designed for the targeted crop. The composition is based on soil tests, crop requirements, and environmental conditions.

3.3 Right time

To optimize N use efficiency, it is essential to apply the correct amount of N in balanced ratios with other essential nutrients and at the appropriate plant growth stages. Early in the growing season, crop nitrogen demand is minimal, but it significantly increases during the late vegetative and early reproductive stages when plants require more nitrogen for growth and development. Applying nitrogen fertilizer just before or during these critical periods ensures that the crop can efficiently utilize the nutrient for maximum yield potential. Due to N susceptibility to leaching, especially in regions with high rainfall or irrigation, it is prudent to split nitrogen applications. This approach minimizes losses and ensures that nitrogen remains available to the crop when it needs it most. Real-time nitrogen application can further enhance precision farming practices. Tools such as Green Seeker™ and SPAD meters, along with LCC, enable farmers to monitor crop N status throughout the season. By adjusting N applications based on real-time sensor readings, farmers can optimize crop nutrition, reduce input costs, and minimize environmental impacts associated with N runoff or emissions, thus promoting sustainable agricultural practices.

3.4 Right method

Using the right method for nitrogen application is crucial because it ensures efficient nutrient uptake by crops, maximizes yield, and minimizes environmental impact. The correct method reduces nitrogen loss through leaching, volatilization, or runoff, improves soil health, and optimizes resource use, leading to more sustainable and cost-effective agricultural practices.

3.4.1 Integrated nutrient management (INM)

INM combines the use of chemical fertilizers with organic amendments, such as compost and manure, to provide a balanced nutrient supply. This approach not only enhances soil fertility and crop productivity [32] but also improves the efficiency of nutrient use, promoting sustainable agriculture.

3.4.2 Split application

Apply a portion of the nitrogen fertilizer before or during planting to support early crop growth. The remaining nitrogen in split doses during critical growth stages, such as tillering and booting for cereals.

3.4.3 Use of controlled-release fertilizers

Use controlled-release or slow-release nitrogen fertilizers to provide a steady supply of nitrogen over an extended period and minimizing nitrogen losses.

3.4.4 Foliar spray

Foliar spray involves applying fertilizers directly to the leaves of plants. This method allows for quick nutrient uptake and is particularly useful for addressing nutrient deficiencies during critical growth stages. Foliar applications can also reduce the risk of nutrient losses associated with soil-applied fertilizers.

3.4.5 Fertigation

Fertigation is the practice of applying fertilizers through irrigation systems. This technique enables precise control over the timing and amount of nutrient application, ensuring efficient use of fertilizers and minimizing environmental losses.

3.4.6 Mud ball techniques

This technique, involves placing a small amount of urea fertilizer at the center of a mud ball. After the mud ball is dried, it is strategically placed in the rice field at specific stages of crop growth. This method allows the mud ball to act as a slow-release fertilizer, gradually providing nitrogen to the crop over an extended period.

3.4.7 Crop rotation

Rotating nitrogen-demanding crops with leguminous crops (like pulses) that fix atmospheric nitrogen can naturally replenish soil nitrogen levels and reduce reliance on external inputs.

3.4.8 Biofertilizers

Incorporating biofertilizers such as nitrogen-fixing bacteria (e.g., *Rhizobium*) and mycorrhizal fungi enhances nitrogen availability in the soil and improves nutrient uptake efficiency by crops.

3.4.9 Precision irrigation

Implementing precision irrigation techniques alongside fertilization helps synchronize nutrient application with water availability, optimizing nutrient uptake and minimizing losses.

3.4.10 Climate-smart agriculture (CSA) practices

Adopting CSA practices such as agroforestry, agroecology, and sustainable intensification that integrate nitrogen management with climate resilience goals, enhancing overall farm sustainability.

4. Conclusions

Day by day, imbalanced fertilization practices and poor nutrient management not only hinder crop productivity but also diminish nitrogen use efficiency (30–50%), leading to environmental degradation, including ozone layer depletion and water quality issues. Additionally, excessive reliance on imported nitrogenous fertilizers places a burden on the central government's finances. Addressing these challenges is crucial. Advancements in precision nutrient management techniques and technologies offer a holistic solution to optimize nitrogen use in agriculture. Integrating tools like Green Seeker™, LCC, and SPAD meters with site-specific nutrient management (SSNM) enables precise nutrient applications. Improving nitrogen use efficiency (NUE) through slow-release fertilizers, Integrated Nutrient Management (INM), nitrogen splitting, nano urea, and inhibitors, alongside advanced methods such as foliar spray and fertigation, significantly reduces environmental impacts and reliance on imported fertilizers. Decision support systems further facilitate informed nutrient management decisions, promoting sustainable agricultural practices.

Author details

Sandeep Kumar[1*], Tanu Sharma[2], Sanjay Singh Rathore[3] and Vinod Kumar Singh[4]

1 ICAR-National Bureau of Soil Survey & Land Use Planning, Regional Centre, Jorhat, India

2 Sardar Vallabhbhai Patel University of Agriculture and Technology, Meerut, India

3 Division of Agronomy, ICAR-Indian Agricultural Research Institute, New Delhi, India

4 ICAR-Central Research Institute of Dryland Agriculture (CRIDA), Hyderabad, India

*Address all correspondence to: sk0097411@gmail.com

References

[1] Stewart WM. News and Views–A Regional News Letter. Potash and Phosphate Institute (PPI), USA and the Potash and Phosphate Institute of Canada (PPIC); 2002. p. 2

[2] Ladha JK, Pathak H, Krupnik TJ, Six J, van Kessel C. Efficiency of fertilizer nitrogen in cereal production: Retrospects and prospects. Advances in Agronomy. 2005;**87**:85-156

[3] Rao KV. Research Themes-Site–Specific Integrated Nutrient Management for Sustainable Rice Production and Growth. Rajendranagar, Hyderabad, India: Rice Knowledge Management Portal (RKMP), Directorate of Rice Research; 2014

[4] Kumar S, Singh VK, Shekhawat K, Upadhyay PK, Dwivedi BS, Rathore SS, et al. Enhancing productivity, economics and energy efficiency through precision nitrogen and water management in conservation agriculture-based maize (*Zea mays*) in the indo-Gangetic Plains. The Indian Journal of Agricultural Sciences. 2024;**94**(3):333-336

[5] Gulati A, Banerjee P. Rationalising Fertiliser Subsidy in India: Key Issues and Policy Options (No. 307). Working Paper. Indian Council for Research on International Economic Relations; 2015. Available from: https://www.econstor.eu/bitstream/10419/176325/1/icrier-wp-307.pdf

[6] Bhatt R, Kukal SS, Busari MA, Arora S, Yadav M. Sustainability issues on rice–wheat cropping system. International Soil and Water Conservation Research. 2016;**4**(1):64-74

[7] Pande D. Role of agriculture in the economic development of a country-an overview. In: Indian Economy Indian Economy–Prospects and Future. Conference: International Peer Reviewed Research Journal. 2017. Available from: https://www.researchgate.net/publication/361901482_Role_of_Agriculture_in_the_Economic_Development_of_A_Country_-An_Overview

[8] Kumar S, Singh VK, Shekhawat K, Upadhyay PK, Rathore SS, Didawat RK. Real time nitrogen and irrigation management for enhanced productivity and nutrient use efficiency of maize under conservation agriculture. Annals of Agricultural Research. 2022;**43**(2):159-163

[9] Diacono M, Rubino P, Montemurro F. Precision nitrogen management of wheat. A review. Agronomy for Sustainable Development. 2013;**33**:219-241

[10] Singh VK, Govil V, Singh SK, Dwivedi BS, Meena MC, Gupta VK, et al. Precision nutrient management strategies using GIS-based mapping in western Uttar Pradesh. Better Crops-South Asia. 2012;**6**(1):15-18

[11] Inoue Y. Satellite-and drone-based remote sensing of crops and soils for smart farming–a review. Soil Science and Plant Nutrition. 2020;**66**(6):798-810

[12] Ali AM, Thind HS. A framework for refining nitrogen management in dry direct-seeded rice using green seeker™ optical sensor. Computers and Electronics in Agriculture. 2015;**110**:114-120

[13] Kumar S, Singh VK, Upadhyay PK, Shekhawat K, Dwivedi BS, Kumar A. Effect of precision nitrogen and water management on growth, productivity

and water budgeting of maize (*Zea mays*) grown under different crop-establishment techniques. Indian Journal of Agronomy. 2021;**66**(2):234-236

[14] Pampolino MF, Witt C, Pasuquin JM, Johnston A, Fisher MJ. Development approach and evaluation of the nutrient expert software for nutrient management in cereal crops. Computers and Electronics in Agriculture. 2012;**88**: 103-110

[15] Buresh RJ, Castillo R, van den Berg M, Gabinete G. Nutrient management decision tool for small-scale rice and maize farmers. In: Technical Bulletin 190. Taipei, Taiwan: Food and Fertilizer Technology Center; 2014

[16] Sharma S, Rout KK, Khanda CM, Tripathi R, Shahid M, Nayak A, et al. Field-specific nutrient management using rice crop manager decision support tool in Odisha, India. Field Crops Research. 2019a;**241**:107578

[17] Kumar A, Ram H, Kumar S, Kumar R, Yadav A, Gairola A, et al. A comprehensive review of nano-urea vs. conventional urea. International Journal of Plant and Soil Science. 2023b;**35**(23): 32-40

[18] Kumar S, Didawat RK, Kumar P. Automation in drip irrigation for increasing the water productivity of cereals in India. Indian Farming. 2023a; **73**(09):29-32

[19] Kumar V, Shekhawat K, Singh RK, Rathore SS, Kumar S. Effect of polyhalite on yield, nutrient uptake by wheat and soil properties. Annals of Plant and Soil Research. 2023c;**25**(1):59-63

[20] Wiesler F, Behrens T, Horst WJ. The role of nitrogen-efficient cultivars in sustainable agriculture. The Scientific World Journal. 2001;**1**(S2):61-69

[21] Ramamoorthy B, Velayutham M. Soil Test Crop Response Correlation Work in India. World Soil Resource Report 110.41. 1971. pp. 96-102

[22] Kumar S. Nitrogen and water management of maize grown under conservation agriculture [M.Sc. Thesis]. New Delhi: Division of Agronomy, ICAR-Indian Agricultural Research Institute; 2020

[23] Ghosh M, Swain DK, Jha MK, Tewari VK, Bohra A. Optimizing chlorophyll meter (SPAD) reading to allow efficient nitrogen use in rice and wheat under rice-wheat cropping system in eastern India. Plant Production Science. 2020;**23** (3):270-285

[24] Dass A, Kumar BS, Dhar S, Kumar K. Soil and plant basedplant-based precision N management in rabi maize (*Zea mays* L.) in North-Indian alluvial plains. Annals of Agricultural Research. 2015;**36**(4)

[25] Singh J, Singh V. Chlorophyll meter basedmeter-based precision nitrogen management in spring maize. Journal of Plant Nutrition. 2022;**46**(1):17-27

[26] Singh V, Singh B, Thind HS, Singh Y, Gupta RK, Singh S, et al. Evaluation of leaf colour chart for need-based nitrogen management in rice, maize and wheat in North-Western India. Journal of Research Punjab Agricultural University. 2014;**51**(3–4):239-245

[27] Hasanain M, Singh VK, Rathore SS, Shekhawat K, Singh RK, Dwivedi BS, et al. Site-specific nutrient management under conservation agriculture-based spring wheat in trans-Gangetic Plains of India. Indian Journal of Agricultural Sciences. 2021;**91**(5):757-760

[28] Shyam CS, Rathore SS, Shekhawat K, Singh RK, Padhan SR, Singh VK.

Precision nutrient management in maize (*Zea mays*) for higher productivity and profitability. The Indian Journal of Agricultural Sciences. 2021;**91**(6): 933-935

[29] Abalos D, Jeffery S, Sanz-Cobena A, Guardia G, Vallejo A. Meta-analysis of the effect of urease and nitrification inhibitors on crop productivity and nitrogen use efficiency. Agriculture, Ecosystems and Environment. 2014; **189**:136-144

[30] Upadhyay PK, Dey A, Singh VK, Dwivedi BS, Singh T, Rajanna GA, et al. Conjoint application of nano-urea with conventional fertilizers: An energy efficient and environmentally robust approach for sustainable crop production. PLoS One. 2023;**18**(7): e0284009

[31] Savant NK, Stangel PJ. Deep placement of urea supergranules in transplanted rice: Principles and practices. Fertilizer Research. 1990; **25**:1-83

[32] Didawat RK, Sharma V, Kumar S, Chobhe K, Bandyopadhyay K, Kumar P, et al. Effect of long-term organic practices on yield and nutrients uptake by rice (*Oryza sativa* L) in an acid inceptisol. Annals of Plant and Soil Research. 2022;**24**(4):547-551

Chapter 7

Perspective Chapter: Redesigning Agroecological Practices for Enhanced Resource Use Efficiency in Agroecosystems

Muhammad Talha Aslam, Imran Khan, Muhammad Umer Chattha, Ali Raza, Muhammad Bilal Chattha, Muqarrab Ali, Muhammad Umair Hassan and Iqra Fatima

Abstract

The projected population of 9.1 billion in 2050 will necessitate a rise in food production to accommodate an extra burden of the human population. Consequently, it is imperative to discuss innovative agricultural methods that have the potential to produce a greater quantity of food while being environmentally friendly. In this chapter, we summarized the pros and cons based on diversification in the following agroecological practices. The current agricultural system is poorly integrated with other agricultural activities and processes (pesticide and herbicide spray, bio-fertilizers, selection of crop in rotation, and allelopathy) and found to be notorious for its wasteful use of resources, which has adverse effects on the environment, biodiversity, and the economy. Quantitative analysis suggested to integrate the world's food chain production with sustainability of agroecosystem for improving carbon-to-nitrogen (C:N) ratio, soil properties, water, and nutrient cycling. Redesigning agroecological practices (legume-based rotation, mulching, no/zero tillage, split fertilization, advanced irrigation practices, biological pest control measure, and organic fertilization) synergized the integrated approach, which leads to organic matter recycling, soil-borne insect-pest and pathogenic resistance, crop adaptability, soil biodiversity, and net farm productivity. Therefore, redesigning agroecological practices is inevitable to increase resource use efficiency within agroecosystems. This chapter will be an invaluable resource for those working in academia, government, and the private sector to promote agroecosystem sustainability.

Keywords: crop, agroecological practices, food, agriculture, agroecosystems, sustainability

1. Introduction

The projected world population (9.1 billion) in 2050 would necessitate a substantial escalation (approximately ˃30%) in food production [1]. Researches have stated that the production of extra food with current agricultural practices would be difficult [2]. Whereas, industrial agriculture is not economical for environment and causes a serious destruction in natural habitat and ecosystem [3]. The best agricultural production practices for higher and sustainable food production are hotly debated. Precision farming [4] or genetically modified crops [5] could meet future food needs. Many studies have suggested to promote sustainable agriculture, which is environmentally friendly, socially equitable, and commercially advantageous. Likewise, Muhie [6] suggested to establish food production practices for conventional, integrated, and organic agriculture through sustainable agronomic practices [7]. Some researchers have suggested to no- or reduced-tillage as an organic agriculture practice, as, it increases soil biota activity and soil fertility [8], whereas others have stated natural biological pest control could be an alternative option to promote sustainability in an agroecological system [9]. Natural pest control is one of the imperative agroecological practices that integrate the natural landscape elements into agricultural use to reduce pesticide use [10]. Furthermore, crop rotation and organic nutrient fertilization techniques are also equally significant agroecological practices that were first established in the 1980s [11, 12]. In a broader way, agroecological practices include resource and biodiversity conservation, livestock integration, agroforestry, cover crops, green manure, intercropping, and biological control. However, their advantages and disadvantages still need to be extensively explored by researchers [13]. The term "Agroforestry" refers to the incorporation of trees into cropland and is not promising for achieving a higher crop yield as trees compete with crops for natural resources, endangering the existence of small land-holding farmers. Which cutting-edge agroecological practices could contribute to the production of more sustainably farmed food? Therefore, substitution practices with efficiency increase become inevitable for sustainability in agroecosystems. The term "substitution practices" describes the act of changing an input or a procedure, such as switching from chemical to natural insecticides. The term "efficiency increase" describes methods that boost crop yield while consuming less input (such as water, fertilizer, and herbicides). There is dire need of redesigning the agroecological practices to increase resource use efficiency within agroecosystems [14]. Therefore, this chapter aims to comprehend the agroecological practices with their associated potential and disadvantages. Moreover, we discussed the system diversification-based practices and their contribution to sustainable agriculture's goals of feeding a growing world population, protecting the environment, and ensuring farmers' economic viability.

2. Agronomic practices to promote efficiency of agroecosystem

2.1 Selection of crop, its spatial and temporal distribution

Selecting the appropriate crop and cultivar can improve crop resistance to infections, diseases, and abiotic stressors [15]. Hybrid and traditional plant breeding can be used to enhance agroecological practices, although availability and cost are important considerations. New pathogenic attack often causes to reduce crop resistance temporarily because it is a selection mechanism for pathogen populations [16]. Rotation and spatial allocation combined with crop resistance can lower this risk.

It's also critical to choose crop species or cultivars that encourage the proliferation of soil microorganisms. These beneficial rhizospheric bacteria increase plant growth by enhancing phytohormone synthesis, pathogen defensive mechanism, and nutrient intake. Arbuscular mycorrhizal fungi (AMF) support the sustainability of agroecosystems and crop growth. Plant development and production are influenced by soil characteristics and plant nature [17]. Crop selection and rotation practices play a critical role in agroecosystems development. AMF populations increased by lowering non-mycorrhizal crops like rapeseed and diversifying crop rotations. Gianinazzi et al. [18] indicate the necessity of adopting AMF characteristics in breeding techniques, instead of intensive fertilizer and chemical use. Plant growth-promoting rhizobacteria (PGPR) also support crop development by strengthening the plants against various diseases through high nutrient availability. However, the effectiveness of PGPR is affected by liming, tillage, and organic amendments. Crop species and cultivar also have an impact on these microbial communities [19]. Although nutrient availability is unclear, breeding and biotechnological approaches can manipulate root exudates to improve plant nutrition and stress resistance [20].

Selecting the right crop or cultivar can enhance cropping systems, reduce pesticide use, and be fitted in rotation. Resource use efficiency (RUE) increased by planting a crop with lower demands following an exhaustive crop. In water-scarce regions, crop rotations can increase water usage efficiency [21]. Cover crops lessen the surface runoff and enhance water uptake during heavy rainfall for the following crop.

2.2 Fertilizer management

Crop nitrogen (N) utilization efficiency is increased by splitting N application. Splitting N doses balance the supply and demand of nitrogen for crops [22]. Crop N uptake can be estimated with appropriate application timing [23]. Improved supply-demand alignment would increase productivity and lessen surface and groundwater contamination by N fertilizers. However, because labor costs are rising, estimating crop N demand may become challenging [22, 24].

Bio-fertilization is a method of nutrient delivery to crop by living microorganisms that, when applied to seed, plant surfaces, or soil, colonizes the rhizosphere or plant interior, and promotes growth by increasing the host plant's supply or availability of primary nutrients [25, 26]. AMF, PGPR, and nitrogen-fixing rhizobia are most common examples of bio-fertilizers [27]. Bio-fertilizers significantly lower fertilizer inputs and raise nutrient availability. However, the commercialization of PGPR and AMF inoculants is currently limited, particularly in developing countries [28].

Several experiments have revealed that the inoculation of PGPR or AMF led to plant growth promotion, improved yield, and enhanced uptake of nutrients [29, 30]. Thus, bio-fertilizers significantly reduced the reliance on synthetic fertilizers and mitigated environmental damage [31]. However, this technique requires additional enhancements to determine the interconnections within the soil-plant-microorganism system in diverse climatic regions [32].

Organic fertilization involves replacing inorganic fertilizers with natural alternatives and enhancing the effectiveness of fertilization by enhancing soil fertility. Applying organic fertilizer leads to greater soil biological activity and potentially higher soil mineralization [33]. However, these approaches may have limitations such as increased manpower and energy requirements, as well as challenges in optimizing nitrogen availability and requirement through organic fertilization to meet the specific needs of plants. In addition, acquiring organic fertilizers from sources outside the farm may provide challenges due to their scarcity, high cost, and the potential for additional expenses, such as transportation expenses, etc. Introducing livestock onto the farm may be necessary for incorporating additional organic fertilizers into the cropping system. This would necessitate a complete overhaul of the entire agricultural sector.

2.3 Irrigation management

Water scarcity is one of the integral issues of the twenty-first century due to immense water use in civil, industrial, and agriculture sector. Agriculture demands plenty of water for various water-loving crops, i.e., sugarcane and rice, to obtain their maximum yield potential. However, climate-induced water scarcity and poor irrigation water quality severely damage agricultural productivity; therefore, effective irrigation management is required [34]. Drip irrigation has significant potential to reduce water usage, enhance water utilization efficiency, and effectively align crop water requirements in terms of timing and location. Additionally, it reduces soil salinization. The primary limitations are installation cost and the expenses associated with overseeing and controlling operations [34]. The implementation of both drip irrigation and cover crops by incorporating rows of cover crops between the main crops offers multiple benefits; provide mulch, control temperature, reduce water evaporation, minimize soil erosion, enhance soil organic matter, and increase nitrogen accumulation, particularly if legumes are being intercrop [35].

2.4 Plant protection approaches

Botanical pesticides, often known as natural pesticides or botanicals, offer a promising alternative to synthetic pesticides and their detrimental consequences [36]. However, there is still a lack of knowledge concerning their application on a wider-scale agricultural land [36]. Additionally, few of them induced environmental contamination. Commercially less availability of natural pesticides is due to varying pest control efficacy, country's rules and regulation, and cost [37, 38]. Botanical pesticides vary depending upon their sources, i.e., tree seeds, plant essential oils, pyrethrum extracted from flowers, crude aqueous extracts of plants, and extracts of trees [39, 40]. While these botanical pesticides may not be as effective as other biocontrol methods, they could be useful for traditional agriculture in developing nations and the expanding organic sector in areas where synthetic pesticides are prohibited. This is because many of these pesticides originate from tropical or subtropical plants, which are native to these regions [38, 41].

Furthermore, biopesticides are employed with botanical insecticides. This involves the utilization of bacteria, AMF inoculants, or other fungi that have the ability to regulate harmful species [42]. Biopesticide effects on pests are carried out through several methods, including antibiosis, competition, induction of plant defense systems, inhibition of pathogen survival, and reduction of pathogenicity [43]. However, in practice, the implementation of field methods often does not effectively prevent the growth of pathogens because of the absence of colonization and absence of rhizosphere/endosphere [44].

Biological pest management uses natural enemies in agroecological practices to decrease the pesticide use. Pheromones also work as biological control method and disrupt insect-pest sexual reproduction [45]. Natural and biological pest control lessen water pollution and health concerns. However, the effectiveness of this method necessitates enough knowledge, management, and demand high-cost pest population [46].

In conclusion, all the agroecological practices aim to boost crop output while decreasing input consumption. Among these, crop selection, organic and biological fertilization, drip irrigation, biological pest control, and natural insecticides are most commonly being practiced globally.

3. Redesigning agroecological practices

The term "redesign practices" denotes the necessity of reevaluating the entire agricultural system, or at least a significant portion of it, in order to incorporate the practice concerned.

3.1 Selection of crop, its spatial and temporal distribution

3.1.1 Introducing green manure or cover crops

As an agroecological strategy, cover cropping helps mitigate soil and wind erosion while simultaneously reducing the fertilizer requirement and thus reduce the water contamination from leaching [47]. Crop diversity increased by practicing the cover crops into rotation. Additionally, selecting leguminous crops increases soil biological activity and supplements N for the following crop [48]. It is also a significant source of carbon that enhances soil structure and encourages microbial growth [49, 50]. Constraints with cover crop practices, however, incur additional labor cost and possible pest growth risk (e.g., snails beneath cover crops). While, brassica crops can be introduced to lower the pest growth risk as they possess multiple uses, including as biocontrol, biofumigant, and biocidal agents against specific insects and pathogens, in addition to being cover and trap crops [51, 52].

3.1.2 Temporal distribution of crop

Crop rotation entails controlling the succession of crops in order to maximize beneficial interactions and crop synergies. Since crops are often present in a sequential manner during rotation, interactions between species are primarily indirect and also rely on crop management and growth circumstances. Crop rotation improved soil fertility, which in turn enhanced crop productivity and reduced pests and diseases [53].

First, cropping sequences can be improved to reduce the demand for fertilizer and increase nutrient availability. For instance, adding leguminous plants to the rotation helps fix nitrogen from the atmosphere and serves as a significant supply of readily absorbed nitrogen for crops that come after. Second, cover crops and winter crops increase soil cover, carbon %, and soil nutrient status, which increases soil stability [54]. As crop leftovers, subsequent crop roots enhance soil biological activity and improve soil structural stability. Root exudates and turnover account for up to 40% of microbial carbon absorption [55]. Cover crops and catch crops may be an excellent rotation addition. In addition, they reduce nitrate leaching and improve nutrient usage efficiency. They may also promote winter water infiltration and crop water availability in temperate regions [56]. Third, crop rotations can reduce pest and disease prevalence by diversifying crops and avoiding successive disease host crops [57]. Fourth, crop rotations reduce weeds' infestation effectively. Due of their tendency to rapidly cover land, crops compete with weeds for soil and light. Weed control can also be achieved through crop management, with different weeding interventions at different times of the year.

It is not possible to improve crop rotation management solely by maximizing ecological services. Preserving high farm production and profitability is a key consideration in redesigning the crop rotation system. In order to increase profitability, productivity, and ecological services, this entails allocating resources in an optimal manner [58].

3.1.3 Spatial distribution of crop by intercropping or agroforestry

The presence of two or more crops in the same field at the same time is known as intercropping. These species can be arranged in a variety of spatial configurations; the associated species and arrangement showed imperative interaction (facilitative/competitive) with environmental factors [58]. Row and strip intercropping, in which at least one of the related crops is planted in a row, is also known as mixed intercropping. This strategy entails "full intercropping," in which interactions between related species take place all the way through the crop cycle. In contrast, relay intercropping limits interactions between species by growing two or more crops together for a portion of their life cycles [59].

Although intercropping systems are more difficult to maintain, they may increase productivity, output stability, resilience, and ecological suitability [60]. The main issue in such a system is managing agricultural resource competition [60, 61]. This is because intercropping with leguminous crops increased land productivity, radiation use efficiency, soil organic matter, nitrogen as well as phosphorus contents, and other nutrient uptake [62, 63]. Twice as many earthworms were produced by wheat combined with clover grass than by wheat alone [64]. Because cereal crops are more resource-efficient and competitive than peas in an intercropped system, peas may be able to suppress weed infestation [64]. Intercropping improves the structure and fertility of soil with reduced compaction and penetration resistance [65]. Intercropping also lessens soil erosion and crusting by increasing soil cover [66].

In addition, multiple agroforestry approaches are considered agroecological practices, as they prevent nutrient leaching, conserve soils, promote production diversification, and generate complementing wood for multiple applications [13, 67]. Agroforestry integrates crops and woody plants in Europe [68].

3.2 Selection of crop, its spatial and temporal distribution

Some plant species release chemicals that inhibit weed, insect, and disease growth called allelochemicals and this effect is known as allelopathic effect. The incorporation of allelopathic plants into crop rotations as intercrop or cover crop is a promising agroecological practice to reduce pesticide use and boost crop yields [13]. The degradation of their residues or the release of toxic substances during their life cycle affects target species. Rye, sorghum, and sunflower can be utilized as green manure or cover crops since their root exudates hinder weed seed germination and growth [69]. When cultivated with carrot, onion releases deterrent compounds that limit carrot fly assaults, making it a "push" crop [13]. Allenopathic substances attract the target organism(s), making the plant a "trap" crop [70]. Biofumigation, a delayed allelopathic action, reduces soil-borne pests and fungi, bacteria, and nematodes [71]. Allelopathic plants provide several benefits in temperate agroecosystems; however, less knowledge about biological process, effectiveness, and results varies depending on local conditions [72], and allelopathic crops as pathogen-hosts [73] are few constraints that need to be addressed.

3.3 Tillage management

Reducing or no-tillage (crop sowing without soil disturbance) minimizes energy consumption, wind and water erosion, soil biota activity, stocking organic C (less C mineralization), and other benefits (**Table 1**). Direct seeders use coulter

Management type	Principal	E, S, R, D	Pros and cons	Reference
Crop selection • Crop spatial distribution • Crop temporal succession	Crops with desired characteristics to promote soil-rhizospheric-activities (AMF, PGPR etc).	E, S	Pros: • High yield • Less nutrient requirement • Better pest management • Less pesticide use Cons: • Risk of non- availability of desired varieties • Crop adaptability	[74–77]
Crop fertilization • Split application	Crop fertilization at its critical growth stages	E	Pros: • Less chemical fertilizer requirement • High nutrient availability and uptake • Higher nutrient use efficiency • Reduced fertilizer loss • Less ground water contamination Cons: • More time and labor energy needed to split fertilizer • Knowledge of critical crop stage	[78–81]
• Bio-fertilizer	Fertilization with living microbes	E, S	Pros: • Same as above Cons: • Variable and inconsistent effects. • Low commercialization rate.	[82–87]
• Organic fertilization	Fertilization with organic source	E, S, R	Pros: • Same as above + • Less energy consumption for transportation of chemical fertilizer to field. • Less ground water pollution risk. • High soil microbial activity. Cons: • Require in high quantity • Slow mineral nutrients availability • Less effective than commercial fertilizers	[88–90]

Management type	Principal	E, S, R, D	Pros and cons	Reference
Crop irrigation • Drip irrigation	Use of drip irrigation (without or in combination with cover crops or mulch).	E	Pros: • High water use efficiency • Less evaporation loss with cover crops or mulch. • Less salinization. Cons: • High installation cost, • Costly equipment • Management costs	[13, 91, 92]
Weeds/pest/disease management • Natural pesti-cides/botanical pesticides	Plant based pesticides	S	Pros: • Less reduction in water quality from synthetic pesticides. • Human health safety. • Imperative for organic agriculture. Cons: • Variable pest control efficiency. • Less commercial availability. • Poor research studies-based knowledge availability. • National regulations and registrations. • Cost.	[93, 94]
• Biological pest control	Use of natural enemies and pheromones	S		
Redesign practices • Crop selection/crop spatial distribution, and crop temporal succession/crop rotation	Integrating various (leguminous and cover crops) crops in rotation	S, R	Pros: • Less weed and pest attack • Decreased pesticide usage • Less fertilizer application requirement • Higher soil microbial ac activity. • Less leaching and erosion. • Higher food and land productivity Cons: • Awareness and less knowledge of appropriate intercrop combination. • Market access • Risk of pest development with cover crops (e.g. snails). • Difficulties in destroying cover crops	[13, 95]
Intercropping	Coexistence of two or more crops on the same field at the same time.	E, S, R	Pros: • Improved land utilization. • Less pest and disease attack • Nitrogen addition in soil (legume intercropping) • Less nutrient requirement • Improved soil structure and fertility. Cons: • High labor demand at harvesting and association management. • Lack of advanced technologies for harvesting. • Inter-species competition.	[96–98]

Management type	Principal	E, S, R, D	Pros and cons	Reference
Weed/pest/disease management • Allelopathic plants	Addition of plants with allelopathic effect in crop rotation	S, R	Pros: • Promising weed and pest control • Less soil-borne diseases infestation with bio-fumigation Cons: • Effectiveness depends on environmental conditions • Less scientific knowledge	
Tillage management • No tillage	Direct seeding into standing cover crop or destroyed (mulch)	E, S, R	Pros: • Less energy requirement for seed bed preparation • Soil protection against wind and water erosion. • Less soil compaction • High soil biota • Improved soil OM and C sequestration. • Less space for weed growth • Less herbicide application cost Cons: • Problematic to control weeds from cover crops. • High weeding cost • Competition risk between main crops and cover crops/living mulch.	[99–101]
Reduced tillage	Soil disturbance at reduced rate			

E = efficiency increase, S = substitution, R = redesign, D = diversification practice.

Table 1.
Principal of agroecological crop management, its type and pros and cons.

discs or tines to cut and open furrows for seeding. Reduced tillage causes minimum soil disturbance without inversion (unlike ploughing). In this, the major goal is to limit soil disturbance and keep organic matter (fresh crop leftovers) at the soil surface or first few centimeters (cm). The benefits of these techniques for soil fertility vary by tillage type. Tillage reduction is widely being adopted in temperate regions of the world [102–104].

More anecic earthworms were observed from a field without tillage practices [105], increasing soil porosity and water and root penetration. Lower tillage is found to be effective over no-tillage practice for earthworm abundance in soil [106]. However, no-tillage enhances predators like ground beetles, and pest control may be used to improve the soil [13].

Thus, no- and reduced-tillage are promising, but adoption is limited. To control weeds and sustain yields, traditional agriculture utilizes more chemical fertilizers and pesticides [107]. Reduced tillage in organic farming increases weed control machine traffic, labor time, and energy expenses [108]. Climate and soil conditions can compact soil in temperate climates [109]. All these limits obscure their impact on agricultural output. In Europe, Soane et al. [109] found that winter crop yields with no- or reduced tillage are equivalent to traditional tillage with ploughing.

Redesigning the cropping systems is necessary for the deployment of such methods in order to reduce limitations and boost efficiency. For example, changing the crop selection and crop rotation strategies will require a comprehensive reevaluation of the cropping system in order to improve weed management.

3.4 Management of landscape elements

Management of landscape in modern agroecological practices includes integrating or reintegrating natural or seminatural landscape components like hedges and plant strips into the field or landscape. Landscape elements can provide shelter, overwintering sites, and alternate prey for favorable insects and pest predators to reduce pesticide use. Their natural plant diversity and flowering fascinate pollinators host them outside crop flowering season, improving crop pollination [110]. The field and landscape also prevent surface water contamination, wind erosion, and soil erosion [111].

Increasing biological pest control to reduce pesticide use is a problem for landscape component integration research in agricultural system [112]. Overall, landscape habitat diversity affects herbivores and their natural enemies in agricultural crops. Obanyi et al. [113] found that early season aphid establishment was reduced in landscapes with ample field margins and perennial crops, regardless of farming practices. They further reported that adding blooming plant strips to cropped fields boosts pollen and nectar, which may need natural pest foes to breed, fecund, and live along. This decreases aphid numbers and increases aphidophagous predators.

Although reduced pest population is observed at landscape and natural habitats, the current challenges are to preserve existing landscape elements and reintroduce them to agroecosystems and agricultural landscapes. Wezel et al. [13] reported that natural enemy search success and parasitism rates decreased below 20% non-crop area. As herbivorous pests and their natural enemies vary in dispersal, landscape structure affects both the number and spatial arrangement of suitable habitats [114]. Tscharntke et al. [115] noted that improving biological control requires a landscape perspective and consideration of landscape context-habitat quality interactions. However, precise advice for designing agricultural landscapes that ensure biological management is needed.

4. Promising agroecological practices

4.1 Intensity required to modify the agroecological practices

Many agroecological practices enhance agricultural productivity without harming the environment. However, agroecological practices are limited in most industrialized large-scale agricultural and animal systems. Unlike intense conventional systems, agroecological approaches have been frequently used in hilly or mountainous locations for decades due to their reduced output potential [116]. Applying the agroecological practices in this system requires appropriate crop management or cropping system changes (**Table 1**). Previous literature suggested that splitting the application of nutrients and changing the crop cultivar are easy to accomplish [79]. However, it was found ineffective, whereas the agroecological practices demand considerable modification by redesigning the current cropping system of specific areas. Likewise, direct seeding into living mulch may require a significant system change, including new machinery for seedbed preparation and weed management in mulch-applying soil.

Diversification helps execute several agroecological practices and their principles (**Table 1**). To boost agroecosystem resilience, limit pest outbreaks, and preserve soil biodiversity, agroecological practice-based crop diversification and reforming the cropping system have become vital.

4.2 Promising agroecological practices integrated with current agriculture

Many agroecological practices are being followed in current agriculture. Modern agriculture integrates organic fertilization with introducing cover crops, installing advanced (drip, rain gun, and central pivot system) irrigation methods, and biological control [117]. Implementing legislation on environmental protection could encourage less polluting and less input-dependent practices in agriculture [118]. However, appropriate research on agroecological practices is required to prove their promising role to adopt on a wide scale for long-term environmental and food security concerns [13, 119]. Their unpredictable yields limit direct seeding and intercropping [13, 120]. There are several successful case studies reported on the promising

Country	Duration (years)	Cropping system	Effects	Reference
China	17	Maize-soybean	NTS improved soil micro pores, water connectivity and SOC	[121]
USA	15	Winter wheat	NTS enhanced the total soil nitrogen and SOC in silty loam soil	[122]
China	3	Rice-wheat	SCP augmented 35.18% water-stable large (>2 mm) macroaggregates, 33.52% small (2–0.25 mm) macroaggregates and microaggregates up to 25.10% in the topsoil. Further, no tillage combining with straw return enhanced the carbon preservation capacity	[123]
Spain	5	Barley	Conservation tillage practices (CTP) promoted the available soil water capacity, OM, SWC and SWR in clayey soil	[124]
Australia	14	Soybean-Oat	Higher soil porosity, structural stability, infiltration and less water run-off was recorded under NTS from silt loamy soil	[125]
Nebraska, USA	35	Maize	CTP increased OM, root mass density, soil physical and chemical properties whereas zero tillage resulted in maximum soil NPK and S in available form from silty clay loam texture	[126]
India	5	Rice-wheat	Introducing mungbean in rice-wheat and maize-wheat cropping system decreased the soil penetration resistance, enhanced infiltration rate, and nutrient availability in loamy soil.	[127]
Germany	8	Different crops	CTP resulted improved pore connectivity, better hydraulic connection, SOM in silt texture soil	[128]

NTS = no tillage system, SCP = soil conservation practices, SOC = soil organic carbon, OM = organic matter, SWC = soil water contents, SWR = soil water retention.

Table 2.
Case studies about soil conservation practices and crop residue retention.

role of agroecological practices (**Table 2**). However, researchers showed their concerns about allelopathic plants, bio-fertilizers, and agroforestry, with the most negligible promising results to use in the future [13, 109].

The above-mentioned practices forced agricultural researchers to revisit the core production areas to achieve high food demand, preserve soil biodiversity, and protect from environmental degradation. In conclusion, most agroecological practices need to be integrated better into modern agriculture and have less potential for widespread adoption in the future. However, organic fertilization, reduced tillage, advanced (drip, central pivot system) irrigation methods, biological control, selection of cultivar, and split application of nutrients have medium or high integration levels in agriculture suited best for future potential.

5. Conclusion or perspective beyond agroecological practices

The world needs more food by 2050, even if agroecological practices are expanded and improved, leading to higher agricultural productivity in both quantity and quality. Based on the above literature, the following findings are suggested to meet the food challenge in the near future: First, it's necessary to break the link between increased income and increased use of animal products, among other things, and to see significant changes in food consumption trends. Second, significant investment on infrastructure, research and development (R&D), and crop improvement is needed in order to combat climate change and preserve ecosystems while also increasing crop yields. Third, efforts are inevitable to regulate the food trade, advance agricultural systems, and enhance consumption by proactive policies implemented at various stages globally. Lastly, decreasing different kind of losses (storage, transport, processing, distribution, and consumption) locally and internationally.

Acknowledgements

Our heartfelt gratitude goes out to everyone who has contributed to this book chapter, whether via their skill or unwavering encouragement. We hope to leave a lasting impression on the scientific community, and we look forward to continuing this adventure of research and discovery with you all. We are also grateful to this publishing group for their supportive policy to spread knowledge in the form of this book.

Author details

Muhammad Talha Aslam[1*], Imran Khan[1], Muhammad Umer Chattha[1], Ali Raza[1], Muhammad Bilal Chattha[2], Muqarrab Ali[3], Muhammad Umair Hassan[1] and Iqra Fatima[4]

1 Department of Agronomy, University of Agriculture Faisalabad, Pakistan

2 Faculty of Agricultural Sciences, Deaprtment of Agronomy, University of the Punjab, Lahore, Pakistan

3 Muhammad Nawaz Shareef University of Agriculture, Multan, Pakistan

4 Department of Botany, University of Agriculture Faisalabad, Pakistan

*Address all correspondence to: taslamuaf@gmail.com

References

[1] Mrabet R. Sustainable agriculture for food and nutritional security. In: Sustainable Agriculture and the Environment. Cambridge, Massachusetts: Academic Press; 2023. pp. 25-90

[2] Wakweya RB. Challenges and prospects of adopting climate-smart agricultural practices and technologies: Implications for food security. Journal of Agriculture and Food Research. 2023;**25**:100698. DOI: 10.1016/j.jafr.2023.100698

[3] Adla K, Dejan K, Neira D, Dragana Š. Degradation of ecosystems and loss of ecosystem services. In: One Health. Cambridge, Massachusetts: Academic Press; 2022. pp. 281-327

[4] Rusmayadi G, Mulyanti DR, Alaydrus AZ. Revolutionizing agrotechnology: Meeting global food demand through sustainable and precision farming innovations. West Science Interdisciplinary Studies. 2023;**1**:619-628. DOI: 10.58812/wsis.v1i08.172

[5] Kumar K, Gambhir G, Dass A, Tripathi AK, Singh A, Jha AK, et al. Genetically modified crops: Current status and future prospects. Planta. 2020;**251**:91. DOI: 10.1007/s00425-020-03372-8

[6] Muhie SH. Novel approaches and practices to sustainable agriculture. Journal of Agriculture and Food Research. 2022;**10**:100446. DOI: 10.1016/j.jafr.2022.100446

[7] Durham TC, Mizik T. Comparative economics of conventional, organic, and alternative agricultural production systems. Economies. 2021;**9**:64. DOI: 10.3390/economies9020064

[8] Francaviglia R, Almagro M, Vicente-Vicente JL. Conservation agriculture and soil organic carbon: Principles, processes, practices and policy options. Soil Systems. 2023;**22**(**7**):17. DOI: 10.3390/soilsystems7010017

[9] Imarhiagbe O, Okafor AC, Ikponmwosa BO, Ogwu MC. Sustainable agricultural pest control strategies to boost food and socioecological security: The allelopathic strategy. In: One Health Implications of Agrochemicals and their Sustainable Alternatives. Singapore: Springer Nature, Springer International Publishing; 2023. pp. 637-657. DOI: 10.1007/978-981-99-3439-3_23

[10] Jacquet F, Jeuffroy MH, Jouan J, Le Cadre E, Litrico I, Malausa T, et al. Pesticide-free agriculture as a new paradigm for research. Agronomy for Sustainable Development. 2022;**42**:8. DOI: 10.1007/s13593-021-00742-8

[11] Arshad MA, Rouf S, Abbas RN, Shahbaz Z, Aleem K, Shahbaz H, et al. Navigating synergies: A comprehensive review of agroforestry system and agronomy crops. Haya: The Saudi Journal of Life Sciences. 2024;**9**:97-113. DOI: 10.36348/sjls.2024.v09i04.003

[12] Akanmu AO, Akol AM, Ndolo DO, Kutu FR, Babalola OO. Agroecological techniques: Adoption of safe and sustainable agricultural practices among the smallholder farmers in Africa. Frontiers in Sustainable Food Systems. 2023;**7**:1143061. DOI: 10.3389/fsufs.2023.1143061

[13] Wezel A, Casagrande M, Celette F, Vian JF, Ferrer A, Peigné J. Agroecological practices for sustainable agriculture. A review. Agronomy for

Sustainable Development. 2014;**34**:1-20. DOI: 10.1007/s13593-013-0180-7

[14] de la Riva EG, Ulrich W, Batáry P, Baudry J, Beaumelle L, Bucher R, et al. From functional diversity to human well-being: A conceptual framework for agroecosystem sustainability. Agricultural Systems. 2023;**208**:103659. DOI: 10.1016/j.agsy.2023.103659

[15] Parvathi MS, Antony PD, Kutty MS. Multiple stressors in vegetable production: Insights for trait-based crop improvement in cucurbits. Frontiers in Plant Science. 2022;**13**:861637. DOI: 10.3389/fpls.2022.861637

[16] Tilman D, Cassman KG, Matson PA, Naylor R, Polasky S. Agricultural sustainability and intensive production practices. Nature. 2002;**418**:671-677. DOI: 10.1038/nature01014

[17] Thakur MP, van der Putten WH, Wilschut RA, Veen GC, Kardol P, van Ruijven J, et al. Plant–soil feedbacks and temporal dynamics of plant diversity–productivity relationships. Trends in Ecology & Evolution. 2021;**36**:651-661. DOI: 10.1016/j.tree.2021.03.011

[18] Gianinazzi S, Gollotte A, Binet MN, van Tuinen D, Redecker D, Wipf D. Agroecology: The key role of arbuscular mycorrhizas in ecosystem services. Mycorrhiza. 2010;**3**:14-22. DOI: 10.1007/s00572-010-0333-3

[19] Kari A, Nagymáté Z, Romsics C, Vajna B, Tóth E, Lazanyi-Kovács R, et al. Evaluating the combined effect of biochar and PGPR inoculants on the bacterial community in acidic sandy soil. Applied Soil Ecology. 2021;**160**:103856. DOI: 10.1016/j.apsoil.2020.103856

[20] Afridi MS, Kumar A, Javed MA, Dubey A, de Medeiros FH, Santoyo G. Harnessing root exudates for plant microbiome engineering and stress resistance in plants. Microbiological Research. 2023;**1**:127564. DOI: 10.1016/j.micres.2023.127564

[21] Nie J, Harrison MT, Zhou J, Yang L, Zhao J, Wang X, et al. Productivity and water use efficiency of summer soybean-winter wheat rotation system under limited water supply in the North China Plain. European Journal of Agronomy. 2023;**151**:126959. DOI: 10.1016/j.eja.2023.126959

[22] Cassman KG, Dobermann A, Walters DT. Agroecosystems, nitrogen-use efficiency, and nitrogen management. AMBIO: A Journal of the Human Environment. 2002;**31**:132-140. DOI: 10.1579/0044-7447-31.2.132

[23] Davies B, Coulter JA, Pagliari PH. Timing and rate of nitrogen fertilization influence maize yield and nitrogen use efficiency. PLoS One. 2020;**15**:e0233674. DOI: 10.1371/journal.pone.0233674

[24] Robertson GP, Vitousek PM. Nitrogen in agriculture: Balancing the cost of an essential resource. Annual Review of Environment and Resources. 2009;**34**:97-125. DOI: 10.1146/annurev.environ.032108.105046

[25] Marín CM, Estrada JA, Naranjo JM, Ocampo PA, Vásquez MP. Cyanobacteria and microalgae as potential sources of biofertilizers: A review. Actualidades Biológicas. 2024;**46**:e4606. DOI: 10.17533/udea.acbi/v46n120a06

[26] Vessey JK. Plant growth promoting rhizobacteria as biofertilizers. Plant & Soil. 2003;**255**:571-586. DOI: 10.1023/A:1026037216893

[27] Akram MS, Cheema MA, Waqas M, Bilal M, Saeed M. Role of bio-fertilizers in sustainable agriculture. Mediterranean

Journal of Basic and Applied Sciences. 2020;**3**:17-23. DOI: 10.20944/preprints202003.0262.v1

[28] Madawala HM. Arbuscular mycorrhizal fungi as biofertilizers: Current trends, challenges, and future prospects. Biofertilizers. 2021;**2**:83-93. DOI: 10.1016/B978-0-12-821667-5.00029-4

[29] Pan J, Huang C, Peng F, Zhang W, Luo J, Ma S, et al. Effect of arbuscular mycorrhizal fungi (AMF) and plant growth-promoting bacteria (PGPR) inoculations on *Elaeagnus angustifolia* L. in saline soil. Applied Sciences. 2020;**10**:945. DOI: 10.3390/app10030945

[30] Mohamed I, Eid KE, Abbas MH, Salem AA, Ahmed N, Ali M, et al. Use of plant growth promoting Rhizobacteria (PGPR) and mycorrhizae to improve the growth and nutrient utilization of common bean in a soil infected with white rot fungi. Ecotoxicology and Environmental Safety. 2019;**171**:539-548. DOI: 10.1016/j.ecoenv.2018.12.100

[31] Iqbal A, Shafi MI, Rafique M, Zaman M, Ali I, Jabeen A, et al. Approaches for using bio-fertilizers as a substitute for chemical fertilizers to improve soil health and crop yields in Pakistan. In: Biofertilizers for Sustainable Soil Management. Boca Raton, Florida, United States: CRC Press; 2023. pp. 89-118. ISBN: 9781003286233

[32] Kumar S, Sindhu SS, Kumar R. Biofertilizers: An ecofriendly technology for nutrient recycling and environmental sustainability. Current Research in Microbial Sciences. 2022;**3**:100094. DOI: 10.1016/j.crmicr.2021.100094

[33] Ye L, Zhao X, Bao E, Li J, Zou Z, Cao K. Bio-organic fertilizer with reduced rates of chemical fertilization improves soil fertility and enhances tomato yield and quality. Scientific Reports. 2020;**10**:177. DOI: 10.1038/s41598-019-56954-2

[34] Ayars JE, Zaccaria D, Bali KM, editors. Microirrigation for Crop Production: Design, Operation, and Management. Amsterdam, Netherlands: Elsevier; 2023

[35] Lopes CM, Santos TP, Monteiro A, Rodrígues ML, Costa JM, Chaves MM. Combining cover cropping with deficit irrigation in a Mediterranean low vigor vineyard. Scientia Horticulturae. 2011;**129**:603-612. DOI: 10.1016/j.scienta.2011.04.033

[36] Dalavayi Haritha M, Bala S, Choudhury D. Eco-friendly plant based on botanical pesticides. Plant archives. 2021;**21**:2197-2204. DOI: 10.51470/PLANTARCHIVES.2021.v21.S1.362

[37] Marrone PG. Pesticidal natural products–status and future potential. Pest Management Science. 2019;**75**:2325-2340. DOI: 10.1002/ps.5433

[38] Isman MB. Botanical insecticides: For richer, for poorer. Pest Management Science. 2008;**64**:8-11. DOI: 10.1002/ps.1470

[39] Khursheed A, Rather MA, Jain V, Rasool S, Nazir R, Malik NA, et al. Plant based natural products as potential ecofriendly and safer biopesticides: A comprehensive overview of their advantages over conventional pesticides, limitations and regulatory aspects. Microbial Pathogenesis. 2022;**173**:105854. DOI: 10.1016/j.micpath.2022.105854

[40] Prakash A, Rao J. Botanical Pesticides in Agriculture. Boca Raton, Florida, United States: CRC Press; 2018

[41] Regnault-Roger C, Philogène BJR. Past and current prospects for the use

of botanicals and plant allelochemicals in integrated pest management. Pharmaceutical Biology. 2008;**46**:41-52. DOI: 10.1080/13880200701729794

[42] Reddy DS, Chowdary NM. Botanical biopesticide combination concept—a viable option for pest management in organic farming. Egyptian Journal of Biological Pest Control. 2021;**31**:1-0. DOI: 10.1186/s41938-021-00366-w

[43] Archana HR, Darshan K, Lakshmi MA, Ghoshal T, Bashayal BM, Aggarwal R. Biopesticides: A key player in agro-environmental sustainability. In: Trends of Applied Microbiology for Sustainable Economy. Cambridge, Massachusetts: Academic Press; 2022. pp. 613-653

[44] Lengai GM, Muthomi JW. Biopesticides and their role in sustainable agricultural production. Journal of Biosciences and Medicines. 2018;**6**:7-41

[45] Witzgall P, Kirsch P, Cork A. Sex pheromones and their impact on pest management. Journal of Chemical Ecology. 2010;**36**:80-100. DOI: 10.1007/s10886-009-9737-y

[46] Ewert F, Baatz R, Finger R. Agroecology for a sustainable agriculture and food system: From local solutions to large-scale adoption. Annual Review of Resource Economics. 2023;**15**:351-381. DOI: 10.1146/annurev-resource-102422-090105

[47] Quintarelli V, Radicetti E, Allevato E, Stazi SR, Haider G, Abideen Z, et al. Cover crops for sustainable cropping systems: A review. Agriculture. 2022;**12**:2076. DOI: 10.3390/agriculture12122076

[48] Kebede E. Contribution, utilization, and improvement of legumes-driven biological nitrogen fixation in agricultural systems. Frontiers in Sustainable Food Systems. 2021;**5**:767998. DOI: 10.3389/fsufs.2021.767998

[49] Kumar S, Meena RS, Lal R, Singh Yadav G, Mitran T, Meena BL, et al. Role of legumes in soil carbon sequestration. Legumes for Soil Health and Sustainable Management. 2018;**2**:109-138. DOI: 10.1007/978-981-13-0253-4_4

[50] Hernanz JL, Sánchez-Girón V, Navarrete L. Soil carbon sequestration and stratification in a cereal/leguminous crop rotation with three tillage systems in semiarid conditions. Agriculture, Ecosystems & Environment. 2009;**133**:114-122. DOI: 10.1016/j.agee.2009.05.009

[51] Fourie H, Ahuja P, Lammers J, Daneel M. Brassicacea-based management strategies as an alternative to combat nematode pests: A synopsis. Crop Protection. 2016;**80**:21-41. DOI: 10.1016/j.cropro.2015.10.026

[52] Dutta TK, Khan MR, Phani V. Plant-parasitic nematode management via biofumigation using brassica and non-brassica plants: Current status and future prospects. Current Plant Biology. 2019;**17**:17-32. DOI: 10.1016/j.cpb.2019.02.001

[53] Sumner DR. Crop rotation and plant productivity. In: Handbook of Agricultural Productivity. Boca Raton, Florida, United States: CRC Press; 2018. pp. 273-314

[54] Scavo A, Fontanazza S, Restuccia A, Pesce GR, Abbate C, Mauromicale G. The role of cover crops in improving soil fertility and plant nutritional status in temperate climates. A review. Agronomy for Sustainable Development. 2022;**42**:93. DOI: 10.1007/s13593-022-00825-0

[55] Lei X, Shen Y, Zhao J, Huang J, Wang H, Yu Y, et al. Root exudates

mediate the processes of soil organic carbon input and efflux. Plants. 2023;**12**:630. DOI: 10.3390/plants12030630

[56] Nitu TT, Milu UM, Jahangir MM. Cover crops and soil nitrogen cycling. In: Cover Crops and Sustainable Agriculture, Boca Raton. Florida, United States: CRC Press; 2021. pp. 209-226

[57] Shah KK, Modi B, Pandey HP, Subedi A, Aryal G, Pandey M, et al. Diversified crop rotation: An approach for sustainable agriculture production. Advances in Agriculture. 2021;**1**:8924087. DOI: 10.1155/2021/8924087

[58] Dury J, Schaller N, Garcia F, Reynaud A, Bergez JE. Models to support cropping plan and crop rotation decisions. A review. Agronomy for Sustainable Development. 2012;**32**:567-580. DOI: 10.1007/s13593-011-0037-x

[59] Glaze-Corcoran S, Hashemi M, Sadeghpour A, Jahanzad E, Afshar RK, Liu X, et al. Understanding intercropping to improve agricultural resiliency and environmental sustainability. Advances in Agronomy. 2020;**162**:199-256. DOI: 10.1016/bs.agron.2020.02.004

[60] Maitra S, Hossain A, Brestic M, Skalicky M, Ondrisik P, Gitari H, et al. Intercropping—A low input agricultural strategy for food and environmental security. Agronomy. 2021;**11**:343. DOI: 10.3390/agronomy11020343

[61] Maitra S. Role of intercropping system in agricultural sustainability. Centurion Journal of Multidisciplinary Research. 2018;**8**:77-90

[62] Kumawat A, Bamboriya SD, Meena RS, Yadav D, Kumar A, Kumar S, et al. Legume-based inter-cropping to achieve the crop, soil, and environmental health security. In: Advances in Legumes for Sustainable Intensification. Cambridge, Massachusetts: Academic Press; 2022. pp. 307-328

[63] Aslam MT, Khan I, Chattha MU, Nawaz M, Mustafa A, Athar F, et al. The biofortification of zinc in legumes to alleviate zinc deficiency. In: Legumes Biofortification. Cham: Springer International Publishing; 2023. pp. 327-346. DOI: 10.1007/978-3-031-33957-8_14

[64] Hallam J, Berdeni D, Grayson R, Guest EJ, Holden J, Lappage MG, et al. Effect of earthworms on soil physico-hydraulic and chemical properties, herbage production, and wheat growth on arable land converted to ley. Science of the Total Environment. 2020;**713**:136491. DOI: 10.1016/j.scitotenv.2019.136491

[65] Ozturkmen AR, Ramazanoglu E, Almaca A, Çakmakli M. Effect of intercropping on soil physical and chemical properties in an olive orchard. Applied Ecology & Environmental Research. 2020;**18**:6-14. DOI: 10.15666/aeer/1806_77837793

[66] Ranjan R, Sharma NK, Kumar A, Pramanik M, Mehta H, Ojasvi PR, et al. Evaluation of maize-based intercropping on runoff, soil loss, and yield in foothills of the Indian sub-Himalayas. Experimental Agriculture. 2021;**57**:69-84. DOI: 10.1017/S0014479721000053

[67] Wezel A, Herren BG, Kerr RB, Barrios E, Gonçalves AL, Sinclair F. Agroecological principles and elements and their implications for transitioning to sustainable food systems. A review. Agronomy for Sustainable Development. 2020;**40**:1-3. DOI: 10.1007/S13593-020-00646-Z

[68] Pantera A, Burgess PJ, Mosquera Losada R, Moreno G, López-Díaz ML, Corroyer N, et al. Agroforestry for

high value tree systems in Europe. Agroforestry Systems. 2018;**92**:945-959. DOI: 10.1007/s10457-017-0181-7

[69] Toungos MD, Bulus ZW. Cover crops dual roles: Green manure and maintenance of soil fertility, a review. International Journal of Innovative Agriculture and Biology Research. 2019;7:47-59

[70] Shelton AM, Badenes-Perez FR. Concepts and applications of trap cropping in pest management. Annual Review of Entomology. 2006;**51**:285-308. DOI: 10.1146/annurev.ento.51.110104.150959

[71] Médiène S, Valantin-Morison M, Sarthou J-P, de Tourdonnet S, Gosme M, Bertrand M, et al. Agroecosystem management and biotic interactions: A review. Agronomy for Sustainable Development. 2011;**14**:211-223. DOI: 10.1007/s13593-011-0009-1

[72] Cheng F, Cheng Z. Research progress on the use of plant allelopathy in agriculture and the physiological and ecological mechanisms of allelopathy. Frontiers in Plant Science. 2015;**6**:1020. DOI: 10.3389/fpls.2015.01020/full

[73] Ratnadass A, Fernandes P, Avelino J, Habib R. Plant species diversity for sustainable management of crop pests and diseases in agroecosystems: A review. Agronomy for Sustainable Development. 2012;**32**:273-303. DOI: 10.1007/s13593-011-0022-4

[74] Lu P, Jiang B, Weiner J. Crop spatial uniformity, yield and weed suppression. Advances in Agronomy. 2020;**161**:117-178. DOI: 10.1016/bs.agron.2019.12.003

[75] Leng G, Huang M. Crop yield response to climate change varies with crop spatial distribution pattern. Scientific Reports. 2017;**7**:1463. DOI: s41598-017-01599-2

[76] Chen Z, Xu C, Ji L, Feng J, Li F, Zhou X, et al. Effects of multi-cropping system on temporal and spatial distribution of carbon and nitrogen footprint of major crops in China. Global Ecology and Conservation. 2020;**22**:e00895. DOI: 10.1016/j.gecco.2019.e00895

[77] Liu Q, Sun X, Wu W, Liu Z, Fang G, Yang P. Agroecosystem services: A review of concepts, indicators, assessment methods and future research perspectives. Ecological Indicators. 2022;**142**:109218. DOI: 10.1016/j.ecolind.2022.109218

[78] Yahaya SM, Mahmud AA, Abdullahi M, Haruna A. Recent advances in the chemistry of nitrogen, phosphorus and potassium as fertilizers in soil: A review. Pedosphere. 2023;**33**:385-406. DOI: 10.1016/j.pedsph.2022.07.012

[79] Chen Y, Peng J, Wang J, Fu P, Hou Y, Zhang C, et al. Crop management based on multi-split topdressing enhances grain yield and nitrogen use efficiency in irrigated rice in China. Field Crops Research. 2015;**184**:50-57. DOI: 10.1016/j.fcr.2015.09.006

[80] Ahmed NU, Ferdous Z, Mahmud NU, Hossain A, Zaman MA. Effect of split application of nitrogen fertilizer on the yield and quality of potato (*Solanum tuberosum*). International Journal of Natural and Social Sciences. 2017;**4**:60-66

[81] Zhang G, Zhao D, Liu S, Liao Y, Han J. Can controlled-release urea replace the split application of normal urea in China? A meta-analysis based on crop grain yield and nitrogen use efficiency. Field Crops Research. 2022;**275**:108343. DOI: 10.1016/j.fcr.2021.108343

[82] Trujillo-Tapia MN, Ramírez-Fuentes E. Bio-fertilizer: An alternative to reduce chemical fertilizer in agriculture. Journal of Global Agriculture and Ecology. 2016;**4**:99-103

[83] Ghaly FA, Abd-Elhamied AS, Shalaby NS. Effect of bio-fertilizer, organic and mineral fertilizaters on soybean yield and nutrients uptake under sandy soil conditions. Journal of Soil Sciences and Agricultural Engineering. 2020;**11**:653-660

[84] Patel N, Patel Y, Mankad A. Bio fertilizer: A promising tool for sustainable farming. International Journal of Innovative Research in Science, Engineering and Technology. 2014;**3**:56-69. DOI: 10.15680/IJIRSET.2014.0309007

[85] Itelima JU, Bang WJ, Onyimba IA, Sila MD, Egbere OJ. Bio-fertilizers as key player in enhancing soil fertility and crop productivity: A review. Direct Research Journal of Agriculture and Food Science. 2018;**3**:73-83. Available from: http://hdl.handle.net/123456789/1999

[86] Kumawat KC, Nagpal S, Sharma P. Present scenario of bio-fertilizer production and marketing around the globe. In: Biofertilizers. Sawston, Cambridge: Woodhead Publishing; 2021. pp. 389-413

[87] Aloo BN, Makumba BA, Mbega ER. Status of biofertilizer research, commercialization, and practical applications: A global perspective. In: Biofertilizers. Sawston, Cambridge: Woodhead Publishing; 2021. pp. 191-208

[88] Zhai L, Wang Z, Zhai Y, Zhang L, Zheng M, Yao H, et al. Partial substitution of chemical fertilizer by organic fertilizer benefits grain yield, water use efficiency, and economic return of summer maize. Soil and Tillage Research. 2022;**217**:105287. DOI: 10.1016/j.still.2021.105287

[89] Rashmi I, Roy T, Kartika KS, Pal R, Coumar V, Kala S, et al. Organic and inorganic fertilizer contaminants in agriculture: Impact on soil and water resources. Contaminants in Agriculture: Sources, Impacts and Management. 2020;**4**:3-41. DOI: 10.1007/978-3-030-41552-5_1

[90] Abebe TG, Tamtam MR, Abebe AA, Abtemariam KA, Shigut TG, Dejen YA, et al. Growing use and impacts of chemical fertilizers and assessing alternative organic fertilizer sources in Ethiopia. Applied and Environmental Soil Science. 2022;**1**:4738416. DOI: 10.1155/2022/4738416

[91] Jayant B, Dahiya K, Rukhiyar A, Raj R, Meena RK. A review of the drip irrigation system. Journal of Engineering Research and Application. 2022;**1**:22-29. DOI: 10.55953/JERA.1103

[92] Wang H, Wang N, Quan H, Zhang F, Fan J, Feng H, et al. Yield and water productivity of crops, vegetables and fruits under subsurface drip irrigation: A global meta-analysis. Agricultural Water Management. 2022;**269**:107645. DOI: 10.1016/j.agwat.2022.107645

[93] Ayilara MS, Adeleke BS, Akinola SA, Fayose CA, Adeyemi UT, Gbadegesin LA, et al. Biopesticides as a promising alternative to synthetic pesticides: A case for microbial pesticides, phytopesticides, and nanobiopesticides. Frontiers in Microbiology. 2023;**14**:1040901. DOI: 10.3389/fmicb.2023.1040901

[94] Ngegba PM, Cui G, Khalid MZ, Zhong G. Use of botanical pesticides in agriculture as an alternative to synthetic pesticides. Agriculture. 2022;**12**:600. DOI: 10.3390/agriculture12050600

[95] Molina GA, Vazquez Pugliese DE. Redesign the agroecosystem through biodiversity: Revising concepts and integrating visions. Agroecology and Sustainable Food Systems. 2022;**46**:1550-1580. DOI: 10.1080/21683565.2022.2121952

[96] Stomph T, Dordas C, Baranger A, de Rijk J, Dong B, Evers J, et al. Designing intercrops for high yield, yield stability and efficient use of resources: Are there principles? Advances in Agronomy. 2020;**160**:1-50. DOI: 10.1016/bs.agron.2019.10.002

[97] Blessing DJ, Gu Y, Cao M, Cui Y, Wang X, Asante-Badu B. Overview of the advantages and limitations of maize-soybean intercropping in sustainable agriculture and future prospects: A review. Chilean journal of Agricultural Research. 2022;**82**:177-188. DOI: 10.4067/S0718-58392022000100177

[98] Khanal U, Stott KJ, Armstrong R, Nuttall JG, Henry F, Christy BP, et al. Intercropping—evaluating the advantages to broadacre systems. Agriculture. 2021;**11**:453. DOI: 10.3390/agriculture11050453

[99] Somasundaram J, Sinha NK, Dalal RC, Lal R, Mohanty M, Naorem AK, et al. No-till farming and conservation agriculture in South Asia–issues, challenges, prospects and benefits. Critical Reviews in Plant Sciences. 2020;**39**:236-279. DOI: 10.1080/07352689.2020.1782069

[100] Gondal AH, Zafar H, Yousaf H, Farooq Q, Imran B, Ch MD, et al. Impacts of tillage technologies on soil, plant, environment and its management: A short communication. Indian Journal of Pure and Applied Biosciences. 2021;**9**:76-83. DOI: 10.18782/2582-2845.8682

[101] Du C, Li L, Effah Z. Effects of straw mulching and reduced tillage on crop production and environment: A review. Water. 2022;**14**:2471. DOI: 10.3390/w14162471

[102] Kuhn NJ, Hu Y, Bloemertz L, He J, Li H, Greenwood P. Conservation tillage and sustainable intensification of agriculture: Regional vs. global benefit analysis. Agriculture, Ecosystems & Environment. 2016;**15**(216):155-165. DOI: 10.1016/j.agee.2015.10.001

[103] Mangalassery S, Sjögersten S, Sparkes DL, Sturrock CJ, Craigon J, Mooney SJ. To what extent can zero tillage lead to a reduction in greenhouse gas emissions from temperate soils? Scientific Reports. 2014;**4**:4586

[104] Holland JM. The environmental consequences of adopting conservation tillage in Europe: Reviewing the evidence. Agriculture, Ecosystems & Environment. 2004;**103**:1-25. DOI: 10.1016/j.agee.2003.12.018

[105] Crittenden SJ, Huerta E, De Goede RG, Pulleman MM. Earthworm assemblages as affected by field margin strips and tillage intensity: An on-farm approach. European Journal of Soil Biology. 2015;**66**:49-56. DOI: 10.1016/j.ejsobi.2014.11.007

[106] Briones MJ, Schmidt O. Conventional tillage decreases the abundance and biomass of earthworms and alters their community structure in a global meta-analysis. Global Change Biology. 2017;**23**:4396-4419. DOI: 10.1111/gcb.13744

[107] Aslam MT, Maqbool R, Khan I, Chattha MU, Nawaz M, Shah AN, et al. Efficacy of different pre and post emergence herbicide application on late sown maize crop under variable planting density. International Journal of Plant Production. 2024;**18**:229-238. DOI: 10.1007/s42106-024-00286-3

[108] Peigné J, Ball BC, Roger-Estrade J, David CJ. Is conservation tillage suitable for organic farming? A review. Soil Use and Management. 2007;**23**:129-144. DOI: 10.1111/j.1475-2743.2006.00082.x

[109] Soane BD, Ball BC, Arvidsson J, Basch G, Moreno F, Roger-Estrade J. No-till in northern, western and south-western Europe: A review of problems and opportunities for crop production and the environment. Soil and Tillage Research. 2012;**118**:66-87. DOI: 10.1016/j.still.2011.10.015

[110] Gilpin AM, O'Brien C, Kobel C, Brettell LE, Cook JM, Power SA. Co-flowering plants support diverse pollinator populations and facilitate pollinator visitation to sweet cherry crops. Basic and Applied Ecology. 2022;**63**:36-48. DOI: 10.1016/j.baae.2022.05.005

[111] Fenta AA, Tsunekawa A, Haregeweyn N, Poesen J, Tsubo M, Borrelli P, et al. Land susceptibility to water and wind erosion risks in the East Africa region. Science of the Total Environment. 2020;**703**:135016. DOI: 10.1016/j.scitotenv.2019.135016

[112] Petit S, Deytieux V, Cordeau S. Landscape-scale approaches for enhancing biological pest control in agricultural systems. Environmental Monitoring and Assessment. 2021;**193**:75-81. DOI: 10.1007/s10661-020-08812-2

[113] Obanyi JN, Ogendo JO, Mulwa RM, Nyaanga JG, Cheruiyot EK, Bett PK, et al. Field margins and cropping system influence diversity and abundance of aphid natural enemies in Lablab purpureus. Journal of Applied Entomology. 2023;**147**:439-451. DOI: 10.1111/jen.13125

[114] Boesing AL, Nichols E, Metzger JP. Effects of landscape structure on avian-mediated insect pest control services: A review. Landscape Ecology. 2017;**32**:931-944. DOI: 10.1007/s10980-017-0503-1

[115] Tscharntke T, Bommarco R, Clough Y, Crist TO, Kleijn D, Rand TA, et al. Conservation biological control and enemy diversity on a landscape scale. Biological Control. 2007;**43**:294-309. DOI: 10.1016/j.biocontrol.2007.08.006

[116] Altieri MA, Nicholls CI, Henao A, Lana MA. Agroecology and the design of climate change-resilient farming systems. Agronomy for Sustainable Development. 2015;**35**:869-890. DOI: 10.1007/s13593-015-0285-2

[117] Meena H, Kumar D, Prasad K. Precision agriculture: Present status & scope: A review. Just Agriculture–Multidisciplinary Newsletter. 2021;**1**:213-219

[118] Schader C, Stolze M, Gattinger A. Environmental performance of organic farming. Green Technologies in Food Production and Processing. 2012;**2**:183-210. DOI: 10.1007/978-1-4614-1587-9_8

[119] Tittonell P, Piñeiro G, Garibaldi LA, Dogliotti S, Olff H, Jobbagy EG. Agroecology in large scale farming—A research agenda. Frontiers in Sustainable Food Systems. 2020;**4**:584605. DOI: 10.3389/fsufs.2020.584605

[120] Mosquera-Losada MR, McAdam JH, Romero-Franco R, Santiago-Freijanes JJ, Rigueiro-Rodróguez A. Definitions and components of agroforestry practices in Europe. Agroforestry in Europe: Current Status and Future Prospects. 2009;**2**:3-19. DOI: 10.1007/978-1-4020-8272-6_1

[121] Guo Y, Fan R, Zhang X, Zhang Y, Wu D, McLaughlin N, et al. Tillage-induced effects on SOC through changes in aggregate stability and

soil pore structure. Science of the Total Environment. 2020;**703**:134617. DOI: 10.1016/j.scitotenv.2019.134617

[122] Omara P, Aula L, Eickhoff EM, Dhillon JS, Lynch T, Wehmeyer GB, et al. Influence of no-tillage on soil organic carbon, total soil nitrogen, and winter wheat (*Triticum aestivum* L.) grain yield. International Journal of Agronomy. 2019;**1**:9632969. DOI: 10.1155/2019/9632969

[123] Song K, Zheng X, Lv W, Qin Q, Sun L, Zhang H, et al. Effects of tillage and straw return on water-stable aggregates, carbon stabilization and crop yield in an estuarine alluvial soil. Scientific Reports. 2019;**9**:4586. DOI: 10.1038/s41598-019-40908-9

[124] Bescansa P, Imaz MJ, Virto I, Enrique A, Hoogmoed WB. Soil water retention as affected by tillage and residue management in semiarid Spain. Soil and Tillage Research. 2006;**87**:19-27. DOI: 10.1016/j.still.2005.02.028

[125] So HB, Grabski A, Desborough P. The impact of 14 years of conventional and no-till cultivation on the physical properties and crop yields of a loam soil at Grafton NSW, Australia. Soil and Tillage Research. 2009;**104**:180-184. DOI: 10.1016/j.still.2008.10.017

[126] Alam MK, Islam MM, Salahin N, Hasanuzzaman M. Effect of tillage practices on soil properties and crop productivity in wheat-mungbean-rice cropping system under subtropical climatic conditions. The scientific World Journal. 2014;**1**:437283. DOI: 10.1155/2014/437283

[127] Jat HS, Datta A, Sharma PC, Kumar V, Yadav AK, Choudhary M, et al. Assessing soil properties and nutrient availability under conservation agriculture practices in a reclaimed sodic soil in cereal-based systems of North-West India. Archives of Agronomy and Soil Science. 2018;**64**:531-545. DOI: 10.1080/03650340.2017.1359415

[128] Vogeler I, Rogasik J, Funder U, Panten K, Schnug E. Effect of tillage systems and P-fertilization on soil physical and chemical properties, crop yield and nutrient uptake. Soil and Tillage Research. 2009;**103**:137-143. DOI: 10.1016/j.still.2008.10.004

Chapter 8

Perspective Chapter: Agroforestry Strategies for Integrated Soil and Water Conservation

Dinesh Jinger, Nandha Kumar N, Chhavi Sirohi, Archana Verma, Pankaj Panwar and Rajesh Kaushal

Abstract

Land degradation has a substantial influence on food security, health, and socioeconomic development, highlighting the critical role that land plays as a resource necessary for maintaining life. As a result, agroforestry interventions become essential tactics for resource preservation and improving sustainable production. Many agroforestry techniques, including agri-horticulture, silvipasture, and agri-silviculture systems, have been developed over the course of substantial study in a variety of agroclimatic zones with the goal of reducing land degradation. The United Nations Convention to Combat Desertification has acknowledged these strategies as essential to reaching land degradation neutrality. The benefits of agroforestry techniques for reducing soil erosion and runoff, increasing soil fertility, and enhancing carbon sequestration are explained in this chapter. It is crucial to promote these affordable and sustainable technologies to guarantee their widespread adoption. As a result, putting in place agroforestry systems is essential for healing impacted regions and addressing issues with livelihoods, environmental sustainability, and food security. In order to make sure that stakeholders receive the proper incentives, national policy programs should incorporate the valuation of the advantages of soil protection. Furthermore, future research endeavors should prioritize the development of economically viable agroforestry systems designed to restore degraded lands, enhance water efficiency, and minimize competition between trees and crops.

Keywords: agroforestry, carbon sequestration, ravine lands, soil erosion, soil organic carbon

1. Introduction

Meeting the increasing global demands for agricultural products, along with the need to enhance ecosystem services, is a significant challenge driven by rising populations and incomes [1, 2]. Amid these concerns, safeguarding the world's soils and preventing soil erosion takes center stage [3, 4]. Global soil and water erosion is on the rise, affecting soil quality and carbon storage and leading to biodiversity loss, reduced crop yields, and even poverty [5–8]. It also hampers the achievement of various sustainable development goals [9–11].

Soil erosion is a historical and persistent challenge that has influenced human civilization and the pursuit of a better quality of life. It arises from natural processes and can be exacerbated by socioeconomic developments over time. The materials eroded due to this phenomenon have detrimental consequences, affecting both the immediate environment and areas farther away, impacting plant and animal life. These repercussions can be magnified by the intricate interactions within the ecosystem, involving both interconnections among ecosystem components and internal dynamics [12]. In regions characterized by an increasing population and expanding agricultural production, construction, urban development, and diverse human activities, soil erosion presents a prominent challenge. This issue has been noted in various studies [13, 14]. Among the contributing factors to soil erosion is inadequate land management, which leads to soil degradation and the subsequent surface runoff of water, rather than facilitating proper soil infiltration [15–17].

Soil erosion poses significant threats to plant growth, agricultural productivity, water quality, and recreational spaces, affecting diverse land types as evidenced by various studies [15]. Its primary drivers are water and wind, leading to substantial annual soil loss. Erosion can manifest gradually and unnoticed or as a rapid process, resulting in the loss of valuable topsoil. In agricultural contexts, it translates into reduced crop yield potential, compromised surface water quality, and disrupted drainage systems [16, 18]. On a global scale, soil erosion stands as a significant environmental concern, depleting soil nutrients and causing land degradation while also giving rise to off-site problems such as flooding, water siltation, and pollution [19–21]. Effectively addressing this challenge necessitates comprehensive monitoring and assessment to ensure the sustainable management of natural resources and the environment.

Soil erosion as a substantial contributor to soil degradation; it is well-established that agroforestry practices hold the potential to enhance rainwater infiltration and reduce runoff, effectively mitigating soil erosion. As raindrops make initial contact with the soil surface, the collective action of trees, annual crops, and grasses through their canopy, surface litter, and intricate root systems comes into play. These components act as a physical barrier, effectively decelerating the movement of surface water and diminishing erosive forces, leading to a reduction in soil erosion [22]. Moreover, the application of various agroforestry techniques, including agri-silviculture, agri-horticulture, silvopasture, and alley cropping, plays a pivotal role in addressing soil erosion. These techniques not only help in erosion control but also contribute to enhanced soil fertility, improved water quality, increased biodiversity, and carbon sequestration. Recent literature underscores the importance of agroforestry in developing integrated, diverse, and sustainable land use systems, bridging the gap between agriculture and conservation [23–25]. By integrating trees, shrubs, crops, and livestock, agroforestry provides a cost-effective and efficient alternative to traditional soil conservation methods, ensuring sustainable land productivity [22, 26]. This chapter delves into the significance of agroforestry strategies in comprehensive soil and water conservation, particularly in regions where perennial ground cover is essential, such as in ravines and poor lands.

2. Unearthing the challenge: soil erosion in India and global

Land resources are essential for human well-being and societal progress [27]. However, land degradation has escalated since the twentieth century due to ecological decline, increased food demand, rapid urbanization, industrialization, and unsustainable land use [28]. Approximately 60% of the world's land is now degraded, posing a significant challenge for sustainable land use [29]. Water and wind erosion contributes to this problem, leading to an annual loss of 24 billion tons of fertile soil and a financial loss of $400 billion [30]. Sustainable soil management is gaining global attention, spearheaded by the Global Soil Partnership (GSP) [4]. The FAO reports an alarming annual global soil loss of 75 billion tons, with India experiencing an average loss of 16 tons per hectare [31]. Rapid land degradation has led to a significant economic loss of 3.17 lakh crore ($46.90 billion) [32]. Soil erosion assessment traces back to Myres [33], and the United Nations Environment Programme (UNEP) has monitored global soil degradation trends from the 1980s to the 1990s [34]. UNEP's Global Assessment of Soil Degradation focuses on static land use and land cover data, while India's land use is becoming more susceptible to surface erosion due to population pressure [4]. Currently, 16.7% of land is allocated for crop cultivation, showing a significant increase since 1985 [35]. The Global Assessment of Soil Degradation Survey highlights that soil erosion, a significant contributor to land degradation, is responsible for the degradation of 84% of affected land [36]. This results in an annual global socioeconomic loss of approximately $40 billion [37]. The United Nations' Sustainable Development Goals (SDGs) have set a target to eradicate land degradation by 2030 [38]. Monitoring reports within the European Union indicate that over 201,000 square kilometers of land are at risk of severe erosion (>10 Mg ha^{-1} $year^{-1}$). Soil erosion plays a critical role in monitoring progress toward SDGs related to climate action, zero hunger, and life on land. This pressing issue is recognized and acknowledged by Lal [37].

Despite the widespread occurrence of water erosion, there is a notable scarcity of comprehensive global data regarding the extent and severity of this issue. The Global Assessment of Human-Induced Soil Degradation (GLASOD) study, as outlined in **Table 1**, estimated that approximately 15% of the world's ice-free land area is affected by various forms of land degradation. Within this assessment, accelerated soil erosion, primarily caused by wind and water, accounts for roughly 28 and 56%, respectively. This indicates that the regions impacted by water erosion span approximately 11 million square kilometers, while wind-induced erosion affects approximately 5.5 million square kilometers. Notably, tillage erosion remains undocumented in some affected areas. As soil is a finite resource formed over time, the global significance of the erosion problem has been gaining recognition only recently.

In India, 120.7 million hectares (M ha) of land are affected by degradation, with soil erosion being the main contributor, affecting 71% of degraded land (85.7 M ha) (**Table 2**). Water erosion alone affects 60.7% (73.3 M ha), and wind erosion affects 10.3% (12.4 million hectares). Furthermore, deforestation (2.07 M ha, 2001–2021), intense rainfall (>7.5 mm ha^{-1}), uncontrolled grazing (5.65 M ha), excessive fertilizer usage (32 mt $year^{-1}$), and shifting cultivation (7.6 M ha) compound the issue of land degradation. Presently, 97.8 M ha (29.7% of the total area) face degradation, with states such as Rajasthan, Maharashtra, Gujarat, Karnataka, Ladakh Union Territory, Jharkhand, Odisha, Madhya Pradesh, and Telangana contributing about two-thirds of the nation's degraded land (23.7% of TGA) [40]. India, a country characterized by its riverine systems and monsoon patterns, faces an escalating threat in the form of soil erosion, a pressing issue in several regions [41]. This erosion, attributed to natural and human-induced factors [42], is exacerbated by climate variations and unscientific agricultural practices, such as deforestation and construction [43–45]. The loss of the topsoil layer poses a significant threat to soil fertility and food security [46]. India has

Degradation	World	Asia	West Asia	Africa	Latin America and Caribbean	North America	Australia and Pacific	Europe
Water erosion	1094	440	84	227	169	60	83	115
Wind erosion	548	222	145	187	47	35	16	42
Nutrient depletion	135	15	6	45	72	—	+	3
Salinity	76	53	47	15	4	—	1	4
Contamination	22	2	+	+	+	—	—	19
Physical	79	12	4	18	13	1	2	36
Others	10	3	1	2	1	—	1	2
Sum	1964	747	287	494	306	96	103	218
Source: Bai et al. [39].								

Table 1.
Estimation of human-induced soil degradation by GLASOD.

Degradation	Arable land (M ha)	Open forest (<40% canopy) (M ha))	Source
Water erosion (>10 Mg ha^{-1} $year^{-1}$)	73.27	9.30	Soil Loss Map of India-IISWC, Dehradun
Eolian erosion	12.40	—	Wind Erosion Map of India, CAZRI, Jodhpur
Sub-total	85.67	9.30	
Chemical degradation			
Exclusively salt-affected soils	5.44	—	Salt-affected Soils of India, CSSRI, Karnal; NBSS&LUP, Nagpur; NRSA, Hyderabad, and others
Salt-affected and water-eroded soils	1.20	0.10	
Exclusively acidic soil (pH < 5.5)	5.09	—	Acid Soils of India, CSSRI, Karnal; NBSS&LUP, Nagpur
Acidic (pH < 5.5) and water-eroded soil	5.72	7.13	
Sub-total	17.45	7.23	
Physical degradation			
Mining and industrial waste	0.19		Wasteland Map of India-NRSA, Hyderabad
Waterlogging (permanent surface inundation)	0.88		
Sub-total	1.07		
Total	104.19	16.53	
Grand total (arable land open forest)	120.72		
Source: Jinger et al. [40].			

Table 2.
Detailed information on India's wastelands and degraded land (in M hectares).

managed to meet food production targets despite monsoon challenges, largely due to advanced mechanization, the use of chemical fertilizers, and extensive tillage [33]. However, this heightened mechanization adversely impacts soil structure and diminishes water-holding capacity [31].

Water-induced soil erosion is a leading cause of land degradation in India. The severity of this erosion varies, spanning from moderate (over 10 Mg ha^{-1} $year^{-1}$) to exceedingly severe (exceeding 80 Mg ha^{-1} $year^{-1}$). Globally, water erosion is a significant contributor to land degradation, affecting approximately 2 billion hectares, with a significant presence in tropical regions. This pervasive issue has substantial implications for essential natural resources, particularly soil and water [41]. The worldwide loss of water and sediment due to soil erosion presents a notable environmental concern [42]. Soil erosion is exacerbated by several factors, including high-intensity rainfall [43], steep terrain [44], and the vulnerability of topsoil [45]. In many regions of India's tropical areas, the annual rainfall is concentrated within a short 4–5-month period (June–September). During the subsequent 7–8-month dry season, water scarcity leads to a severe shortage of fodder on farmlands, resulting in heightened grazing pressure on forest and community lands. It is worth noting that nearly one-third of India's fodder requirements are met through the utilization of forest resources for grazing and harvested fodder [46]. Land degradation in croplands and grasslands is predominantly accelerated by inappropriate land use [31, 33] and mismanagement [32]. The runoff water generated by water erosion poses a substantial and far-reaching threat to soil quality, resulting in the depletion of organic carbon, disturbances in nitrogen balance, soil compaction, reduced soil biodiversity, and contamination from pesticides and heavy metals.

3. Impacts of erosion: assessing the consequences on water and soil erosion

3.1 Impact on nutrient mobilization

The examination of erosion processes provides valuable insights into the intricate environmental dynamics of our planet. Notably, water erosion emerges as a dominant force, contributing to an annual sediment flux of roughly 28 Pg. In addition, tillage and wind erosion introduce approximately 5 and 2 Pg, respectively, leading to a total sediment flux of roughly 35 ± 10 Pg per year. This erosional activity also results in an agricultural carbon erosion flux of around 0.5 ± 0.15 Pg C annually, with an estimated 0.08 ± 0.02 Pg C transported to river systems through water erosion. When it comes to nitrogen dynamics, soil erosion is a significant factor, mobilizing roughly 23–42 Tg of nitrogen every year, a magnitude similar to the nitrogen applied to agricultural land *via* chemical fertilizers, the nitrogen removed by harvested crops, and the estimated riverine fluxes of particulate nitrogen. In terms of phosphorus, we estimate that soil erosion leads to the movement of 2.1–3.9 Tg of organic phosphorus and 12.5–22.5 Tg of inorganic phosphorus on an annual basis. While global mean phosphorus fluxes are notably lower than the 40 Pg stored in soils worldwide, they are similar in magnitude to crop uptake and fertilizer phosphorus additions to agricultural land. However, in specific regions, erosion-induced phosphorus fluxes surpass phosphorus additions, posing challenges for soil fertility and food production. This comprehensive evaluation highlights the intricate global interplay of erosion processes and their far-reaching implications for soil and nutrient dynamics [45, 47].

3.2 Impact on carbon cycle

Soil erosion encompasses the processes of soil mobilization, transport, and deposition. A holistic understanding of its effects on the carbon cycle requires considering all three phases. When soil material is mobilized, it disrupts soil structure, potentially leading to a significant increase in soil organic carbon (SOC) mineralization, resulting in the loss of over 20% of the total SOC as carbon dioxide. Regarding the influence of transport on SOC mineralization, it is essential to distinguish between SOC deposited locally after relatively short-distance transport (<500 m) *via* water or tillage within a brief timeframe (<1 day) and the fate of SOC transported to rivers. Field observations and erosion-deposition simulations based on 137Cs inventories indicate that the additional SOC mineralization during soil transport over land is of limited significance [48].

Recent field observations have also showed that SOC losses from soil redeposited after a short transport phase are relatively minor, constituting less than 2.5% of eroded SOC and, therefore, having a negligible impact on the global carbon budget. Understanding erosion's implications for the carbon cycle necessitates a comprehensive examination of both short-term and long-term impacts [49]. Contemporary research suggests that erosion plays a dual role in influencing carbon dynamics. When soil structure is disrupted during erosion, it can trigger the immediate release of carbon dioxide. Over extended timeframes, the increased carbon emissions are associated with eroded soils losing their capacity to support plant growth, leading to reduced carbon inputs from plant and root matter. Surprisingly, erosion can also contribute to carbon sequestration. As soil erosion integrates carbon-poor subsoil into the plow layer, newly exposed mineral surfaces may bind organic matter, offering potential for expanding soil carbon inventories [50].

The process of erosion, characterized by the removal of soil, also has the effect of bringing subsoil and parent material closer to the surface. Both empirical and theoretical evidence are increasingly pointing to a connection between erosion and heightened rates of chemical weathering in areas with silicate-rich parent material under steady-state conditions. This is significant because the weathering of silicate minerals consumes carbon dioxide, suggesting a potential link between erosion-induced weathering and the consumption of carbon dioxide [48]. However, it is expected that this carbon dioxide flux is relatively small. On the contrary, in regions with calcareous parent materials, accelerated weathering could lead to the release of carbon dioxide into the atmosphere. For instance, in the Canadian prairies, it has been estimated that approximately 10% of the carbonates acidified during erosion might be emitted as carbon dioxide, resulting in an estimated loss of carbon ranging from 0.12 to 1.2 Mg ha^{-1} $year^{-1}$ [51].

3.3 Impact on nutrient cycles

Current research on erosion's impact on nitrogen and phosphorus cycling has primarily concentrated on assessing how these nutrients are mobilized and delivered to aquatic ecosystems. However, there is limited understanding of how erosion influences nitrogen and phosphorus cycling in terrestrial environments. Soil organic matter serves as a significant reservoir for nitrogen and phosphorus. Therefore, increased soil carbon mineralization due to erosion-induced soil mobilization leads to a proportional rise in dissolved nitrogen and phosphorus, making them more readily available to organisms than particulate or organic forms. Conversely, the burial and preservation of deposited carbon contribute to the stability of organic nitrogen. Nitrogen's stability in depositional settings may be high, primarily influenced by the rate of carbon mineralization. This phenomenon may account for the consistent C:N ratios observed in surface soils within specific ecological contexts, as well as the similarities between contemporary and ancient soil C:N values. In eroding environments, the dynamic replacement of carbon also promotes nitrogen stabilization. Nevertheless, nitrogen can also influence carbon cycling, particularly in environments where nitrogen availability directly limits biomass production and, consequently, dynamic carbon replacement [48].

Erosion significantly reduces soil phosphorus levels over time, not just through physical removal but also due to the exposure of low-phosphorus subsoil. This decline is influenced by erosion rates and chemical weathering. The soil's phosphorus composition changes, shifting from various forms to predominantly organic and occluded form [52]. Depositional areas receive a significant phosphorus influx, as seen in Hawaii, where dust deposition alleviates phosphorus limitations in aging soils [53]. In regions dominated by erosion with limited nitrogen and phosphorus inputs, primary production exponentially declines as erosion intensifies [54]. This is driven by nutrient loss, soil structure degradation, and reduced water availability due to thinner soils. Water erosion enriches sediment with carbon, nitrogen, and phosphorus relative to the parent soil during mobilization and deposition [54]. This leads to changes in the relative abundance of these elements in soils, with water and wind erosion causing greater enrichment than tillage erosion. Loss of carbon, nitrogen, and phosphorus from mobilization sites may initiate a feedback loop, reducing plant productivity and increasing erosion vulnerability. Soil resistance to erosion is closely tied to organic matter and vegetation cover [54]. Conversely, deposition sites experience higher nutrient and carbon contents, boosting primary productivity and enhancing soil fertility, plant growth, and erosion resistance [16].

4. Agroforestry's real-world effects on biodiversity

Agroforestry offers many advantages associated with the conservation of biodiversity (BD) and is found to have a major role in BD conservation and protection, according to recent reviews. Studies show that a variety of aboveground vegetation types and their interactions have a significant impact on belowground biodiversity; belowground communities are enhanced by systems that combine above- and belowground features. Comparing practices such as alley cropping and silvi-pasture to monocropping, the latter shows lower leaching losses and higher plant growth, along with superior soil attributes such as improved soil porosity, water dynamics, and nutrient cycling efficiency. The temperature of the surface soil rises when shade trees are removed, which has an impact on the soil communities, decomposition rates, humidity, and nutritional status. In Missouri, Agroforestry buffers outperform monocrop regions in terms of soil thermal characteristics and resistance to high temperatures. Additionally, there were differences in the microclimate characteristics of crop and buffer areas. The buffers had higher soil temperature, humidity, and wind speeds than the crop areas. Numerous studies conducted globally have demonstrated advantageous soil characteristics, microclimates, and nutritional condition, hence endorsing a wider range of agroforestry management approaches. These results suggest that specific species combinations at appropriate spacing and configurations can contribute to the improvement and conservation of BD by fostering a favorable microclimate and favorable soil properties [55].

5. Agroforestry approaches to mitigate soil erosion: measures and techniques

Soil is the most crucial resource in agriculture for meeting the varied food needs of the world's rapidly expanding population. It is an essential component of ecosystems for providing the main ecosystem services. But according to recent reports, increased soil erosion poses a serious risk to soil, and approximately every year, 24 billion tons (BT) of rich soil are lost only as a result of erosion caused by water [55]. Erosion strongly affects soil quality by decreasing the available soil nutrients, which subsequently affects the plant growth, yield, and carbon stock potential [56]. The idea of soil conservation has evolved during the last 20 years in a number of ways. Soil conservation was once associated with soil erosion, and efforts to limit it were conducted independently of other land management activities. By the 1970s, the phrase had come to refer to not just keeping the soil in its proper location but also preserving or even improving its production. Different agroforestry system in different parts of the country have been mentioned (**Figures 1–6**). Here are some key strategies and practices within agroforestry systems that help prevent soil erosion:

Establish windbreaks and shelterbelts: planting rows of trees and shrubs along field boundaries or across the prevailing wind direction helps reduce wind erosion. These vegetative barriers slow down the wind and minimize the impact of wind-driven soil erosion.

Create riparian buffers: in areas adjacent to water bodies like streams, rivers, or ponds, plant trees and shrubs as riparian buffers. These buffer zones filter runoff water, reduce soil erosion, and improve water quality.

Integrate contour farming: implement contour farming, where crops are planted along contour lines of the land, thereby reducing the speed of surface runoff and minimizing soil erosion.

Terracing: construct terraces on sloping lands to create flat surfaces that trap water and soil, reducing soil erosion. Plant trees or shrubs on terraces to further stabilize the soil.

Cover cropping: intercrop or rotate main crops with cover crops like legumes or grasses. Cover crops protect the soil from erosion, enhance soil structure, and add organic matter.

Figure 1.
Sapota-based agri-horticulture (sapota + castor + cowpea) system on bench terrace in Mahi ravine land of Central Gujarat, India.

DOI: http://dx.doi.org/10.5772/intechopen.1005772

Figure 2.
Melia dubia-based silvo-aromatic (Melia dubia + lemon grass) system in gully bed of Mahi ravine land of Central Gujarat, India.

Figure 3.
Dragon fruit-based horti-silviculture (Dragon fruit + Melia dubia) system in Mahi ravine land of Central Gujarat.

Figure 4.
Silvo-aromatic system (Melia dubia + lemon grass) in degraded lands of Shivalik hills of Chandigarh.

Figure 5.
Silvo-aromatic system (Melia dubia + lemon grass) in degraded lands of Dehradun under Western Himalaya.

Figure 6.
Silvi-agriculture system (Melia dubia + green gram) in red soil of Southern Karnataka, India.

Agroforestry alley cropping: in alley cropping systems, rows of trees or shrubs are planted with crop alleys. The tree canopy and the root systems help protect the soil from erosion and provide shade to the crops.

Silvopasture: integrate trees with livestock grazing areas. Trees provide shade, reduce trampling of the soil by livestock, and contribute to soil stabilization.

Minimize soil disturbance: implement no-till or reduced-till farming practices to reduce soil disturbance during planting and harvesting. This preserves the soil structure and reduces the risk of erosion.

Maintain ground cover: keep the soil covered with crop residues or mulch, which reduces the impact of raindrops and minimizes water erosion. Mulching can be a crucial practice in agroforestry systems.

Proper agroforestry design: plan and design agroforestry systems to ensure that tree and shrub species are compatible with crops and provide adequate soil protection. The arrangement and spacing of trees should be carefully considered.

Soil fertility management: maintain proper soil fertility through nutrient management and organic matter addition. Healthy soils with adequate nutrients are less susceptible to erosion.

Manage livestock access: if livestock are part of the agroforestry system, manage their access to sensitive areas and prevent overgrazing, which can lead to soil compaction and erosion.

Regular monitoring and maintenance: continuously monitor the agroforestry system and address any issues promptly, such as damaged windbreaks or deteriorating terraces. Agroforestry practices can vary depending on the specific ecosystem and agricultural goals. Combining these techniques can provide a holistic approach to preventing soil erosion while enhancing biodiversity, improving soil health, and increasing overall agricultural sustainability.

6. Prevention of soil erosion

Agroforestry is a valuable approach to prevent soil erosion while also promoting sustainable land use and agricultural practices (**Figure 7**). Here are some studies indicate the strategies within agroforestry systems that helps to prevent soil erosion: A study reported that a single bamboo plant can bind up to 6 m of soil [57]. However, in Jiulongjang and Dayingjang rivers in China, it was noted that each clump can protect 12 m of river embankment, also yielding shoots and bamboo timber [58]. Wiersum [59] found that different agroforestry systems cause the lowest soil erosion. Compared to pure cropping systems, agroforestry produced the following results: 50% less erosion, 75% higher infiltration, 57% lower runoff, 22% greater macro-aggregation, and 30% higher mean weight diameter [60]. In Brazil, Beliveau et al. [61] compared erosion in short cycle cropping systems, 2-year-old agroforestry systems, and forest systems. They discovered that while soil particle loss, total Hg, and cation mobilized in agroforestry systems were similar to those in forests, they were significantly lower in short cycle cropping systems. The ability of agroforestry systems to mitigate soil erosion varies depending on their kind and management style. Only a small percentage of the current agroforestry systems are able to prevent farmland from eroding, according to an Indonesian study. This suggests that more agroforestry is required to reduce soil erosion and preserve watershed quality while promoting sustainable agriculture [62]. Purwaningsih et al. [63] recommended use agroforestry with proper species and arrangement as technique to control landslide reactivation on volcanic foot slopes in Java. Agroforestry functions in two ways to control erosion of soil: supplementary and directly; by supplementary function,

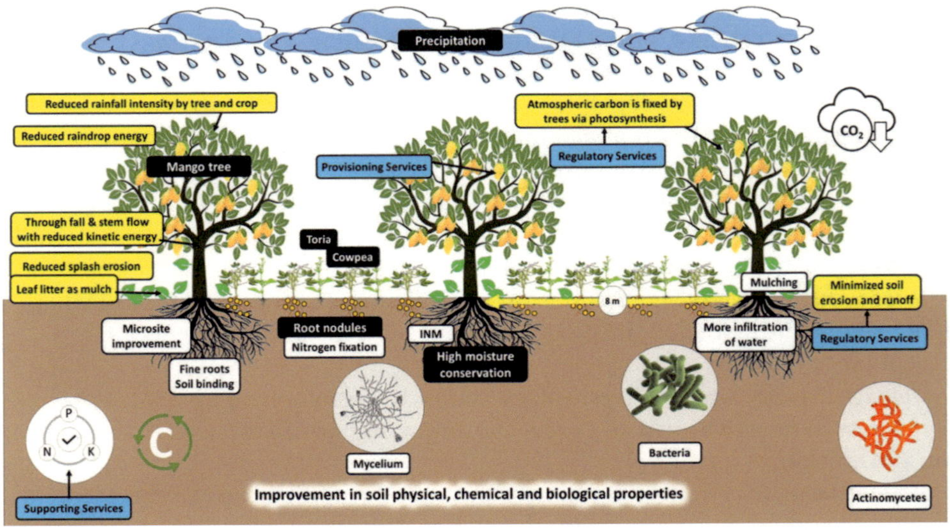

Figure 7.
Ecosystem services received from agroforestry system (Source: Rathore et al. [2]).

plants stabilize soil structure and improve texture and enhance soil productivity, while vegetation directly by itself reduces erosion by reducing impact of rainfall.

7. Bioengineering measures

For soil and water conservation, structural and mechanical methods are frequently employed; nevertheless, they have drawbacks, including a limited lifespan, soil disturbance, high cost, labor intensity, and technical requirements. On the other hand, using plants, bioengineering techniques provide an affordable way to increase output while preserving natural resources [64]. These techniques, which have been successfully shown in Dehradun, use plant materials to alleviate erosion problems. At Nalotanala, for example, a 4 ha area was completely restored and revegetated in less than 10 years, greatly lowering the sediment load from 320 to 5.5 Mg ha^{-1} $year^{-1}$ over a 30-year period. In addition, when therapy ended, the flow of dry weather rose from 100 to 250 days. The impact of bioengineering techniques on the Doon Valley's Shahastradhara mining watershed was evaluated by Juyal et al. [65]. Over a 20-year period, they discovered that afforestation raised vegetation cover from 10 to 90%. After treatment, bulk density dropped from 1.63 to 1.45 Mg m^{-3}, pH dropped from 8.1 to 7.4, and slope was cut in half. Soil organic carbon increased from 0.13 to 0.45%. Main drainage channels also became perennial, and in November and February, respectively, water flow reached 265 and 100 cubic meters per day.

8. Soil health improvement through agroforestry practices

Under both sequential and simultaneous agroforestry, the incorporation of trees into croplands can aid in maintaining the physico-biochemical qualities. In addition to supplying soil organic matter, tree litter and prunings increase soil fertility by releasing nutrients into the soil through mineralization. The amount and quality of tree litter or prunings, the local climate, the kind of soil, and the growing activity of microorganisms in the root zone all play a role. The roots of trees typically extend considerably beyond the crop zone, allowing them to absorb nutrients from the lower soil layers and boost nutrient and water consumption efficiency while posing minimal threat to crops. Improved environmental conditions provided by shade trees for crops and silvo-pastoral systems stimulated the activity of earthworms, termites, and other soil engineers, leading to an increase in the development of both tiny and large faunal aggregates [66]. The soil fertility enhancement varies with the species and the quantity of litter added to the soil **Table 3**. Various workers have reported that trees directly or indirectly influence the soil physic-chemical-biochemical properties. Rivest et al. [70] found that windbreaks increased the soil organic C, total N, and C:N ratio while reducing the soil pH, relative to adjacent crop fields. Meta-analysis by Muchane et al. [60] revealed that agroforestry can reduce soil erosion rates by 50%, enhances soil organic carbon by 21%, raises nitrogen by 13%, enhances available phosphorous by 11%, and alleviates pH of soil by 2% than pure cropping. This might be due to reduced runoff, high infiltration, increased macro-aggregation, and more stable soil structure in agroforest. The physical and chemical characteristics of the soil under coconut monocropping and 5-year-old and 20-year-old Gliricidia-coconut-based mixed systems were studied for 3 years by Raveendra et al. [71]. They discovered that the 20-year-old Gliricidia-coconut-based mixed systems had 22%

Species	Soil fertility enhancement	References
Populus deltoides	SOC: 2.9–4.8 Mg ha^{-1}	[67]
Acacia nilotica	SOM: 1.4 g kg^{-1} soil	[68]
S. cannabina	209 kg N ha^{-1}	[69]

Table 3.
Soil fertility enhancement through trees.

higher organic matter content, 20% higher available P content, and 69% higher total exchangeable K content than the coconut monocrop. This study demonstrated the potential of Gliricidia-coconut-based mixed system for rehabilitation of degraded coconut-growing soils in Sri Lanka.

Biological nitrogen fixation (BNF) constitutes a key nutrient input to agro ecosystems. The contribution of leguminous trees to building up N in degraded soils through BNF is well recognized as an important component of the ecosystem service of nutrient cycling [72]. There are significant differences in estimates of BNF in trees, ranging from high rates up to 472 kg N ha^{-1} year $^{-1}$ in *L. leucocephala*, *Gliricidia sepium*, *C. calothyrsus* to low rates <50 kg N ha^{-1} $year^{-1}$ in *Acacia melanoxylon* and *A. holoserica* [73], whereas roots of Casuarina species fix nitrogen up to 350 kg N ha^{-1} $year^{-1}$. Improvement of soil organic carbon by 36.5% and total nitrogen stocks by 33.2% at 0–10 cm soil layer was observed in alley cropping of rubber trees with *Coffea liberica* as compared to rubber monoculture [74]. In agroforestry system, BNF trees contribute toward complementary interaction, which makes agroforestry system sustainable for nutrient requirements.

Leguminous trees contribute to soil nitrogen enrichment through the process of N-fixation, which occurs when nitrogen-rich biomass is accumulated in the soil through root decomposition, litter fall, and prunings. In addition to being leguminous, some nonleguminous shade plants, such as rambutan and durian, improved the soil's carbon and nitrogen content when added to cocoa agroforestry [75]. Humid and subhumid tropical regions are at risk of losing biodiversity due to soil degradation, and the majority of these soils have inadequate nutrient reserves, are prone to erosion, are acidic, and have other related toxicities. Most of these obstacles may be greatly alleviated by agroforestry, which would also guarantee increased food production, better health and nutrition, and the preservation of natural resources [76]. Stocker et al. [76] also estimated the agroforestry improves several soil physical attributes over a short time, by diversification of tree root systems and accumulation of plant residues. Piza et al. [77] compared decomposition and release of nutrients in different land use systems in Columbia and found that litter decomposition and nutrient release in coffee agroforests were similar to a nearby forest.

9. Effect of agroforestry on erosion factors

Agroforestry is the intentional cultivation of trees, crops, and/or animals in interdependent combinations for a range of purposes. Although agroforestry has been used for a long time all over the world, it was not until the late 1970s that the phrase and concept gained widespread recognition as a viable choice for sustainable land management. These days, a wide range of land-use systems in various contexts are included by the phrase [78]. Agroforestry, which bridges the gap between agriculture and forestry, is currently viewed as a potential method of land use, particularly in emerging tropical and subtropical nations [79].

The idea that agroforestry systems might lessen rainfall erosivity is one that is commonly accepted. In some agroforestry scenarios, it might not hold true, though. When there is a tall, broad-leaved canopy, the kinetic energy of raindrops falling can

Degraded land	Location	Agroforestry system	Impact	Reference
Degraded sloping land	Northeast India	Hedgerow cropping	There was a 94 and 78% decrease in soil loss and runoff, respectively. *Michelia oblonga* enhances soil physical behavior and SOC significantly due to its continuous leaf litter and root exudates.	[81]
	Dehradun	Silvi-pasture system	Zero soil loss, with eucalyptus providing additional revenue of about Rs. 4000 per hectare per year just from commercial grass.	[82]
	Shivalik hills	Horti-pasture system	4.9–30.7 cm of water and 862–2818 kg ha^{-1} of soil were preserved in the *Emblica officinalis* + *Chrysopogon fulvus* horti-pasture system.	[83]
	Karnataka	Ley farming (Vegetative barriers)	Plants such as *Cenchrus ciliaris* and *Cymbopogon martini* reduced soil loss by 16% and discharge by 38%.	[84]
	Kashmir	Silvi-agriculture	Lower temperatures and decreased soil erosion combined with higher agricultural yields.	[85]
Gullied and ravine lands	Gujarat	Agri-horticultural	They showed that sapota with bench terraces and trenches decreased runoff by 16–34% and soil loss by 15–25%.	[86]
	Gujarat	Cowpea, Castor, and Sapota in an Agroforestry System	Compared to a single tree plantation, there was a reduction in total soil loss and runoff of 37.7 and 19.1%, and an increase in system productivity of 81.9%.	[22]
	Uttar Pradesh	Agri-horticulture	Increased yields of fuel wood, wheat, pearl millet, and ber were recorded.	[87]
	Rajasthan	Alley cropping	In alley cropping systems based on Leucaena, higher yields, land equivalent ratios, and soil organic carbon levels were observed.	[88]
Shifting cultivation land	Northeast India	Nitrogen-fixing, fast-growing multipurpose tree plantations	SOC increased by 96.2%, aggregate stability increased by 24.0%, porosity increased by 10.9%, accessible soil moisture increased by 33.2%, bulk density decreased by 15.9%, and erosion ratio decreased by 39.5%.	[81]
	Odisha	Alley cropping	There was a 23–32% decrease in runoff and a 49–52% decrease in soil loss.	[89]
	East India	The grass filter strip and hedgerow of *Gliricidia sepium*	There was a 32% decrease in runoff and a 35% decrease in soil loss.	[90]

Table 4.
Agroforestry practices with potential for control of soil erosion.

be increased. Raindrops aggregate to form larger droplets that, when descending from a height of roughly 30 meters above the ground, can reach a high speed and significantly erode splash zones due to their contact. Although field measurements from such agroforestry systems have been sparse, it is probable that a low and dense canopy will lower erosivity.

When wind is a significant contributor to soil erosion, agroforestry techniques such as windbreaks and shelterbelts are crucial for managing the erosion. Protecting fields, houses, canals, and other places from wind and blowing sand is done by planting windbreaks, which are small strips of trees, shrubs, and/or grasses. Protecting agricultural fields from tidal wave flooding is the purpose of shelterbelts, a type of windbreak that are long, multiple rows of trees and bushes, commonly along sea shores. In the semiarid tropics [80] as well as in semiarid temperate parts of North America, Europe, and Asia, windbreaks have long been used to protect crops and soil from wind and wind erosion. When properly designed and maintained, a windbreak reduces the velocity of the wind and thus its ability to carry and deposit soil and sand (**Table 4**). It can improve the microclimate in a given protected area by decreasing evaporation from the soil and plants. In many cases, windbreaks have been shown to increase the productivity of the crops they protect. Additionally, windbreaks can provide a wide range of useful products, from poles and fuel wood, to fruit, fodder, fiber, and mulch [78].

10. Agroforestry approaches to mitigate water erosion: measures and techniques

Water erosion poses a significant threat to soil health and agricultural productivity, leading to the degradation of arable land and the loss of essential nutrients. In recent years, agroforestry has emerged as a sustainable and effective approach to mitigate water erosion, offering a holistic solution that integrates the cultivation of trees and crops. This portion explores various measures and techniques within agroforestry systems that prove instrumental in combating the detrimental effects of water erosion. Integrating cover crops and living mulches into agroforestry systems is another effective strategy. The roots of cover crops bind soil particles together, preventing them from being washed away by water runoff. Leguminous cover crops, such as clover and vetch, offer the added benefit of fixing nitrogen in the soil, enhancing overall soil fertility. These living mulches also provide a protective cover, reducing the impact of raindrops on the soil surface and minimizing soil compaction. Uttar Pradesh implemented an agri-horticulture system, resulting in higher yields of Ber, pearl millet, wheat, and fuel wood, as reported by Prakash et al. [87]. Another instance in Northeast India involved multipurpose tree plantation, specifically fast-growing nitrogen-fixing trees, in fallow areas. This approach led to a 15.9% reduction in bulk density, a 39.5% decrease in erosion ratio, and significant increases in soil organic carbon (96.2%), aggregate stability (24.0%), porosity (10.9%), and available soil moisture (33.2%), as observed by Saha et al. [81]. Contour planting is a key technique in agroforestry that helps mitigate water erosion. By planting along the contours of the land, the flow of water is slowed down, giving it more time to infiltrate the soil. This reduces the risk of surface runoff and soil erosion. Agroforestry systems that incorporate contour planting often include a mix of deep-rooted trees and shrubs alongside crops. The roots of these trees play a crucial role in stabilizing the soil structure. Buffer strips comprising trees, shrubs, and other perennial vegetation can be strategically established between agricultural fields and water bodies. These buffer strips act as a buffer zone, filtering and slowing down surface runoff before it reaches water bodies. This not only prevents soil erosion but also helps in retaining nutrients and reducing the influx of sediment into water sources. Native species are often preferred for buffer strips to enhance biodiversity and ecosystem services. In Dehradun, a silvipasture system was adopted, demonstrating no soil loss and yielding an annual return of about Rs. 4000 ha^{-1} $year^{-1}$ from commercial grass alone, in addition to returns from Eucalyptus, according to Sharda and Venkateswarlu [82]. The horti-pasture system in the Shivalik hills, incorporating *Emblica officinalis* and *Chrysopogon fulvus*, proved effective in water and soil conservation, saving 4.9–30.7 cm of water and 862–2818 kg ha^{-1} of soil, as documented by Prasad et al. [87]. In Karnataka, Ley farming with vegetative barriers using *Cenchrus ciliaris* and *Cymbopogon martini* resulted in a 38% reduction in runoff and a 16% decrease in soil loss, according to Ramajayam et al. [84]. In Kashmir, the adoption of silvi-agriculture not only reduced scorching heat but also mitigated soil erosion and increased crop production, as highlighted by Mughal and Makaya [85]. Gujarat saw success with an agri-horticulture system incorporating soil moisture conservation practices, revealing that sapota with trenches and bench terraces reduced runoff by 16–34% and soil loss by 15–25%, as indicated by Kumar et al. [88]. Additionally, an agroforestry system involving cowpea, castor, and sapota in Gujarat, as studied by Jinger et al. [22], exhibited a 37.7% reduction in total soil loss, a 19.1% decrease in runoff, and an 81.9% increase in system productivity compared to sole tree plantation. Alley cropping, where rows of trees

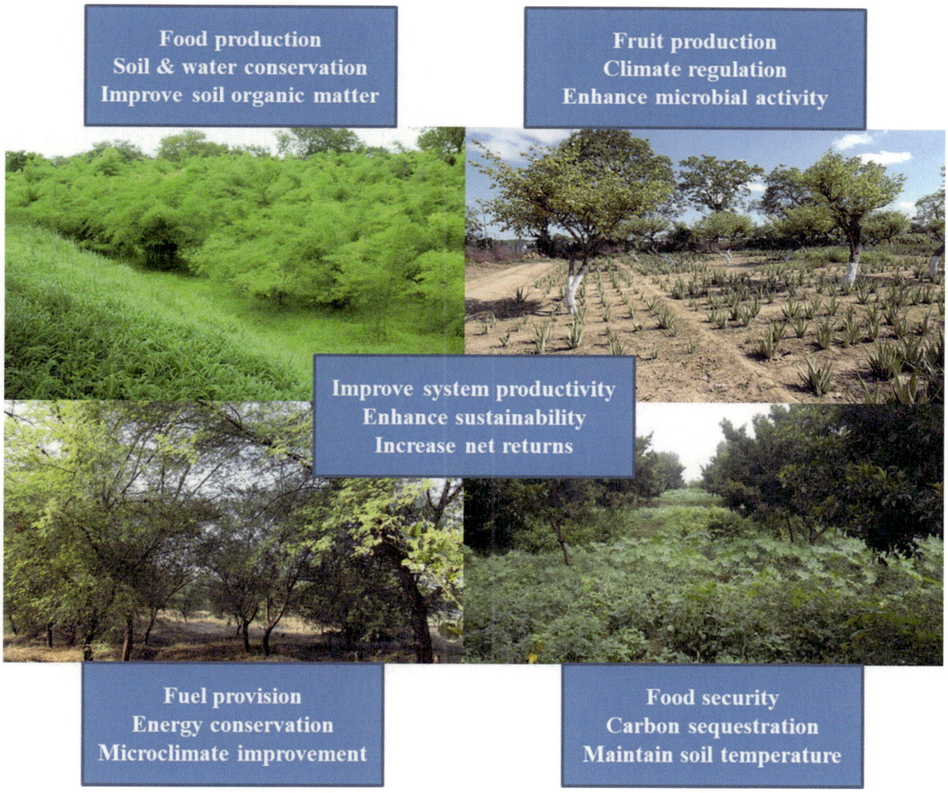

Figure 8.
Manifold benefits of agroforestry system.

are planted between rows of crops, is an agroforestry technique that promotes water conservation and reduces erosion. The tree rows serve as windbreaks and contribute organic matter to the soil through leaf litter, enhancing soil structure. Additionally, the shading effect of the trees helps in moderating soil temperature and moisture, creating a more favorable environment for crops. In Northeast India, hedgerow cropping was implemented, resulting in a substantial reduction of soil loss and runoff by 94 and 78%, respectively, as reported by Saha et al. [81]. Rajasthan adopted alley cropping, leading to higher yields, a favorable land equivalent ratio, and increased soil organic carbon in Leucaena-based systems, according to Dhyani et al. [88]. In Odisha, the implementation of alley cropping resulted in a noteworthy reduction of 23–32% in runoff and 49–52% in soil loss, as reported by Adhikary et al. [89]. Finally, in East India, the use of *Gliricidia sepium* hedgerow and grass filter strip effectively reduced runoff by 32% and soil loss by 35%, according to Lenka et al. [90]. Agroforestry presents a multifaceted approach to address water erosion challenges in agriculture. The combination of windbreaks, cover crops, contour planting, buffer strips, and alley cropping creates a resilient and sustainable system that not only mitigates erosion but also enhances overall soil health and productivity. As we move forward in our quest for sustainable agricultural practices, agroforestry stands out as a promising solution to combat water erosion and promote the long-term sustainability of our agricultural landscapes (**Figure 8**).

11. Conclusion

Many agroforestry techniques, including agri-silviculture, silvi-pastroral, and agri-horticulture, have demonstrated a high potential for generating both financial and environmental gains from these lands. India has a large number of species that could be used for agroforestry; all that is needed is to fully realize their potential. Farmers can significantly boost their revenue by widely cultivating and adopting high-value trees. On an average, soil loss and runoff could be reduced by 45–55 and 55–65%, respectively, by different agroforestry system. Moreover, SOC and system productivity could be enhanced by 100–150 and 30–40%, respectively, under different agroforestry system. More studies are required to develop agroforestry practices that can be widely accepted by farmers and stakeholders. Understanding fundamental processes and implementing technology well would be crucial for successfully managing and repairing soil eroded land through agroforestry. Therefore, to reduce soil erosion, it is necessary to develop appropriate agroforestry practices, policies, and action plans for the greater promotion and adoption of agroforestry in India.

Author details

Dinesh Jinger[1*], Nandha Kumar N[1], Chhavi Sirohi[2], Archana Verma[3], Pankaj Panwar[4] and Rajesh Kaushal[5]

1 ICAR-Indian Institute of Soil and Water Conservation, Research Centre, Anand, Gujarat, India

2 Choudhary Charan Singh Haryana Agricultural University, Hisar, Haryana, India

3 ICAR-Central Arid Zone Research Institute, Jodhpur, Rajasthan, India

4 ICAR-Indian Institute of Soil and Water Conservation, Research Centre, Chandigarh, India

5 ICAR-Indian Institute of Soil and Water Conservation, Dehradun, Uttarakhand, India

*Address all correspondence to: dineshjinger28@gmail.com

References

[1] Foley JA, Ramankutty N, Brauman KA, Cassidy ES, Gerber JS, Johnston M, et al. Solutions for a cultivated planet. Nature. 2011;**478**(7369):337-342. DOI: 10.1038/nature10452

[2] Rathore AC, Singh C, Jayaprakash J, Gupta AK, Doharey VK, Jinger D, et al. Impact of conservation practices on soil quality and ecosystem services under diverse horticulture land use system. Frontiers in Forests and Global Change. 2023;**6**:1289325. DOI: 10.3389/ffgc.2023.1289325

[3] Montanarella L. Agricultural policy: Govern our soils. Nature. 2015;**528**(7580):32-33. DOI: 10.1038/528032a

[4] Borrelli P, Robinson DA, Fleischer LR, Lugato E, Ballabio C, Alewell C, et al. An assessment of the global impact of 21st century land use change on soil erosion. Nature Communications. 2017;**8**(1):1-13. DOI: 10.1038/s41467-017-02142-7

[5] Adhikari K, Hartemink AE. Linking soils to ecosystem services—A global review. Geoderma. 2016;**262**:101-111. DOI: 10.1016/j.geoderma.2015.08.009

[6] Borrelli P, Robinson DA, Panagos P, Lugato E, Yang JE, Alewell C, et al. Land use and climate change impacts on global soil erosion by water (2015-2070). Proceedings of the National Academy of Sciences. 2020;**117**(36):21994-22001. DOI: 10.1073/pnas.2001403117

[7] UNCCD. The global land outlook. In: Global Indicator Framework for the Sustainable Development Goals and Targets of the 2030 Agenda for Sustainable Development. Bonn, Germany: UNSD; 2017. Available from: https://www.unccd.int/resources/global-land-outlook/overview [Accessed on March, 2024]

[8] Uthappa AR, Devakumar AS, Das B, Mahajan GR, Chavan SB, Jinger D, et al. Comparative analysis of soil quality indexing techniques for various tree based land use systems in semi-arid India. Frontiers in Forests and Global Change. 2024;**6**:1322660. DOI: 10.3389/ffgc.2023.1322660

[9] Banwart S. Save our soils. Nature. 2011;**474**(7350):151-152. DOI: 10.1038/474151a

[10] Quine TA, Van Oost K. Insights into the future of soil erosion. Proceedings of the National Academy of Sciences. 2020;**117**(38):23205-23207. DOI: 10.1073/pnas.2017314117

[11] Yin C, Zhao W, Pereira P. Soil conservation service underpins sustainable development goals. Global Ecology and Conservation. 2022;**33**:e01974. DOI: 10.1016/j.gecco.2021.e01974

[12] Issaka S, Ashraf MA. Impact of soil erosion and degradation on water quality: A review. Geology, Ecology, and Landscapes. 2017;**1**(1):1-11. DOI: 10.1080/24749508.2017.1301053

[13] Singhal V, Jinger D, Rathore AC, Pal R, Samal I, Bhoi TK, et al. COVID-19, deforestation, and green economy. Front. Frontiers in Forests and Global Change. 2024;**6**:1305779. DOI: 10.3389/ffgc.2023.1305779

[14] Leh M, Bajwa S, Chaubey I. Impact of land use change on erosion risk: An integrated remote sensing, geographic information system and modeling

methodology. Land Degradation and Development. 2011;**24**(5):409-421. DOI: 10.1002/ldr.1137

[15] Kumar D, Ranjan R, Meena MK, Yadav RS, Gupta G, Jinger D, et al. Exploring conservation agricultural practices in Bundelkhand region, Central India. In: Jayaraman S, Dalal RC, Patra AK, Chaudhari SK, editors. Conservation Agriculture: A Sustainable Approach for Soil Health and Food Security. Berlin, Germany: Springer; 2021. pp. 195-222. DOI: 10.1007/978-981-16-0827-8_9

[16] Montgomery DR, Huang MY-F, Huang AY-L. Regional soil erosion in response to land use and increased typhoon frequency and intensity, Taiwan. Quaternary Research. 2014;**81**(1):15-20. DOI: 10.1016/j.yqres.2013.10.005

[17] Rathore AC, Singh C, Islam S, Gupta AK, Patra S, Singhal V, et al. Long-term conservation practice in litchi (*Litchi chinensis L.*) cultivation improves the crop productivity and soil health of degraded lands. Land Degradation and Development. 2024;**6**:1-12. DOI: 10.1002/ldr.5077

[18] Munodawafa A. The effect of rainfall characteristics and tillage on sheet erosion and maize grain yield in semiarid conditions and granitic sandy soils of Zimbabwe. Applied and Environmental Soil Science. 2012;**2012**:1-8. DOI: 10.1155/2012/243815

[19] Al-Wadaey A, Ziadat F. A participatory GIS approach to identify critical land degradation areas and prioritize soil conservation for mountainous olive groves. Journal of Mountain Science. 2014;**11**(3):782-791. DOI: 10.1007/s11629-013-2827-x

[20] Dahal G, Holcomb J, Socci D. Surfactant-oxidant co-application for soil and groundwater remediation. Remediation. 2016;**26**(2):101-108. DOI: 10.1002/rem.21461

[21] Jinger D, Khatri P, Kumari K, Kumar D, Dinesh D. Agroforestry-based ecosystem services for livelihood resilience. Food Scientific Report. 2022b;**3**:50-55

[22] Jinger D, Kumar R, Kakade V, Dinesh D, Singh G, Pande VC, et al. Agroforestry system for controlling soil erosion and enhancing system productivity in ravine lands of Western India under climate change scenarios. Environmental Monitoring and Assessment. 2022;**194**:267. DOI: 10.1007/s10661-022-09910-z

[23] Chavan SB, Dhillon RS, Sirohi C, Uthappa AR, Jinger D, Jatav HS, et al. Carbon sequestration potential of commercial agroforestry Systems in Indo-Gangetic Plains of India: Poplar and eucalyptus-based agroforestry systems. Forests. 2023;**14**:559. DOI: 10.3390/f14030559

[24] Fahad S, Chavan SB, Chichaghare AR, Uthappa AR, Kumar M, Kakade V, et al. Agroforestry systems for soil health improvement and maintenance. Sustainability. 2022;**14**:14877. DOI: 10.3390/su142214877

[25] Chavan SB, Dhillon RS, Sirohi C, Keerthika A, Kumari S, Bharadwaj KK, et al. Enhancing farm income through boundary plantation of poplar (*Populus deltoids*): An economic analysis. Sustainability. 2022;**14**:8663. DOI: 10.3390/su14148663

[26] Chavan SB, Dhillon RS, Sirohi C, Saleh IA, Uthappa AR, Keerthika A, et al. Optimizing planting geometries in eucalyptus-based food production systems for enhanced yield and carbon sequestration in North Western

India. Frontiers in Sustainable Food System. 2024;**8**:1386035. DOI: 10.3389/fsufs.2024.1386035

[27] Xie H, Zhang Y, Wu Z, Lv T. A bibliometric analysis on land degradation: Current status, development, and future directions. Land. 2020;**9**(1):28. DOI: 10.3390/land9010028

[28] Hammad A, Tumeizi A. Land degradation: Socioeconomic and environmental causes and consequences in the Eastern Mediterranean. Land Degradation and Development. 2010;**23**(3):216-226. DOI: 10.1002/ldr.1069

[29] Pimentel D. Soil erosion: A food and environmental threat. Environment, Development and Sustainability. 2006;**8**(1):119-137. DOI: 10.1007/s10668-005-1262-8

[30] Global Soil Partnership. Endorses Guidelines on Sustainable Soil Management. 2017. Available from: http://www.fao.org/global-soil-partnership/resources/highlights/detail/en/c/416516/ [Accessed: January 8, 2023].

[31] Kassam A, Friedrich T, Shaxson F, Pretty J. The spread of conservation agriculture: Justification, sustainability and uptake. International Journal of Agricultural Sustainability. 2009;7(4):292-320. DOI: 10.3763/ijas.2009.0477

[32] Mandal D, Sharda VN. Appraisal of soil erosion risk in the eastern Himalayan region of India for soil conservation planning. Land Degradation and Development. 2011;**24**(5):430-437. DOI: 10.1002/ldr.1139

[33] Wuepper D, Borrelli P, Finger R. Countries and the global rate of soil erosion. Nature Sustainability. 2019;**3**(1):51-55. DOI: 10.1038/s41893-019-0438-4

[34] Myers N, Kent J. Gaia: An atlas of planet management. The Journal of Academic Librarianship. 1993;**19**(3):200. DOI: 10.1016/0099-1333(93)90650-t

[35] Xiong M, Sun R, Chen L. Global analysis of support practices in USLE-based soil erosion modeling. Progress in Physical Geography: Earth and Environment. 2019;**43**(3):391-409. DOI: 10.1177/0309133319832016

[36] FAO. State of the World's Forests: Forests and Agriculture: Land-Use Challenges and Opportunities. Italy: Food and Agriculture Organization; 2016. DOI: 10.1146

[37] Lal R. Soil erosion by wind and water: Problems and prospects. In: Soil Erosion Research Methods. Netherland: Routledge; 2017. pp. 1-10

[38] Oldeman LR, Hakkeling R, Sombroek WG. World Map of the Status of Human-Induced Soil Degradation: An Explanatory Note. Florida: International Soil Reference and Information Centre; 2017

[39] Bai ZG, Dent DL, Olsson L, Schaepman ME. Proxy global assessment of land degradation. Soil Use and Management. 2008;**24**(3):223-234. DOI: 10.1111/j.1475-2743.2008.00169.x

[40] Jinger D, Kaushal R, Kumar R, Paramesh V, Verma A, Shukla M, et al. Degraded land rehabilitation through agroforestry in India: Achievements, current understanding, and future prospectives. Frontiers in Ecology and Evolution. 2023;**11**:1-27. DOI: 10.3389/fevo.2023.1088796

[41] Jat ML, Chakraborty D, Ladha JK, Rana DS, Gathala MK,

McDonald A, et al. Conservation agriculture for sustainable intensification in South Asia. Nature Sustainability. 2020;**3**(4):336-343. DOI: 10.1038/s41893-020-0500-2

[42] Sidle RC, Ziegler AD, Negishi JN, Nik AR, Siew R, Turkelboom F. Erosion processes in steep terrain—Truths, myths, and uncertainties related to forest management in Southeast Asia. Forest Ecology and Management. 2006;**224**(1-2):199-225. DOI: 10.1016/j.foreco.2005.12.019

[43] Krishnamurthy L, Krishnamurthy V. Influence of PDO on South Asian summer monsoon and monsoon–ENSO relation. Climate Dynamics. 2013;**42**(9-10):2397-2410. DOI: 10.1007/s00382-013-1856-z

[44] Singh D, Tsiang M, Rajaratnam B, Diffenbaugh NS. Observed changes in extreme wet and dry spells during the South Asian summer monsoon season. Nature Climate Change. 2014;**4**(6):456-461. DOI: 10.1038/nclimate2208

[45] Turner AG, Annamalai H. Climate change and the South Asian summer monsoon. Nature Climate Change. 2012;**2**(8):587-595. DOI: 10.1038/nclimate1495

[46] Schulze ED, Freibauer A. Carbon unlocked from soils. Nature. 2005;**437**(7056):205-206. DOI: 10.1038/437205a

[47] Seitzinger SP, Harrison JA, Dumont E, Beusen AHW, Bouwman AF. Sources and delivery of carbon, nitrogen, and phosphorus to the coastal zone: An overview of global nutrient export from watersheds (NEWS) models and their application. Global Biogeochemical Cycles. 2005;**19**(4):1-11. DOI: 10.1029/2005gb002606

[48] Seutloali KE, Beckedahl HR. Understanding the factors influencing rill erosion on roadcuts in the South Eastern region of South Africa. Solid Earth. 2015;**6**(2):633-641. DOI: 10.5194/se-6-633-2015

[49] Berhe AA, Harte J, Harden JW, Torn MS. The significance of the erosion-induced terrestrial carbon sink. Bioscience. 2007;**57**(4):337-346. DOI: 10.1641/b570408

[50] Cole JJ, Prairie YT, Caraco NF, McDowell WH, Tranvik LJ, Striegl RG, et al. Plumbing the global carbon cycle: Integrating inland waters into the terrestrial carbon budget. Ecosystems. 2007;**10**(1):172-185. DOI: 10.1007/s10021-006-9013-8

[51] Gabet EJ, Mudd SM. A theoretical model coupling chemical weathering rates with denudation rates. Geology. 2009;**37**(2):151-154. DOI: 10.1130/g25270a.1

[52] Stocking MA. Tropical soils and food security: The next 50 years. Science. American Association for the Advancement of Science. 2003;**302**(5649):1356-1359. DOI: 10.1126/science.1088579

[53] Filippelli GM. The global phosphorus cycle: Past, present, and future. Elements. 2008;**4**(2):89-95. DOI: 10.2113

[54] Quinton JN, Govers G, Van Oost K, Bardgett RD. The impact of agricultural soil erosion on biogeochemical cycling. Nature Geoscience. 2010;**3**(5):311-314. DOI: 10.1038/ngeo838

[55] FAO. Food and Agriculture Organization. The State of the World's Land and Water Resources for Food and Agriculture (SOLAW)–Managing Systems at Risk. FAO, United Nations: Rome and London; 2011

[56] Bashagaluke JB, Logah V, Opoku A, Sarkodie-Addo J, Quansah C. Soil nutrient loss through erosion: Impact of different cropping systems and soil amendments in Ghana. PLoS One. 2018;**13**(12):e0208250. DOI: 10.1371/journal.pone.0208250

[57] Anonymous. Healing Degraded Land. INBAR Magazine. 1997;**5**(3):40-45

[58] Maoyi F, Banik RL. Bamboo production system and their management. In: People and the Environment: Propagation and Management, Rao VR, Sastry CB, Widjaja E, editors. Proceedings of 5th International Bamboo Workshop and 4th International Bamboo Congress. Antwerp, Belgium: World Bamboo Organization; 1996. pp. 18-33

[59] Wiersum KF. Forest gardens as an 'intermediate' land-use system in the nature–culture continuum: Characteristics and future potential. Agroforestry Systems. 2004;**61-62** (1-3):123-134. DOI: 10.1023/b:agfo.0000028994.54710.44

[60] Muchane MN, Sileshi GW, Gripenberg S, Jonsson M, Pumariño L, Barrios E. Agroforestry boosts soil health in the humid and sub-humid tropics: A meta-analysis. Agriculture, Ecosystems and Environment. 2020;**295**:106899. DOI: 10.1016/j.agee.2020.106899

[61] Béliveau A, Lucotte M, Davidson R, Paquet S, Mertens F, Passos CJ, et al. Reduction of soil erosion and mercury losses in agroforestry systems compared to forests and cultivated fields in the Brazilian Amazon. Journal of Environmental Managementt. 2017;**203**:522-532. DOI: 10.1016/j.jenvman.2017.07.037

[62] Rachman LM, Hidayat Y, Tarigan SD, Sitorus SP, Fitri R, Ain AQ. The effect of agroforestry system on reducing soil erosion in upstream Ciliwung watershed. IOP Conference Series: Earth and Environmental Science. 2020;**556**(1):012010. DOI: 10.1088/1755-1315/556/1/012010

[63] Purwaningsih R, Sartohadi J, Setiawan MA. Trees and crops arrangement in the agroforestry system based on slope units to control landslide reactivation on volcanic foot slopes in Java, Indonesia. Land. 2020;**9**(9):327. DOI: 10.3390/land9090327

[64] Sharda VN, Sikka AK, Juyal GP. Participatory Integrated Watershed Management—A Field Manual Dehradun. Dehradun: Central Soil Water Conservation Research and Training Institute; 2006

[65] Juyal GP, Katiyar VS, Dhadwal KS, Joshie P, Arya RK. Mined Area Rehabilitation in Himalayas–Shahastradhara Experience Dehradun. Dehradun: Central Soil and Water Conservation Research and Training Institute; 2007. p. 104

[66] Rodríguez L, Suárez JC, Rodriguez W, Artunduaga KJ, Lavelle P. Agroforestry systems impact soil macroaggregation and enhance carbon storage in Colombian deforested Amazonia. Geoderma. 2021;**384**:114810. DOI: 10.1016/j.geoderma.2020.114810

[67] Gupta N, Kukal SS, Bawa SS, Dhaliwal GS. Soil organic carbon and aggregation under poplar-based agroforestry system in relation to tree age and soil type. Agroforestry Systems. 2009;**76**(1):27-35. DOI: 10.1007/s10457-009-9219-9

[68] Dagar JC, Singh G, Singh NT. Evaluation of forest and forest trees used for rehabilitation of semiarid alkali soils in India. Arid Land Research Management. 2001;**15**:115-133

[69] Pandey AK, Gupta VK, Solanki KR. Productivity of neem-based agroforestry system in semi-arid region of India, range manage. Agroforestry. 2010;**31**(2):144-149

[70] Rivest M, Whalen JK, Rivest D. Variation of soil microbial and earthworm communities along an agricultural transect with tree windbreak. Agroforestry Systems. Netherland: Elsevier; 2019;**94**(5):1639-1649. DOI: 10.1007/s10457-019-00476-3

[71] Raveendra SAST, Nissanka SP, Somasundaram D, Atapattu AJ, Mensah S. Coconut-gliricidia mixed cropping systems improve soil nutrients in dry and wet regions of Sri Lanka. Agroforestry Systems. 2021;**95**(2):307-319. DOI: 10.1007/s10457-020-00587-2

[72] Ram A, Dev I, Uthappa AR, Kumar D, Kumar N, Chaturvedi OP, et al. Reactive nitrogen in agroforestry systems of India. In: The Indian Nitrogen Assessment. Netherland: Elsevier; 2017. pp. 207-218

[73] Giller KE. Nitrogen Fixation in Tropical Cropping Systems. 2nd ed. Wallingford: CAB International Publishing; 2001

[74] Li X, Ge T, Chen Z, Wang S, Ou X, Wu Y, et al. Enhancement of soil carbon and nitrogen stocks by abiotic and microbial pathways in three rubber-based agroforestry systems in Southwest China. Land Degradation and Development. 2020;**31**(16):2507-2515. DOI: 10.1002/ldr.3625

[75] Wartenberg AC, Blaser WJ, Roshetko JM, Van Noordwijk M, Six J. Soil fertility and Theobroma cacao growth and productivity under commonly intercropped shade-tree species in Sulawesi, Indonesia. Plant and Soil. 2019;**453**(1-2):87-104. DOI: 10.1007/s11104-018-03921-x

[76] Stöcker CM, Bamberg AL, Stumpf L, Monteiro AB, Cardoso JH, ACR d L. Short-term soil physical quality improvements promoted by an agroforestry system. Agroforestry Systems. 2020;**94**(5):2053-2064. DOI: 10.1007/s10457-020-00524-3

[77] Piza PA, Suárez JC, Andrade HJ. Litter decomposition and nutrient release in different land use located in Valle del Cauca (Colombia). Agroforestry Systems. 2021;**95**(2):257-267. DOI: 10.1007/s10457-020-00583-6

[78] Nair PKR. Agroforestry Systems in the Tropics. Netherlands: Springer; 1989

[79] Nair PKR. The Prospects for Agroforestry in the Tropics. Washington: International Crops Research Institute for the Semi-Arid Tropics; 1990

[80] Vandenbeldt RJ, editor. Faidherbia Alb Ida in the West African Semi-Arid Tropics. Hyderabad, India and Nairobi, Kenya: ICRISAT and ICRAF; 1992

[81] Saha R, Chaudhary RS, Somasundaram J. Soil health management under hill agroecosystem of North East India. Applied and Environmental Soil Science. 2012;**2012**:1-9. DOI: 10.1155/2012/696174

[82] Sharda VN, Venkateswarlu B. Crop diversification and alternate land use systems in watershed management. In: Wani SP, Mula Rosana P, Sreedevi TK, Raju KV, editors. Proceedings of the Comprehensive Assessment of Watershed Programs in India. Hyderabad: International Crops Research Institute for the Semi-Arid Tropics; 2007. pp. 23-27

[83] Prasad R, Yadav RP, Panwar P. Aonla based agroforestry systems for soil and water conservation in Shivaliks. In: Panwar P, Tiwari A

K, Dadhwal KS, editors. Agroforestry System for Resource Conservation Livelihood Security in Lower Himalayas. New Delhi: New India Publishing Agency; 2007. pp. 171-178

[84] Ramajayam D, Mishra PK, Nalatwadmath SK, Mondal B, Adhikari RN, aand Murthy, B. K. N. Evaluation of different grass species for growth performance and soil conservation in vertisols of Karnataka. Indian Journal of Soil Conservation. 2007;**35**:54-57

[85] Mughal AH, Makaya AS. Suitable agroforestry models for the degraded and wastelands of Srinagar district of Kashnir. In: Khan MA, editor. Environment Biodiversity and Conservation. New Delhi: APH Publication Corporation; 2000. pp. 493-500

[86] Kumar R, Bhardwaj AK, Rao BK, Vishwakarma AK, Bhatnagar PR, Patra S, et al. Development of degraded ravine lands of Western India using Sapota (*Achras zapota*) plantation with terracing vs. trenching-on-slope-based conservation measures. Land Degradation and Development. 2020;**32**(1):101-111. DOI: 10.1002/ldr.3691

[87] Prakash O, Dubey RK, Dubey SK, Deshwal JS. Evaluation of Ber (*Zizyphus mauritiana Lamk.*) based agri-horti systems under various tree root management practices on reclaimed Yamuna ravines. Indian Journal of Soil and Conservation. 2011;**39**:59-62

[88] Dhyani SK, Ajit HA, Jabeen N, Uma. Agroforestry: An integrated system for conservation of natural resources in Northern India. Indian Journal of Forestry. 2011;**34**(2):121-130. DOI: 10.54207/bsmps1000-2011-949428

[89] Adhikary PP, Hombegowda HC, Barman D, Jakhar P, Madhu M. Soil erosion control and carbon sequestration in shifting cultivated degraded highlands of Eastern India: Performance of two contour hedgerow systems. Agroforestry Systems. 2016;**91**(4):757-771. DOI: 10.1007/s10457-016-9958-3

[90] Lenka NK, Dass A, Sudhishri S, Patnaik US. Soil carbon sequestration and erosion control potential of hedgerows and grass filter strips in sloping agricultural lands of Eastern India. Agriculture, Ecosystems and Environment. 2012;**158**:31-40. DOI: 10.1016/j.agee.2012.05.017

Section 3

Water and Irrigation Management

Chapter 9

Agricultural Drainage Systems in Irrigated Ecologies

Mohamed Joseph Sesay, Felix K. Abagale and Shaibu Abdul-Ganiyu

Abstract

The production of crops ultimately depends on the health of the soil, a natural resource that is negatively impacted by poor drainage management, particularly in irrigated rice ecologies, due to drainage problems. Ineffective drainage management often has negative effects on farm economics because it can lead to land abandonment, which makes the land unproductive and causes large production depressions. Saline and wet conditions due to drainage problems can have a significant impact on crop productivity, making soils difficult to work with and severely limiting crop choice, intensification, and diversification. In order to retain soil moisture, leach salts beyond plant roots, improve crop productivity, and provide enough root zone aeration on poorly drained soils, an efficient artificial drainage system is essential. This makes irrigated land sustainable for the long term. This chapter focused on agricultural drainage systems in irrigated ecologies. A brief background on agricultural drainage systems was critically examined, looking at definitions, types, agricultural drainage in the global and African context, and benefits of drainage systems in agricultural production. The chapter also takes into account the factors affecting drainage systems, the influence of irrigation on drainage practices, drainage coefficients, and problems relating to drainage management in irrigated ecologies.

Keywords: drainage, irrigation, waterlogging, drainage coefficients, salinization

1. Introduction

In many soils, sustainable agronomic production depends on the drainage of surplus soil water. Drainage can increase agricultural yields, lessen production unpredictability from year to year, and create field conditions that are manageable for field operations during planting or harvest. Crop productivity can be affected by drainage system design and management, which can affect the environment. Given the significant impacts that drainage has on ecosystem services and the exponential growth in the area of drained croplands, it is imperative that efficient drainage system design and management receive more attention. Even while irrigation improves 300 Mha of agriculture compared to 130–200 Mha of artificial drainage, an additional 450 Mha of cropland might gain from better drainage [1]. Furthermore, due to changes in agricultural patterns, climate change, and the end of design life, current drainage systems are being enlarged and intensified. Crop yields, soil organic carbon (SOC) stocks, greenhouse gas (GHG) emissions, nutrient losses to water resources, and cropping systems' resilience to climate change will all be impacted by the extension and intensification of drainage systems [1].

According to estimates from [2, 3], irrigation is used on between 306 and 367 Mha (million hectares) of arable and cultivated land, while artificial drainage is used on between 186 and 200 Mha. A greater range of other ecosystem services, such as supporting services (like good soil), regulatory services (like flood management), and cultural services (like recreational or spiritual benefits), can be offered in addition to irrigation and drainage, which assist the economic development of rural regions.

Modern drainage design guidelines in Iowa, USA, for instance, prescribe a drainage coefficient of 1.27–2.54 cm/day; however, an analysis of Iowa's key drainage infrastructure showed that 95% of drainage basins have a coefficient of less than 0.95 cm/day. These modern drainage systems are insufficient for the Midwest of the United States [1].

An overview of agricultural drainage is provided in this chapter, which also covers its definition, types, global and African contexts, and the advantages of drainage systems for agricultural output. The chapter also takes into consideration the factors affecting drainage system, the influence of irrigation on drainage practices, physico-chemical and hydrological properties of soil in relation to drainage, drainage system performance indicators, drainage coefficients including the various methods of its determination and problems relating to drainage management in irrigated ecologies. Several perspectives are drawn and conclusions made on some aspects of proper drainage system operation and management.

2. Overview of drainage on agricultural lands

2.1 Definition of agricultural drainage

The act of draining something literally refers to the removal of a liquid. Agricultural drainage is the process of removing excess water out of affected agricultural areas and disposing of it by ditch construction, open channels, or vertical drains. Land drainage makes it possible to increase farming productivity and repurpose unproductive areas like marshes and flooded areas [4].

Drainage is the act of eliminating excess water and dissolved salts from the soil's surface and subsurface in order to promote plant growth [5]. Agricultural land drainage, according to Gurovich and Oyarce [6], is a collection of technical techniques and hydraulic structures that enable the removal of excess water and/or salt from the soil volume occupied by the roots of crops. This creates an environment that is suitably oxygenated for roots to develop normally while maintaining appropriate water and air relative proportions in accordance with crop physiological needs. Eventually, this promotes soil sustainability for crop productivity.

Systems for draining water from agricultural lands are a combination of subsurface and surface drainage systems. Surface drainage takes excess water out of the soil's top layer in agricultural settings. Subsurface or tile drainage lowers the water table by removing excess water from the plant's root zone [7].

2.2 Types of drainage systems

The two (2) primary drainage systems that can be applied in an agricultural area are surface and subsurface drainage. These can either be regular or checked

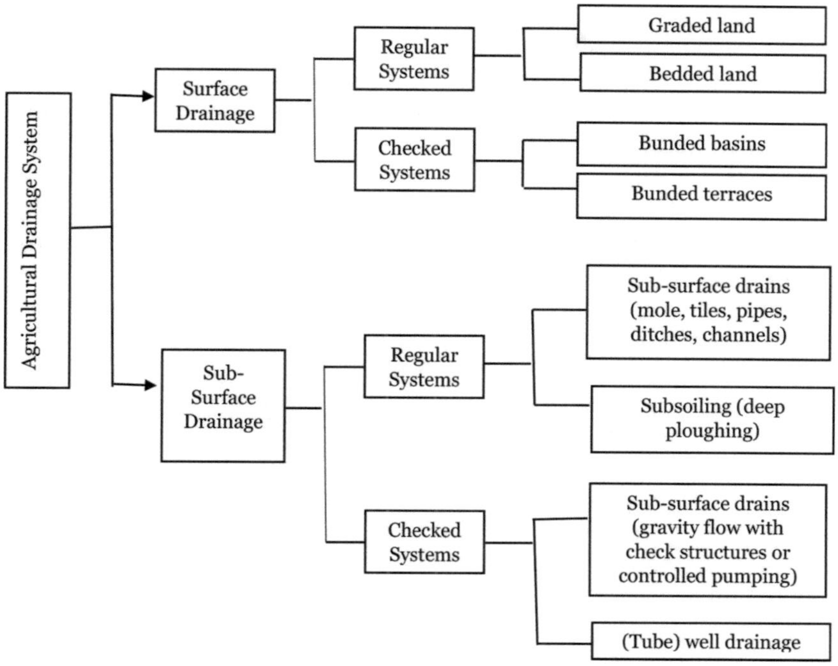

Figure 1.
Types of agricultural drainage systems.

systems (**Figure 1**). According to King et al. [8], a field's drainage needs are determined by various factors, including the type of crop being grown, topography, soil properties, and appropriate outlets. The process of building open channels and drains to remove excess water from the uppermost portion of the soil is known as surface drainage. According to Nangia et al. [9], surface drainage is the most traditional and straightforward technique available to farmers. According to Schultz et al. [10], surface drainage is the process of using canals and/or land shapes to remove surplus water from the surface layer of the land.

Depending on how serious the drainage condition is, perforated plastic pipes of varying sizes, spacing, and depths are installed using subsurface or tile drainage. If the flow occurs only occasionally throughout the year, the drainage depth is often less than 1 m; if the flow occurs continuously throughout the year, the drainage depth increases [11]. In contrast to surface drainage, Kamiri et al. [12] noted that subsurface drainage pipelines are buried and necessitate that water percolate through the soil before collecting, which promotes infiltration within the soil strata.

According to Xian et al. [13], subsurface drainage lowers the groundwater table and creates favorable conditions within the root zone, whereas surface drainage removes excess water from the soil's outer layer before it enters the root zone. For this reason, much of the world uses subsurface drainage to get rid of extra water in plant roots [14].

2.3 Agricultural drainage in the global context

Poor drainage causes millions of valuable agricultural areas to be lost every year on a global scale [15]. According to FAO [16], between 0.25 and 0.5 million hectares of important agricultural land are degraded, which lowers crop production potential. Salinization and waterlogging are two of the main issues with inadequate drainage.

Groundwater flow, excessive irrigation, precipitation, and overland flow are some of the causes of excess water in the soil. Among these, [17] noted that irrigation is a major factor in waterlogging (the rising of the soil's water table) and salinization (the accumulation of salts in the upper soil layers) in semi-arid climates when rainfall is insufficient to meet crop water requirements.

Drainage issues are prevalent in agricultural areas, particularly in cases where natural sources of self-draining are insufficient. Extreme drainage issues can lead to land abandonment; the abandoned areas of ancient Chaldea, the Tigris Valley, and the Euphrates Rivers in Mesopotamia are prime examples [18]. According to McCarty et al. [19], an irrigation system should ideally precisely provide the water required for plant growth and eliminate any excess to create the perfect soil-water conditions for both plant development and productive, sustainable agriculture.

The need for more intense drainage to maintain cost-effective agricultural output in the United States arose when crop production practices changed, leading to fewer producers, larger fields, and larger equipment. The increased agricultural production's off-site environmental effects became noticeable, bringing agricultural water management techniques front and center in the environmental protection arena. In order to reduce the amount of drainage needed to achieve the leaching requirement, various irrigation management techniques were created in response to competition for water resources in irrigated producing areas.

Numerous investigations have been carried out regarding the ability of drainage systems in rice paddies in Japan and monsoon Asia to prevent flooding. According to this research, drainage systems in lowland paddies function as buffers and boost the systems' total capacity for removing excess water or storing water when combined with irrigation canals and reservoirs [20].

The planning of water development, drainage, and other related tasks led to the founding of a public corporate entity in Israel in 1952. The Israel Water Law was passed in 1959 with the goals of establishing water as a public resource, regulating its use, distribution, and prevention of contamination, as well as preserving water [21]. A number of initiatives have been implemented nationwide with the goal of increasing the operational productivity of water dissemination systems, recovering wastelands, running off interceptions, and core distribution system growth. The government made significant efforts to drain marshes and other flooded areas throughout the winter, which resulted in the redemption of numerous agricultural lands. The next step was to regulate drainage basins by extending regional activities to a larger scale. The focus was on implementing effective drainage management techniques to alleviate poorly drained soils, prevent flooding, drain swamps, and divert runoff water from agricultural land [21].

Recent research on the effects of flooding and waterlogging in paddy fields has focused on the damage that high water temperatures in the midst of heat stress can do to rice plants, as well as the decline in rice yield and quality [22]. The surrounding temperature, water velocity, ponding water depth, and paddy plant stage all have an impact on the temperature of the water [23]. In Finland, research on the effects of sub-irrigation and controlled drainage in peat soils is currently gaining attention. According to preliminary findings, N and dissolved P leach more readily from subterranean drains when the peat layer is thicker above them [24, 25].

2.4 Agricultural drainage in the African context

The long-term sustainability of irrigated agriculture in Africa is seriously threatened by poor soil drainage and the salinization that results from it. Despite the significant contributions made by governments and other actors, issues with policy, institutional capacity, technology, infrastructure, and markets impede the management of water in irrigated agriculture in Africa, with Ethiopia serving as an example [26]. Sustainable irrigation systems face greater difficulties/challenges, particularly in regions with limited water resources where insufficient drainage systems result in waterlogging and salinity of the soil in vast areas of cropland and other challenges including gaps in capacity building, lack of understanding of the multifunction of drainage, and effects of climate change (**Figure 2**) [27]. Although irrigated agriculture has enormous potential benefits, actual achievements in many irrigated zones of the nation are far short of potential because of ineffective water management techniques that result in salinization, waterlogging, and other related issues [28].

The primary sources of salt are naturally occurring saline seeps and shallow groundwater tables. Secondary salinization is further facilitated by inadequate drainage and ineffective irrigation water management techniques [29]. Due to salinity and sodicity problems in irrigated ecologies, the Awash basin for instance, which makes up roughly one-third of Ethiopia's total irrigated area, irrigation projects that were poorly planned and unsupported by improved management technologies had seriously degraded the land [30, 31].

In North Africa, drainage is a strategy used to safeguard irrigation development projects by getting rid of excess water and salts that irrigation water brings in, as well as helping farmers deal with little and/or erratic rainfall [32]. The ability of the soil to function as a storage space is enhanced by drainage, particularly in situations where operational control is possible to keep the water table elevated. Controlled drainage has been shown to increase irrigation efficiency, according to studies on water balance conducted in Egypt.

Figure 2.
Challenges of agricultural drainage in Africa.

The semi-arid country of South Africa is vulnerable to waterlogging and salinization brought on by irrigation [33]. Waterlogging and salinization affect between 15% and 18% of South Africa's irrigated land, and the expense of drainage exacerbates the issue [34]. Reinders et al. [33] have highlighted the seriousness of drainage issues, pointing out that some South African agricultural estates have been deemed worthless, unproductive, and ultimately abandoned as a result of these issues.

According to Namara et al. [35], around 1 Mha of land in West Africa is under irrigation, of which 60% is set up for full-control irrigation and 40% is in lowlands. There is an estimated 9.1 Mha of irrigation potential, with 55% of that potential located in only three countries: Nigeria, Ghana, and Sierra Leone. Poor drainage, waterlogging, eutrophication, soil salinity, and acidity are frequent issues with Agricultural Water Management (AWM) programs, according to a report by African Union [36]. It further said that if left unchecked, this might lead to irrigation system failure, rendering land unfit for growing crops and hence costing money in lost revenue.

2.5 Benefits of drainage on crop production

Soils are natural resources that are necessary for agricultural production. According to FAO [16], 250,000–500,000 hectares of valuable agricultural soils are lost yearly due to poor drainage, which causes salinity and waterlogging, with over 34 million hectares being salinized. Salinity and waterlogging have negative effects on farm economics because they can lead to land abandonment, which makes the land unproductive and causes large production depressions. Saline and wet conditions can significantly limit crop choice, intensification, and diversification. They can also negatively impact agricultural output, making soils difficult to work with.

According to Singh [37], maintaining soil moisture, enhancing crop productivity, and facilitating adequate root zone aeration on poorly drained soils all depend on an efficient artificial drainage system (**Figure 3**). One way to prevent flooding and save lives and property is through drainage. Strong drainage networks act as a buffer against heavy rainfall in certain areas. In humid tropical regions, irrigation has changed the hydrology of the soils, and natural events like monsoon flooding and waterlogging have made the issue worse. Uncontrolled floods can cause periodic devastation to resources and significant loss of life; Bangladesh and India are two main examples.

According to FAO [38], 5 million people die each year from water-related diseases like malaria, schistosomiasis, lymphatic filariasis, guinea worm infection, water-borne gastrointestinal diseases that are oral and caused by feces, as well as diseases linked to the spread of pesticides and pesticide residues in drainage water (non-communicable). Inadequately maintained drains and silted-up drains from stagnant water on poorly drained soil increase the likelihood and spread of infections, resulting in human misery, medical expenses, and incalculable costs associated with unavailability or insufficiency of labor forces.

Four advantages of tile drainage are categorized by Fraser and Fleming [39]. These include (1) enhanced drainage may result in changes to land use; (2) intensive land use may raise crop yields and quality, increase fertilizer use, and improve crop rotations; (3) enhanced drainage may lower production costs; and (4), the use of tile drainage would enhance resource allocation on the farm by cutting down on the amount of time needed to cover the field, potentially leading to the employment of smaller, more energy-efficient tractors. Reducing the number of damp regions or

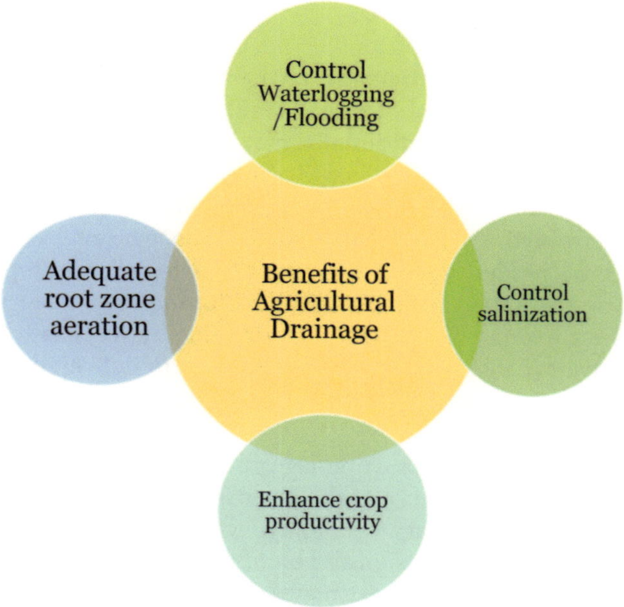

Figure 3.
Benefits of agricultural drainage.

damp fields inside an agricultural land allows for more versatility in crop arrangement within crop fields and may potentially alter the nature of farming activities [39].

According to Muma et al. [40], installing a drainage system in any agricultural area has a number of effects, some of which are direct and some of which are indirect. The removal of water from agricultural fields and a reduction in the amount of water retained in or on the soil are the immediate effects. But the main goals of drainage are usually not the direct impacts. Improved aeration of the root zone for increased crop yield is one of the indirect consequences [41]. Other indirect advantages include better fertilizer use, decreased weed growth, decreased denitrification, and less restriction on crop selection. Long-term sustainability of irrigated land is achieved through the leaching of salts through drainage, which stops additional salinization of the rootzone [42].

According to Alakukku et al. [43], drainage also increases the amount of land that is accessible, facilitates workable soil, enhances soil structure, increases land carrying capacity, and increases the activity of microfauna by keeping the soil drier, all of which support efficient crop growth. In addition, Fraser and Fleming [39] claimed that peak flow volumes decreased in streams connected to artificially drained land and that overall runoff spread out more over time, perhaps resulting in a decrease in surface runoff, based on their analysis of the environmental benefits of tile drainage.

3. Factors affecting drainage systems

When irrigation is added, an area's natural conditions can vary over time and may need a drainage system. Consideration must be given to issues related to the soil and hydrology of the area where drainage must be built in order to predict the effects of the changes that will take place. The following are a few of the primary factors.

3.1 Drainage requirement

Design and construction requirements are part of the drainage requirements, and if these are carefully followed, they will result in irrigated ecologies having dependable drainage systems. Additionally, man's potential agricultural use of the physical aspect and conditions of lands will affect the final drainage requirement [44]. FAO [5] emphasized that while designing a drainage system, the drainage need must be determined. This is the water that needs to be extracted from a specific area within a predetermined amount of time in order to stop surface or groundwater levels from rising unintentionally. Eliminating the surplus that can be drained offers two advantages: it lessens waterlogging by artificially preserving a sufficiently deep water table, and it ensures that any salt delivered by irrigation does not reach a concentration that is harmful to the crops by removing adequate water from the crop root zone [5].

The amount of data required for a drainage problem varies depending on the specific problem and the importance of the investigations or reports being prepared. The basic data must be representative enough to enable the selection of an appropriate drainage plan from which to design and build a drainage system that is functionally sound. Reasonably accurate cost estimates must be established for the purposes at hand. Inadequate or inaccurate data presents a serious risk when assessing drainage requirements and cost projections [44].

3.2 Water table

The term "water table" refers to the top layer of the groundwater. It is the point at which atmospheric pressure and groundwater pressure are equal. Over time, the water table changes. Following irrigation or precipitation, it rises sharply before progressively falling due to water entering the drainage system.

According to Gholamhosseinian et al. [45], controlling the water table improves the soil environment for vegetative growth, manages water for irrigation and drainage, improves water quality, makes better use of rainfall, lowers the demand for water for irrigation, lowers freshwater runoff to saline nursery areas, and encourages leaching of saline and alkali soil. A water table near the base of the rootzone may alter the way soil moisture is extracted, influencing deep percolation and drainage needs [44].

3.3 Depth to the water table

The water table's depth is ascertained using piezometers or observation wells. A plastic pipe buried in the ground that has a small diameter (> Ø 12 mm) is called an observation well [5]. In order to keep small particles like silt and clay from clogging the perforations and preventing the flow of water, holes are occasionally filled in the bottom of the pipe along a length of the pipe where variations in the water table are anticipated. When thick clay soils are present, an auger hole may be adequate and pipe is not required.

Because a groundwater system is made up of numerous components that recharge and outflow, the water table is always shifting. Any drainage research must take into account the highest and lowest mean positions of the water table as well as the mean depths of the water table within a hydrological year [44].

3.4 Dissolved salts in the groundwater

All groundwater contains dissolved salts. The geological setting, the groundwater's source, and its flow all influence the kinds of salts that would generally be present. A further factor in the groundwater's salinity content is irrigation. In addition to supplying salts to the soil, it dissolves salts in the root zone. The concentrations of salts in water from the rootzone of irrigated soil are generally many times higher than in the irrigation water that was initially supplied. Although there are many factors that lead to the formation of saline soils, Ritzema [44] proposed that the consumption of capillary groundwater and irrigation water containing salts is the primary cause of saline soil development in most cases.

3.5 Hydraulic conductivity

The K-value, also known as hydraulic conductivity, is a measurement of the soil's capacity to transfer water. FAO [5] reported that the K-values of various soil types vary significantly and are mostly influenced by their texture. The ability of soil to flow water through it is determined by a property called hydraulic conductivity. Determining and comprehending this characteristic is essential for identifying and fixing the majority of subsurface drainage problems.

3.6 Topography

Gravity most times is used to remove excess water; thus, understanding the geography of an area with drainage issues is essential. Because small variations in the land surface elevation are crucial, physical characteristics, both natural and man-made, that will impact the design of the drainage system must be displayed on the topographic map [5]. Topographic maps are crucial for drainage studies, and their significance affects the overall layout that should be planned for the placement of the outlet, sub-outlet, and collector drains. Even before drainage planning and design stages come around, topographic factors are important to be examined.

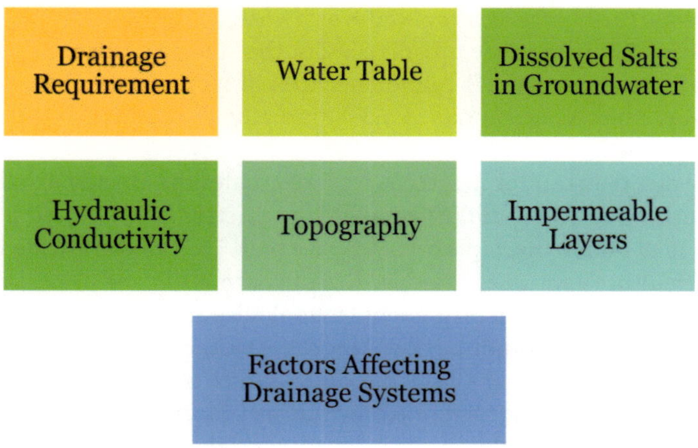

Figure 4.
Factors affecting drainage systems.

3.7 Impermeable layers

Soils are typically not consistently homogeneous or uniform in the vertical direction. A layer of impermeable soil will always exist below the earth's surface. When the impermeable layer is deep and groundwater only partially fills the porous upper layer, the water table is free to rise and fall. When this occurs, the groundwater is referred to as phreatic, unconfined, or below the water table (**Figure 4**) [5].

4. Influence of irrigation on drainage practices

Applying water to crops in a timely and controlled manner to promote or maintain higher yields is known as irrigation. According to Malakar et al. [46], irrigation is the process of applying water using an irrigation system during the growing season. It also involves applying water during crucial phases of crop production, such as field preparation, pre-irrigation, weed control, harvesting, and leaching salts from the root zone.

Irrigated agriculture is still one of the biggest and most wasteful users of water, which is gradually running out globally. It has also been identified as the most consumptive use of freshwater [47]. There is a significant chance that the population will grow and that more people will go without enough food, particularly in impoverished and water-scarce places. To achieve food security, agriculture must boost food output by an estimated 60% by 2050. A large amount of irrigation will be used to meet this need. Mehta et al. [48] predicted that irrigation will become even more reliant on inferior water sources and face more rivalry from other industries (like tourism or industry) for water resources in the drive to raise agricultural production.

Water resources have been severely degraded as a result of a number of factors, including the effects of global climate change, changes in land use, agricultural and urban expansion, and overexploitation brought on by economic development [49, 50]. It is anticipated that as water resources are overused and deteriorate, so will the demand for the services they provide. Globally, agricultural ecosystems are the primary providers of food and the primary consumers of water resources. Depending on the region's economic growth and climate, these ecosystems can utilize anywhere from 60% to 90% of the water available [51]. According to Hedley et al. [52], there are 275 million hectares of irrigated cropland worldwide, and this number is growing at a rate of 1.3% annually. This made up only 23% of the land that was farmed. According to Hedley et al. [52], there are 275 million hectares of irrigated cropland worldwide, and this number is growing at a rate of 1.3% annually. This made up only 23% of the land that was farmed.

The high water demand for irrigated rice exacerbates the already existing scarcity of water. According to Tuong and Bouman [53] prediction, there would be a lack of water for about 18 million hectares of paddy rice by 2025. The worldwide issue of water scarcity in agriculture continues to jeopardize the output of paddy rice. Investments in irrigation and other water management technologies are crucial for agricultural growth since they increase productivity. According to Domènech [54], irrigation is a crucial part of the contemporary input package—which also includes enhanced tillage, fertilizer, and seed—that is needed to achieve crop yields high enough to satisfy the world's expanding food demand. Beyond boosting production, irrigation interventions can also enhance nutritional outcomes by increasing access to food sources and promoting a better diet in terms of both quantity and quality. They can also help to lower poverty.

Understanding the existing level of knowledge and investigating the impacts of controlled drainage and irrigation depths on crop production are therefore essential. This knowledge will support maintaining appropriate food and soil quality, as well as ensuring sufficient agricultural production to meet rising demand.

4.1 Types of irrigation

The type of irrigation practiced in a particular agroecological zone can have a major contributing effect on the drainage system of that particular area. In addition to increasing agricultural productivity and food security, irrigation is a land management method that frequently entails permanent and irreversible land alterations [4]. Numerous irrigation techniques have been created in contemporary agriculture to satisfy the needs of particular crops or regions. At first, there are three primary systems for grouping irrigation techniques: surface, drip, and sprinkler [55]. More contemporary sources, such as Refs. [56, 57], have attempted to generalize irrigation techniques globally, typically classifying them into four primary delivery types.

1. *Surface:* The water is delivered to the field through hydrants, and it is spread over the whole area (furrow/basin irrigation) or from its borders (border irrigation) [56].

2. *Sprinkler:* The water is delivered as raindrops that precipitate over the agricultural area. There are several variations of this method that can depend on the height (over/under the foliage) and diameter of the application, and rate [56].

3. *Micro:* The so-called "localized" irrigation method, in which water is applied directly to a small area of soil surrounding the plants. Moreover, the water is applied with little pressure and at a slow rate. But this technique is mostly used in permanently installed systems, like vineyards and trees [56].

4. *Sub-irrigation:* According to Bjorneberg [56], this technique involves applying water to alter the height of the water table and has great promise for regions with significant drainage needs. In greenhouses and nursery plant production, this technique is typically employed in the horticulture sector. But since irrigation water is required for crop germination, it is not utilized in arid or semi-arid regions [56].

In the border strip method, the farm is divided into many strips each strip measuring 3–20 m wide and 100–400 m long. These portions are separated by low boundaries. The strips are flat between borders even if they slope naturally throughout their length. However, the initial land preparation required by this technique is costly. Water from the supply canal is channeled slowly toward the downstream end of these strips, where it soaks into and irrigates the land. When the supply is turned off, water flows from the upstream to the downstream end [58].

Surface irrigation using the check/basin method involves putting irrigation water on a level or almost level region that is encircled by dikes all around. The land is divided into multiple almost level sections, all surrounded by levees. These plots receive water from the farmer's watercourse in a rotation. This method is effective on a wide range of soil types, including heavy and very permeable soils. The farmer may easily regulate the amount of water sprayed on his field. It is feasible to lessen the quantity of water lost to deep infiltration and surface runoff, enabling adequate irrigation of the whole farm [58].

Although surface irrigation methods are the most often used worldwide for rice cultivation, their low efficiency and homogeneity can be due to inadequate water management and design [59]. Surface irrigation methods including furrow, border, and basin irrigation are all characterized by inefficient irrigation. According to Mitiku et al. [60], this leads to water waste, salinization, and contamination of surface and groundwater resources.

Because paddy fields have high water levels, drainage management is usually a problem and requires a lot of labor. Farmers usually use manual equipment to control the input and output flow. However, Liu et al. [61] suggested that effective water management practices could allow flood irrigation to continue.

Many approaches in water saving have been tested to increase rice production including direct seeded rice (DSR) [62], alternate wetting and drying (AWD) [63], aerobic rice [64], system of rice intensification [65], saturated soil culture [53], intermittent dry spells [66], and a lot more.

The technique known as alternate wetting and drying (AWD) includes letting the surface water in rice fields evaporate before irrigating them once more as a substitute for flooding them, as in traditional rice production. Keeping soil conditions intermittently aerobic is the core of AWD. This limits the biological process of methanogenesis in the soil, which lowers the amount of methane gas released into the atmosphere and lowers the amount of water needed for rice cultivation.

Rice fields that adopt water-saving strategies have seen an increase in paddy yield as well as increased carbon sequestration, resource (labor, energy, water, and time) savings, and a decrease in greenhouse gas (GHG) emissions [67]. The potential

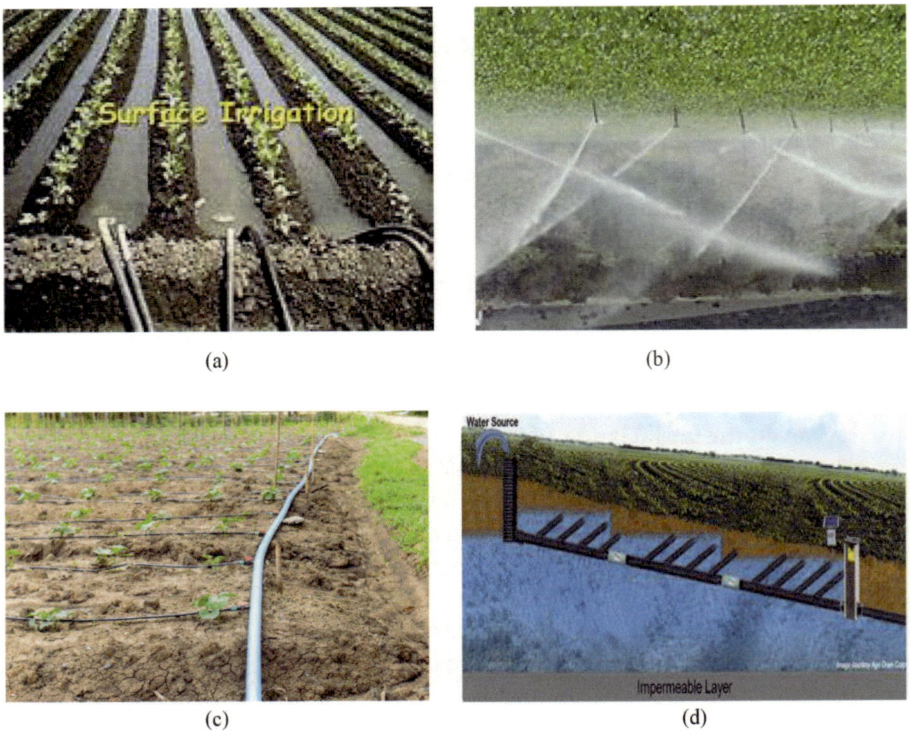

Figure 5.
(a) Surface irrigation. (b) Sprinkler irrigation. (c) Micro (drip) irrigation. (d) Sub-irrigation

benefits of the AWD system include requiring 23–33% less water [68], reducing the anthropogenic GHG emissions by 45–90% [69], improving water use efficiency (WUE) [70], and maintaining or even increasing the grain yield. The intermittent irrigation with AWD in rice has been effective in decreasing insect pest (92%) and disease (100%) infestation [71] (**Figure 5**).

4.2 Physicochemical and hydrological properties of soil in relation to drainage

When cultivating crops like rice, maize, millet, etc., the relationship between the environment, nutrient dynamics, and soil characteristics is complex in any setting [72]. According to Mamun et al. [73], the physical characteristics of soil affect root penetration, water movement into and through soils, and oxygen availability. Delgado and Gomez [74] claim that in addition to providing support, soils also act as a tank for storing water and nutrients.

It has been proposed that land use, soil texture, and proper drainage are important factors determining infiltration. According to Fischer et al. [75], values of infiltration within different locations for different crops may differ from the reported values in various literature due to the significance of vegetation and the geographical variances of places that have been documented, as shown in **Table 1**. In drainage studies, an understanding of hydrological features like infiltration is becoming more crucial since infiltration enables the soil to temporarily retain water, making it accessible to plants. In level locations, ponding may occur from an infiltration rate that is too slow, whereas in sloping areas, surface runoff and erosion may arise from an inadequate drainage system [77].

According to Habtamu [78], environmental conditions, continuous agriculture, uncontrolled drainage, and a lack of agricultural inputs all exacerbate the deterioration of soil physicochemical properties. The physicochemical characteristics of soil and water are influenced by parent rocks, vegetation, altitude, drainage, and human activity [79].

One of the biggest issues with agricultural production is the overuse of pesticides and fertilizers, especially when there is inadequate drainage. Such waterbodies' physicochemical characteristics will likely alter, making the waters unfavorable to the surrounding environment [80].

Numerous factors, including organic matter, the kind and quantity of clay, pH, the current redox conditions, and routes, affect the capacity for elements in soils and sediments to become mobilized or immobilized and to redistribute [79].

4.3 Drainage system performance indicators

A system's performance must be evaluated using a variety of indicators linked to criteria and target levels. To classify an indication as good or not, criteria are

Soil/crop	Average (mm/h)	Minimum (mm/h)	Maximum (mm/h)
Clay	4	<0.1	10
Sand	26	2	250
Rice	4	0.004	18
Maize	204	12	925

Table 1.
Summarized literature values of stable infiltration rates, based on soils (clay and sand) and crops (rice and maize) (Source: Adapted from [76]).

predetermined ranges of values (e.g., the salinity of soil might be classified as non-saline, somewhat saline, saline, or severely saline). A nominal value and a departure from the permitted value range surrounding the nominal value are used to define the target levels. The difference between an indicator and a parameter was highlighted by Murray-Rust and Snellen [81]. An indicator is a variable that can be measured and that changes over time, whereas a parameter is a variable that is measured at a certain time.

According to Bos and Nugteren [82], indicators should be structured as similarly or identically as possible to allow for the ease of cross-national comparison of performance evaluation studies. An important factor to take into account when choosing an indicator is its nature; it can be used to designate a specific activity or to aggregate or convert a set of underlying activities. Performance assessment (PA), according to Smedema and Vlotman [83], is the process of evaluating the drainage system's operation in comparison to pre-established design requirements. It can be done directly or indirectly. According to their research, PA is comprised of two components: an examination of the cause and an assessment of the drainage system's functionality. Bos [84] studied the performance indicators for drainage system's general features as:

$$\text{Sustainability of irrigation area}=\frac{\text{Current irrigation area}}{\text{Initial total irrigation area}} \tag{1}$$

$$\text{Relative groundwater depth}=\frac{\text{Actual groundwater depth}}{\text{Critical groundwater depth}} \tag{2}$$

$$\text{Relative salinity (EC) ratio}=\frac{\text{Actual EC value}}{\text{Critical EC value}} \tag{3}$$

No	Indicator	Author(s)
1	Water table as function of time	[85, 87]
2	Water table draw-down rate	[85, 87]
3	Relative groundwater depth (ICID)	[85, 87]
4	Water table depth as function of the area	[85, 87, 88]
5	Change in hydraulic conductivity over time	[85, 89]
6	Sustainability of drained area	[85, 87]
7	Ratio of design discharge versus actual discharge over time	[85, 87]
8	Relative change in soil salinity	[85, 87]
9	Relative change in soil alkalinity (SAR or ESP)	[85, 87]
10	Infiltration rate over time	[85, 87, 90]
11	Relative yield change	[85, 87]
12	Relative cropped area change	[85, 87]
13	Changes in cropping pattern	[85, 87, 88]
14	Workability	[85, 87]
15	Waterlogging index	[85, 87, 90]

Table 2.
Drainage performance assessment indicators.

Any primary parameter's critical value quantifies a physical process wherein a chemical's concentration limits crop yield or impairs health if it is surpassed. For example, irrigation water's salinity has a critical value that, if exceeded, lowers crop yield [85].

One of the metrics used to evaluate the effectiveness of the drainage system is the degree of waterlogging. Sieben [86] used the SEW30 index to evaluate Australia's agricultural drainage infrastructure. A day's worth of values (in centimeters) for which the water table is closer than 30 cm to the soil surface is added up to form the SEW30. According to the SEW30 indicator, the intensity of waterlogging rises linearly when the water table rises above 30 cm. It can be challenging to compile the existing literature because various authors refer to the same indicator using different terminology. On the other hand, several writers' performance assessment metrics are shown in **Table 2**.

5. Drainage coefficients

Determining the drainage coefficient is crucial since it plays a major role in determining a drainage system's capacity. As drainage coefficient values increase, more advantages are achieved at the same time as higher costs do as well. It is necessary to optimize the anticipated benefits in relation to the expenses when selecting a coefficient for a drainage system. Standard procedures have been adopted by several countries to determine q-values, and standard drainage coefficient values have been accepted for use in various scenarios. The narrow range of 5–10 mm/day is where drainage coefficients, which are utilized in subsurface drainage design, predominantly change for controlling the water table under conditions of rainfall recharge.

Under these circumstances, a technical-economic optimum for pipe drainage has been determined to be characterized by these values. Prior to determining drainage coefficients for subsurface drainage system regulation of the water table, groundwater balance calculations and salt balance studies are required [91]. According to Murty and Jha [92], one of the key design parameters for hydraulic systems in agricultural drainage is the concept of drainage coefficients. The rate at which open channels and drains in surface drainage networks should remove water from a drainage area varies depending on the drainage area's size, characteristics, irrigation frequency and intensity, type of crops grown there, and level of waterlogging protection needed.

5.1 Methods of determining drainage coefficients

Drainage coefficients have been determined using a variety of techniques, such as Simplified Hydrologic Accounting Method, Cypress Creek Formula, and Boston Society Formula are a few of the techniques.

5.1.1 Simplified Hydrologic Accounting Method

Raadsma and Schilze [93] used the Simplified Hydrologic Accounting approach to estimate the number of hours required to remove excess water by analyzing rainfall data and crop tolerance statistics. It is necessary to assess the rainfall data for duration and frequency. When planning drainage, it is important to consider the length and frequency of rainfall during a particular farming season. The surplus rainfall is calculated and the channel's storage capacity is taken into consideration. It is possible to determine the necessary drainage system capacity by knowing the drainable surplus [92].

The process uses a graphical method to calculate the drainage coefficient using the depth-duration-frequency rainfall curve. Because of the maximum saturated conditions of the soil surface and atmosphere, the losses resulting from evapotranspiration, infiltration, and interception are disregarded in this process. A crop may have a minimum amount of surface ponding, and any water that exceeds that must be drained within the crop's tolerance period [44].

5.1.2 Cypress Creek Formula

Data from an enormous number of American watersheds have been used to establish an empirical relationship between the catchment region and the outer end of the drainage basin which relates to the frequency of drainage design. Famously known as the Cypress Creek equation, this relationship is thought to be effective at a mean watershed slope of 0.45%. Consequently, Eq. 4, which relates the runoff rate with the watershed area, was obtained and cited by United States Department of Agriculture (USDA) and Natural Resources Conservation Service (NRCS) [77] as follows:

$$Q=C\times A^{P} \tag{4}$$

where Q = runoff rate (m^3/s), A = area of the watershed or agricultural land (km^2), P = 5/6 (approximate average value) and C = coefficient (0.2098 + 0.0074Y), where Y is the direct runoff volume (mm) estimated by the CN method.

Since its inception, the SCS-CN method has been applied to a wide range of regions, land uses, and climatic conditions. It has also been used for a wide range of applications outside of its original scope, such as integration into long-term, daily time-step hydrological models and runoff estimation in large-scale river basins. For more than 60 years, it has also been the focus of numerous evaluations on both theoretical and practical grounds, as well as numerous attempts at improvement, adaptation, and modification [94].

Combining the water balance equation and proportional equality hypothesis, the NRCS-CN method is represented by the equation given by Mockus [95]:

$$Q=\frac{(P-0.2S)^{2}}{(P+0.8S)} \tag{5}$$

S is the retention parameter. This parameter changes spatially with changes in soils, landuse and slope and temporally due to in soil water content changes and is defined mathematically by Mockus [95]:

$$S=\left(\frac{25{,}400}{CN}\right)-254 \tag{6}$$

CN is a dimensionless number and a function of landuse antecedent soil moisture content and other factors affecting runoff and retention of watershed [96].

5.1.3 Boston Society Formula

The Boston Society of Civil Engineers suggested the peak discharge for surface drains design, and this was reported by Uppal and Sehgal [97] as:

$$Q = C \times A^{0.5} \tag{7}$$

where Q = peak discharge in cusec (ft^3/s), C = coefficient (dimensionless) and A = peak catchment area in square miles.

5.2 Drainage coefficients for irrigated areas

According to FAO [98], an irrigated field's drainage coefficients can be determined by examining the soil's properties and water balance. Peak evapotranspiration (PeakET) should be calculated manually or with the use of an appropriate model, such as CropWat. Then, laboratory analyses of bulk density, field capacity, and permanent wilting point should be performed. The proper link between soil depth, bulk density, field capacity, and permanent wilting point would be used to estimate the soil's readily available moisture (RAM). The link between the permissible soil moisture depletion, root zone depth, field capacity, and permanent wilting point should then be used to compute net irrigation.

The shortest irrigation interval is calculated from the equation:

$$SII = \frac{RAM}{ET_C} \tag{8}$$

where SII = shortest irrigation interval (days), RAM = readily available moisture (mm) and Peak ET_c = peak evapotranspiration (mm day^{-1}).

Deep percolation and seepage are computed as percentages of water losses.

Leaching requirements are typically disregarded because deep percolation in surface irrigation systems is larger than leaching requirements, hence computations only take the former into account. It is reasonable to suppose that any extra water that deeply percolates into the soil can be used for leaching [98]. In the event of precipitation throughout the growing season, the total amount should be noted and incorporated into the surface drainage coefficient simulation for irrigated ecosystems.

Total water input into drains should be equal to the sum of deep percolation, seepage and rainfall (if any) as given in Eq. 9

$$TW_d = D_p + S + R \tag{9}$$

where TW_d = total water into drains (mm), D_p = deep percolation (mm day^{-1}), S = seepage (mm day^{-1}) and R = rainfall (if any) (mm).

Drainage coefficient would then be equal to total water input into drains divided by shortest irrigation interval as given by the following equation:

$$DC = \frac{TW_d}{SII} \tag{10}$$

where DC = drainage coefficient (mm day^{-1}), TW_d = total water input into drains (mm), and SII = shortest irrigation interval (days).

6. Problems of drainage in irrigated ecologies

Drainage issues are prevalent in agricultural areas, particularly in cases where natural sources of self-draining are insufficient. Extreme drainage issues can lead to land abandonment; the abandoned areas of ancient Chaldea, the Tigris Valley, and the Euphrates Rivers in Mesopotamia are prime examples [18]. According to FAO [16], an irrigation system should ideally precisely provide the water required for plant growth and eliminate any excess to create the perfect soil-water conditions for both plant development and productive, sustainable agriculture.

According to FAO [16], between 0.25 and 0.5 million hectares of important agricultural land are degraded, which lowers crop production potential. Salinization and waterlogging are two of the main issues with inadequate drainage. Groundwater flow, excessive irrigation, precipitation, and overland flow are some of the causes of excess water in the soil. Among these, [17] noted that irrigation is a major factor in waterlogging (the rising of the soil's water table) and salinization (the accumulation of salts in the upper soil layers) in semi-arid climates when rainfall is insufficient to meet crop water requirements.

While there are enormous potential benefits to irrigated agriculture, the actual accomplishments in many irrigated zones within the country are significantly less than the potential due to inefficient water management strategies leading to waterlogging, salinization, and other related problems [28]. Challenges are more extensive in sustainable irrigation systems, especially in water-scarce areas, in which large areas of crop production are affected by inadequate drainage systems, causing soil salinity and waterlogging [27].

The primary sources of salt are naturally occurring saline seeps and shallow groundwater tables. Secondary salinization is further facilitated by inadequate drainage and ineffective irrigation water management techniques [29]. Due to salinity and sodicity problems in the Awash basin, which makes up roughly one-third of Ethiopia's total irrigated area, irrigation projects that were poorly planned and unsupported by improved management technologies had seriously degraded the land [31].

According to Namara et al. [35], around 1 Mha of land in West Africa is under irrigation, of which 60% is set up for full-control irrigation and 40% is in lowlands. There is an estimated 9.1 Mha of irrigation potential, with 55% of that potential located in only three countries: Nigeria, Ghana, and Sierra Leone. Poor drainage, waterlogging, eutrophication, soil salinity, and acidity are frequent issues with Agricultural Water Management (AWM) programs, according to a report by the African Union [36]. It further said that if left unchecked, this might lead to irrigation system failure, rendering land unfit for growing crops and hence costing money in lost revenue.

Improving the efficiency of the water management, agronomic practices and improving drainage systems in irrigated ecologies will go a long way in alleviating the problems and consequences of water scarcity, thereby increasing farmers' socioeconomic and sustainable livelihoods.

One of the biggest obstacles to agricultural drainage in Africa in the future is climate change; its effects on semi-arid and arid regions must be taken into account while developing new design standards for agricultural drainage systems.

For technical and financial analysis with little funding, advanced technologies like remote sensing and Geographic Information System (GIS) can be highly helpful and valuable decision-making tools at the pre-investigation stage of drainage installation.

There is a need to have sensitization/education programs of the local farmers and agricultural producers on the benefits of drainage systems. Therefore, a key element for future development and management of drainage system, especially in Africa, is the establishment/strengthening of farmer associations and cooperatives.

To enhance sustainable agriculture in irrigated ecologies, proper drainage system remain a priority. This can be done through the reinvention of irrigation ensuring full consideration of agricultural drainage and its effect on the environment.

7. Conclusion

This chapter focused on agricultural drainage systems in irrigated ecologies. A brief background on agricultural drainage systems was critically examined looking at definition, types, agricultural drainage in the global and African context and benefits of drainage systems in agricultural production.

Drainage is seen to have affected several parts of the world and has resulted to the abandonment of land in certain instances. However, when drainage systems in irrigated ecologies are properly designed, they come with a lot of benefits including providing good rootzone aeration, maintain soil moisture, and improving crop production. On a broader level, it leads to intensive utilization of land, flexibility in crop usage, better fertilizer usage, and the prevention of soils from further salinization.

In implementing drainage project, a lot of factors needs to be considered. These include drainage requirement, water table depth, dissolved salts in the groundwater, hydraulic conductivity, topography of the field and presence of impermeable layers.

The type of irrigation practiced also has an influence on the drainage system practices. Adoption of efficient irrigation and drainage techniques is essential for effective management of farm lands. Several indicators could be used to examine the performance of a drainage system. One of the metrics used to evaluate the effectiveness of a drainage system is the degree of waterlogging. Others include water table as a function of time, salinity index, ratio of design discharge versus actual discharge etc.

The effect of climate change such as extreme events of floods and droughts are posing serious challenges on the implementation of drainage systems. Therefore, impact of climate change on drainage planning, design, construction, operation and management must be carefully considered in drainage system implementation.

In order to ensure sustainability of a drainage system, there is need to emphasize on policy formulation in most countries to pay critical attention to drainage system performance. Also, design criteria such as determination of drainage coefficients, soil physical and hydraulic properties are of essence since they play a major role in determining the system's capacity.

Acknowledgements

This publication was made possible through support provided by the West African Centre for Water, Irrigation and Sustainable Agriculture (WACWISA), University for Development Studies, Ghana with funding support from the Government of Ghana and The World Bank through the African Centres of Excellence for Development Impact (ACE Impact) initiative with Grant No. IDA Credit 6389.

Conflict of interest

The authors declare no conflict of interest.

Author details

Mohamed Joseph Sesay[1,2,3*], Felix K. Abagale[1,2] and Shaibu Abdul-Ganiyu[1,2]

1 West African Centre for Water, Irrigation and Sustainable Agriculture, University for Development Studies, Tamale, Ghana

2 School of Engineering, University for Development Studies, Tamale, Ghana

3 Faculty of Agriculture and Food Sciences, University of Makeni, Sierra Leone

*Address all correspondence to: sesaymohamedj127@gmail.com

References

[1] Castellano MJ, Archontoulis SV, Helmers MJ, Poffenbarger HJ, Six J. Sustainable intensification of agricultural drainage. Nature Sustainability. 2019;**2**(10):914-921. DOI: 10.1038/s41893-019-0393-0

[2] Food and Agriculture Organization of the United Nations (FAO). The State of the World's Land and Water Resources for Food and Agriculture—Systems at Breaking Point. Synthesis Report. Rome, Italy: Food and Agricultural Organization; 2021

[3] van der Laan M, Virtanen S, Matsuno Y, Castelli G, Fayrap A, Cresswell R. Ecosystem services of irrigated and controlled drainage agricultural systems: A contemporary global perspective. Irrigation and Drainage. 2024:1-15. DOI: 10.1002/ird.2974

[4] Valipour M, Krasilnikof J, Yannopoulos S, Kumar R, Deng J. The Evolution of Agricultural Drainage from the Earliest Times to the Present. Sustainability. 2020;**1**:416. DOI: 10.3390/su12010416

[5] FAO. Drainage of Irrigation Lands. Rome, Italy: Food and Agriculture Organization of the United Nations; 1996

[6] Gurovich L, Oyarce P. New approaches to agricultural land drainage: A review. Irrigation and Drainage Systems Engineering. 2015;**04**(02). DOI: 10.4172/2168-9768.1000135

[7] Ritzema H. In: Ritzema H, Abdel-Dayem S, editors. Drainage to Sustain Irrigated Agriculture in North-Africa. Wageningen, The Netherlands: Alterra; 2014

[8] King KW, Fausey NR, Williams MR. Effect of subsurface drainage on streamflow in an agricultural headwater watershed. Journal of Hydrology. 2014;**519**:438-445. DOI: 10.1016/j.jhydrol.2014.07.035

[9] Nangia V, Sunohara MD, Topp E, Gregorich EG, Drury CF, Gottschall N, et al. Measuring and modelling the effects of drainage water management on soil greenhouse gas fluxes from corn and soybean fields. Journal of Environmental Management. 2013;**129**:652-664

[10] Schultz B, Zimmer D, Vlotman WF. Drainage under increasing and changing requirements. Irrigation and Drainage. 2007;**56**:S3-S22. DOI: 10.1002/ird.372

[11] Kleinman PJA, Smith DR, Bolster CH, Easton ZM. Phosphorus fate, management and modelling in artificially drained systems. Journal of Environmental Quality. 2015;**44**:460-466

[12] Kamiri H, Kreye C, Becker M. Dynamics of agricultural use differentially affect soil properties and crop response in East African Wetlands. Wetlands Ecology and Management. 2013;**21**:417-431

[13] Xian C, Qi Z, Tan CS, Zhang TQ. Modeling hourly sub-surface drainage using steady-state and transient methods. Journal of Hydrology. 2017;**550**:516-526

[14] Kennedy CD, Batallie C, Liu Z, Ale S, VanDeVelde J, Roswell CR, et al. Dynamics of nitrate and chloride during storm events in agricultural catchments with different sub-surface drainage intensity (Indiana, USA). Journal of Hydrology. 2012;**466467**:1-10. DOI: 10.1016/j.jhydrol.2012.05.002

[15] FAO. Soil Organic Carbon: The Hidden Potential. Rome, Italy: Food and

Agriculture Organization of the United Nations; 2017. ISBN: 978-92-5-109681-9

[16] FAO. The State of the World's Land and Water Resources for Food and Agriculture (SOLAW)—Managing Systems at Risk. Rome and London: Food and Agriculture Organization of the United Nations, Earthscan; 2011

[17] Reinders F, Senzanje A, Oosthuizen H. Efficient subsurface drainage for improved efficacy. In: 22nd International Congress on Irrigation and Drainage; 2014 September, 1-11. 2014

[18] Malota M. An Assessment of Shallow Water Tables and the Development of Appropriate Drainage Design Criteria for Sugarcane in Pongola, South Africa. South Africa: University of KwaZulu-Natal; 2012

[19] McCarty LB, Hubbard LR, Quisenberry V. Applied Soil Physical Properties, Drainage, and Irrigation Strategies. Switzerland: Springer International Publishing; 2016. DOI: 10.1007/978-3-319-24226-2

[20] Hatcho N, Yamasaki K, Hirofumi O, Kimura M, Matsuno Y. Estimation of the function of a paddy field for reduction of flood risk. In: Yadav B, Mohanty MP, Pandey A, Singh VP, Singh RD, editors. Sustainability of Water Resources. Water Science and Technology Library. Vol. 116. Cham: Springer; 2022. DOI: 10.1007/978-3-031-13467-8_11

[21] Shevah Y. Irrigation and agriculture experience and options in Israel. WCD Thematic Review. 1999

[22] Nishida K, Yoshida S, Shiozawa S. Numerical model to predict water temperature distribution in a paddy rice field. Agricultural Water Management. 2021;**245**:106553. DOI: 10.1016/j.agwat.2020.106553

[23] Xie W, Kimura M, Iida T, Kubo N. Simulation of water temperature in paddy fields by a heat balance model using plant growth status parameter with interpolated weather data from weather stations. Paddy and Water Environment. 2021;**19**(1):35-54. DOI: 10.1007/s10333-020-00818-3

[24] Pham T, Yli-Halla M, Marttila H, Lötjönen T, Liimatainen M, Kekkonen M. Leaching of nitrogen, phosphorus and other solutes from a controlled drainage cultivated peatland in Ruukki, Finland. Science of the Total Environment. 2023;**904**:166769. DOI: 10.1016/j.scitotenv.2023.166769

[25] Yli-Halla M, Lötjönen T, Kekkonen J, Virtanen S, Marttila H, Liimatainen M. Thickness of peat influences the leaching of substances and greenhouse gas emissions from a cultivated organic soil. The Science of the Total Environment. 2022;**806**(1):150499. DOI: 10.1016/j.scitotenv.2021.150499

[26] Awulachew SB, Erkossa T, Namara R. Irrigation Potential in Ethiopia: Constraints and Opportunities for Enhancing the System. Addis Ababa, Ethiopia: IWMI; 2010

[27] Wichelns D, Qadir M. Achieving sustainable irrigation requires effective management of salts, soil salinity, and shallow groundwater. Agricultural Water Management. 2015;**157**:31-38. DOI: 10.1016/j.agwat.2014.08.016

[28] Hordofa T, Menkir M, Bekele S, Erkossa T. Irrigation and rain-fed crop production system in Ethiopia. In: Impact of Irrigation on Poverty and Environment in Ethiopia. Colombo, Sri Lanka: International Water Management Institute; 2008. pp. 27-36. Available from: https://publications.iwmi.org/pdf/H044065.pdf

[29] Abebe TF, Alamirew T, Abegaz F. Appraisal and mapping of soil salinity problem in amibara irrigation farms, Middle Awash Basin, Ethiopia. International Journal of Innovation and Scientific Research. 2015;**13**:298-314

[30] Abraham L, Roehrig J, Chekol D. Climate change impact on Lake Ziway watershed water availability. Catchment and Lake Research. (LARS 2007); 2007. pp. 18-23

[31] Gebrehiwot KA. A review on waterlogging, salinization and drainage in Ethiopian irrigated agriculture. Sustainable Water Resources Management. 2018;**4**(1):55-62. DOI: 10.1007/s40899-017-0121-8

[32] Ritzema H. Performance Assessment of Subsurface Drainage Systems Case Studies from Egypt and Pakistan. Wageningen, The Netherlands: ILRI, Wageningen University and Research Centre; 2015

[33] Reinders FB, Oosthuizen H, Senzanje A. Development of technical and financial norms and standards for drainage of irrigated lands: Report to the Water Research Commission by Agricultural Research Council, Institute for Agricultural Engineering. 2016

[34] Malota M, Senzanje A. Modelling mid-span watertable depth and drainage discharge dynamics using DRAINMOD 6.1 in a sugarcane field in Pongola, South Africa. Water SA. 2015;**41**(3):325-334. DOI: 10.4314/wsa.v41i3.04

[35] Namara RE, Horowitz L, Kolavalli S, Kranjac-Berisavljevic G, Dawuni BN, Barry B, et al. Typology of irrigation systems in Ghana. In: IWMI Working Paper 142. Colombo, Sri Lanka: International Water Management Institute; 2010. DOI: 10.5337/2011.200. 27 p

[36] African Union. Framework for Irrigation Development and Agricultural Water Management in Africa. Addis Ababa Ethiopia: African Union; 2020

[37] Singh A. Managing the salinization and drainage problems of irrigated areas through remote sensing and GIS techniques. Ecological Indicators. 2018;**89**:584-589

[38] FAO. Quality Control of Wastewater for Irrigated Crop Production. Rome, Italy: Food and Agriculture Organization; 1997

[39] Fraser H, Fleming R. Environmental Benefits of Tile Drainage—Literature Review. Vol. 23. Ontario, Canada: University of Guelph; 2001

[40] Muma M, Rousseau AN, Gumiere SJ. Modeling of subsurface agricultural drainage using two hydrological models with different conceptual approaches as well as dimensions and spatial scales. Canadian Water Resources Journal. 2016;**1784**(October):1-16. DOI: 10.1080/07011784.2016.1231014

[41] Nousiainen AO, Björklöf K, Sagarkar S, Nielsen JL, Kapley A, Jørgensen KS. Bioremediation strategies for removal of residual atrazine in the boreal groundwater zone. Applied Microbiology and Biotechnology. 2015;**99**(23):10249-10259. DOI: 10.1007/s00253-015-6828-2

[42] Jafari-talukolaee M, Ritzema H, Darzinaftchali A, Shahnazari A. Subsurface drainage to enable the cultivation of winter crops in consolidated paddy fields in Northern Iran. Sustainability. 2016;**8**:1-19. DOI: 10.3390/su8030249

[43] Alakukku L, Nuutinen V, Ketoja E, Koivusalo H, Paasonen-Kivekäs M. Soil macroporosity in relation to subsurface

drain location on a sloping clay field in humid climatic conditions. Soil and Tillage Research. 2010;**106**(2):275-284. DOI: 10.1016/j.still.2009

[44] Ritzema HP. Drainage Principles and Applications, Publication 16. 2nd ed. Wageningen, The Netherlands: International Institute for Land Reclamation and Improvement (ILRI); 1994. ISBN: 90-70754-3-39

[45] Gholamhosseinian A, Bashtian M, Sepehr A. Soil quality: Concepts, importance, indicators, and measurement. In: Rakshit AS, Ghosh S, Vasenev V, Pathak H, Rajput VD, editors. Soils in Urban Ecosystem. Singapore: Springer Nature Singapore Pte Ltd.; 2022. DOI: 10.1007/978-981-16-8914-7_8

[46] Malakar A, Snow DD, Ray C. Irrigation water quality—A contemporary perspective. Water (Switzerland). 2019;**11**(7):1-24. DOI: 10.3390/w11071482

[47] Ingrao C, Strippoli R, Lagioia G, Huisingh D. Water scarcity in agriculture: An overview of causes, impacts and approaches for reducing the risks. Heliyon. 2023;**9**(8):e18507. DOI: 10.1016/j.heliyon.2023.e18507

[48] Mehta P, Siebert S, Kummu M, Deng Q, Ali T, Marston L, et al. Half of twenty-first century global irrigation expansion has been in water-stressed regions. Nature. Mar 2024;**2**:254-261. DOI: 10.1038/s44221-024-00206-9

[49] Nhemachena C, Nhamo L, Matchaya G, Nhemachena CR, Muchara B, Karuaihe ST, et al. Climate change impacts on water and agriculture sectors in Southern Africa: Threats and opportunities for sustainable development. Water. 2020;**12**:1-17. DOI: 10.3390/w12102673

[50] Kalfas D, Kalogiannidis S, Papaevangelou O, Chatzitheodoridis F. Assessing the connection between land use planning, water resources, and global climate change. Water. 2024;**16**(333):1-18. DOI: 10.3390/w16020333

[51] Velasco-muñoz JF, Aznar-sanchez AJ, Batlles-delafuente A, Fidelibus M. Sustainable irrigation in agriculture: An analysis of global research. Water. 2019;**11**(1758):1-26. DOI: 10.3390/w11091758

[52] Hedley C, Knox J, Raine S, Smith R. Water: Advanced irrigation technologies. In: Encyclopedia of Agriculture and Food Systems. Vol. 5. 2014. pp. 378-406. DOI: 10.1016/B978-0-444-52512-3.00087-5

[53] Tuong TP, Bouman B. Rice production in water-scarce environments. In: Kijne JW, Barker R, Molden D, editors. Water Productivity in Agriculture: Limits and Opportunities for Improvement. Manila, Philippines: CABI Publishing; 2003. pp. 53-67

[54] Domènech L. Improving irrigation access to combat food insecurity and undernutrition: A review. Global Food Security. 2015;**6**:24-33. DOI: 10.1016/j.gfs.2015.09.001

[55] Brouwer C, Prins K, Kay M, Heibloem M. Irrigation water management: Training manual No. 5: Irrigation methods. In: Agricultural and Food Sciences, Environmental Science. Rome: Food and Agriculture Organization of the United Nations; 1985

[56] Bjorneberg DL. Irrigation methods. In: Reference Module in Earth Systems and Environmental Science. Oxford, UK: Elsevier Inc.; 2013. DOI: 10.1016/b978-0-12-409548-9.05195-2

[57] Irmak S. Subirrigation in Agricultural Fields. Nebraska, USA: University of Nebraska; 2020

[58] Sable R, Kolekar S, Gawde A, Takle S, Pednekar A. A review on different irrigation methods. International Journal of Applied Agricultural Research. 2019;**14** (0973-2683):49-60

[59] Kaur G, Singh G, Motavalli PP, Nelson KA, Orlowski JM, Golden BR. Impacts and management strategies for crop production in waterlogged or flooded soils: A review. Agronomy Journal. 2020;**112**:1475-1501. DOI: 10.1002/agj2.20093

[60] Mitiku A, Deribew KT, Moisa MB, Worku K. Spatial evaluation of surface water irrigation potential areas to improve rural crop productivity in the Gomma district, southwestern Ethiopia. Cogent Food & Agriculture. 2024;**10**(1):1-12. DOI: 10.1080/23311932.2024.2328424

[61] Liu X, Liu J, Huang C, Liu H, Meng Y, Chen H, et al. The impacts of irrigation methods and regimes on the water and nitrogen utilization efficiency in subsoiling wheat fields. Agricultural Water Management. 2024, 2024;**295**:108765. DOI: 10.1016/j.agwat.2024.108765

[62] Farooq M, Siddique KHM, Rehman H, Aziz T, Lee DJ, Wahid A. Rice direct seeding: Experiences, challenges and opportunities. Soil and Tillage Research. 2011;**111**(2):87-98. DOI: 10.1016/j.still.2010.10.008

[63] Lampayan RM, Rejesus RM, Singleton GR, Bouman BAM. Adoption and economics of alternate wetting and drying water management for irrigated lowland rice. Field Crops Research. 2015;**170**:95-108. DOI: 10.1016/j.fcr.2014.10.013

[64] Ishfaq M, Farooq M, Zulfiqar U, Hussain S, Akbar N, Nawaz A, et al. Alternate wetting and drying: A water-saving and ecofriendly rice production system. Agricultural Water Management. 2020;**241**:106363. DOI: 10.1016/j.agwat.2020.106363

[65] Toungos MD. System of rice intensification. International Journal of Innovative Agriculture & Biology Research. 2019;**6**(July):2318-2318. DOI: 10.1007/978-94-024-1179-9_301491

[66] Feng L, Bouman BAM, Tuong TP, Cabangon RJ, Li Y, Lu G, et al. Exploring options to grow rice using less water in northern China using a modelling approach. I. Field experiments and model evaluation. Agricultural Water Management. 2007;**88**(1-3):1-13. DOI: 10.1016/j.agwat.2006.10.006

[67] Yang J, Zhou Q, Zhang J. Moderate wetting and drying increases rice yield and reduces water use, grain arsenic level, and methane emission. Crop Journal. 2017;**5**(2):151-158. DOI: 10.1016/j.cj.2016.06.002

[68] Carrijo DR, Lundy ME, Linquist BA. Rice yields and water use under alternate wetting and drying irrigation: A meta-analysis. Field Crops Research. 2017;**203**:173-180. DOI: 10.1016/j.fcr.2016.12.002

[69] Linquist B, Snyder R, Anderson F, Espino L, Inglese G, Marras S, et al. Water balances and evapotranspiration in water- and dry-seeded rice systems. Irrigation Science. 2015;**33**(5):375-385. DOI: 10.1007/s00271-015-0474-4

[70] Djaman K, Mel VC, Diop L, Sow A, El-Namaky R, Manneh B, et al. Effects of alternatewetting and drying irrigation

regime and nitrogen fertilizer on yield and nitrogen use efficiency of irrigated rice in the Sahel. Water (Switzerland). 2018;**10**(6). DOI: 10.3390/w10060711

[71] Chapagain T, Yamaji E. The effects of irrigation method, age of seedling and spacing on crop performance, productivity and water-wise rice production in Japan. Paddy and Water Environment. 2010;**8**(1):81-90. DOI: 10.1007/s10333-009-0187-5

[72] Ololade IA, Ajayi IR, Gbadamosi AE, Mohammed OZ, Sunday AG. A study on effects of soil physico-chemical properties on cocoa production in Ondo State. Modern Applied Science. 2010;**4**(5):35-43. DOI: 10.5539/mas.v4n5p35

[73] Mamun SA, Rahman F, Yeasmin F, Islam MA. Suitability of the physical properties of soil for crop production: A study in Tangail Sadar. Journal of Environmental Science and Natural Resources. 2011;**4**(2):121-125

[74] Delgado A, Gomez JA. The soil physical, chemical and biological properties. In: Villalobos FJ, Fereres E, editors. Principles of Agronomy for Sustainable Agriculture. Spain: Springer International Publishing; 2016. DOI: 10.1007/978-3-319-46116-8

[75] Fischer C, Roscher C, Jensen B, Eisenhauer N, Baade J, Attinger S, et al. How do earthworms, soil texture and plant composition affect infiltration along an experimental plant diversity gradient in grassland? PLoS ONE. 2014;**9**(6):e98987. DOI: 10.1371/journal.pone.0098987

[76] Söderberg MH. Measuring soil infiltration rates in cultivated land—A case study of Ifakara, Tanzania [unpublished Master's thesis]. Stockholm, Sweden: Department of Physical Geography—Stockholm University; 2015

[77] United States Department of Agriculture (USDA), Natural Resources Conservation Service (NRCS). Watertable control. In: Part 624 of National Engineering Handbook on Hydrology. USA: USDA; 2004

[78] Habtamu A. Land use/land cover change and impacts of Jatropha on soil fertility: The case of Mieso and Bati Districts, Ethiopia [M.Sc. thesis]. Ethiopia: Haramaya University; 2011. pp. 11-30

[79] Manga VE, Neba GN, Suh CE. Environmental geochemistry on mine tailing soils in the artisanal gold mining district of Betare-Oya, Cameroon. Environmental Pollution. 2017;**6**(1):52-61

[80] Tening AS, Chuyong GB, Asongwe GA, Fonge BA, Lifongo LL, Mvondo-Ze AD, et al. Contribution of some water bodies and the role of soils in the physico-chemical enrichment of the Douala-Edea Mangrove Ecosystem, Cameroon. African Journal of Environmental Science and Technology. 2013;7(5):336-349. DOI: 10.5897/AJEST12.081. ISSN: 1996-0786

[81] Murray-Rust DH, Snellen WB. Irrigation System Performance Assessment and Diagnosis, Joint IIMI/ILRI/IHEE Publication. Colombo, Sri Lanka: International Irrigation Management Institute; 1994

[82] Bos MG, Nugteren J. On Irrigation Efficiencies, ILRI Publication No. 19. 2nd ed. Wageningen, The Netherlands: International Institute for Land Reclamation and Improvement (ILRI); 1990. 90-70260-875

[83] Smedema LK, Vlotman WF, editors. Workshop on Performance Assessment

of Drainage Systems. Cairo, Egypt: Drainage Research Institute (DRI) and Egyptian Public Authority for Drainage Projects (EPADP); 1996

[84] Bos MG. Performance indicators for irrigation and drainage. In: Workshop on the Evaluation of the Performance of Sub-Surface Drainage Systems, 16th ICID. Congress. Cairo, Egypt: ICID-Cemagref; 1996

[85] Bos MG, Burton MA, Molden DJ. Irrigation and Drainage Performance Assessment Practical Guidelines. Oxfordshire, Cambridge, MA, UK: CABI Publishing; 2005. 0-85199-967-0

[86] Sieben WH. Relationships between drainage condition and crop yieldfor young loam soils in the Nordoostpolder. Van Zee tot Land. 1964;**40**:107-123

[87] Vincent B, Voltman WF, Zimmer D, Hornbuckle JW, Christen E, Vincent B, editors. Performance assessment and potential indicators for drainage systems. In: CSIRO Land and Water Science Report 62/07. Griffith, Australia: CSIRO Land and Water Science; 2007. 1834-6618

[88] Molden DJ, Sakthivadivel R, Perry CJ, de Fraiture C, Kloezen W. Indicators for comparing performance of irrigated agricultural systems. In: International Water Management Institute, Research Report 20. Colombo, Sri Lanka: IWMI; 1998. 26 p. DOI: 10.3910/2009.028

[89] Ijir TA, Burton MA. Performance assessment of the Wurno irrigation scheme, Nigeria. ICID Journal. 1998;**47**:31-46

[90] Garces C. A methodology to evaluate the performance of irrigation systems: Application to Philippine National Systems [unpublished PhD thesis]. Ithaca, New York: Cornell University; 1983

[91] Ochs WJ, Bishay BG. Drainage Guidelines, World Bank Technical Paper No. 195. Washington, DC: The World Bank; 1992. 0253-7494

[92] Murty VVN, Jha MK. Land and Water Management Engineering. 6th ed. Ludhiana, India: Kalyani Publishers; 2011

[93] Raadsma S, Schilze FE. Surface field drainage systems. In: Drainage Principles and Applications. Vol. 4. Wageningen, The Netherlands: International Institute for Land Reclamation and Improvement; 1974

[94] Soulis KX. Soil conservation service curve number (SCS-CN) method: Current applications, remaining challenges and future perspectives. Water. 2021;**13**:192

[95] Mockus V. Estimation of total (and Peak Rates of) surface runoff for individual storms. In: Interim Survey Report, Grand (Neosho) River Watershed. USDA Soil Conservation Service, USA; 1949

[96] Shi WH, Huang MB, Gongadze K, Wu LH. A modified SCS-CN method incorporating storm duration and antecedent soil moisture estimation for runoff prediction. Water Resources Management. 2017;**8**:1-15. DOI: 10.1007/s11269-017-1610-0

[97] Uppal HL, Sehgal SR. Drainage Requirements of the Punjab State, International Commission on Irrigation and Drainage (ICID), Bulletin 23-31. New Delhi, India: ICID; 1965

[98] FAO. Crop Water Requirements and Irrigation Scheduling. Irrigation Manual Module 4. Harare, Zimbabwe: Food and Agriculture Organization of the United Nations; 2002

Chapter 10

Agrometeorological Aspects of Sugarcane

Alexandrius de Moraes Barbosa

Abstract

Sugarcane is an important raw material for the production of sugar and ethanol. Bioelectricity and biogas can also be produced from its byproducts. The productive potential of sugarcane is a result of the interaction among genotype, phenotype, and environment. Considering the various agrometeorological elements that influence crop development, solar radiation, air temperature, and soil water balance stand out. The combination of these factors can promote or inhibit the development of sugarcane, given that the crop has different agrometeorological needs at each stage of development. Therefore, understanding the effects of agrometeorological elements, as well as the interaction between these elements and their relationships with the sugarcane development phases, is of great importance for agricultural planning, aiming to maximize the yield of stalks and sugar, mainly in a future climate change.

Keywords: development phases, phenological stages, solar radiation, air temperature, water balance

1. Introduction

Sugarcane belongs to the *Poaceae* family and the genus *Saccharum* L. The current modern commercial cultivars are hybrids resulting from genetic improvement and cross-breeding of several species (mainly *S. officinarum* and *S. spontaneum*). Sugarcane is a plant with a C4 physiological cycle, and its main growing regions are located in tropical environments due to the agrometeorological conditions that are favorable for the development of the crop.

The development and yield of sugarcane is the result of the interaction among genotype, phenotype, and environment. The environment is made up of several agrometeorological elements that act simultaneously on the plant (**Figure 1**). Considering the various elements, three stand out, such as solar radiation, air temperature, and water balance in the soil.

Air temperature and solar radiation, associated with the genotypic and phenotypic characteristics of the cultivar and the concentration of CO_2 in the atmosphere, are the defining factors of the potential productivity of sugarcane (**Figure 2**). Potential productivity (Yp) is understood as the yield of a crop grown under conditions in which the development of the plant is not limited by the lack of water, nutrients, and also by

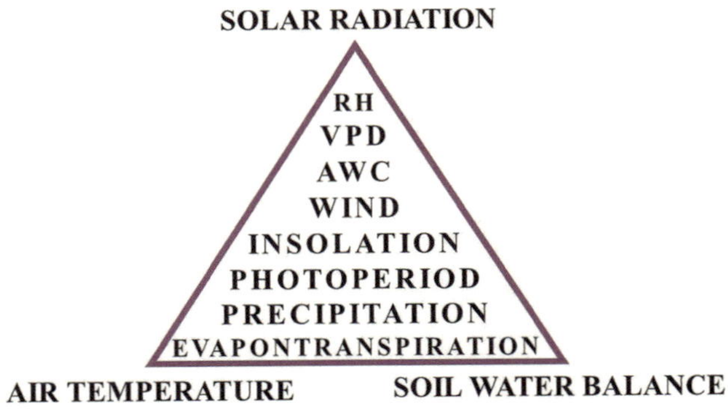

Figure 1.
Agrometeorological elements that influence the development of sugarcane.

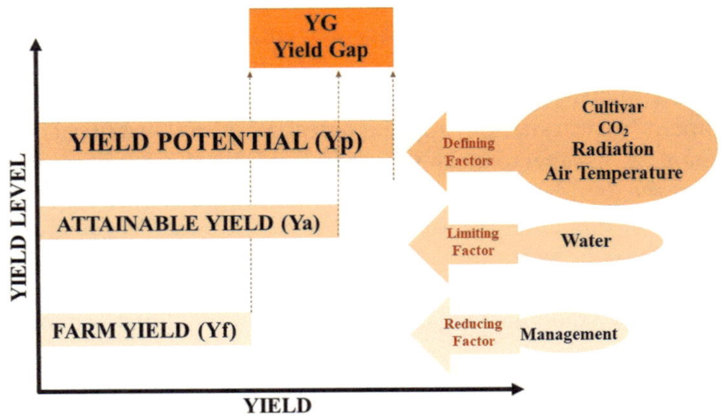

Figure 2.
Different levels of sugarcane yield potential. Adapted from Van Ittersum et al. [1].

biotic stresses. Its yield is usually influenced only by the crop and the environmental conditions (solar radiation, air temperature, and photoperiod) [1, 2].

In rainfed (non-irrigated) crops, productivity is limited by water, also called attainable productivity (Yw), equivalent to the productive potential limited by water. Average productivity (Yf) is defined as the current yield achieved by the producer in the field [1–3].

The difference between potential productivity or water-limited productivity and the productivity achieved by producers has commonly been called the “Yield Gap” (YG) (2,4). Through Yield Gap analysis, it is possible to identify the main factors that affect agricultural productivity, providing important information for agricultural management, research, and public policies [1, 4, 5].

Monteiro and Sentelhas [6] observed that the average potential yield of sugarcane in Brazil is 186.4 Mg ha^{-1}. The average Yield Gap related to water is 100.8 Mg ha^{-1}, and the Yield Gap related to management is 32.4 Mg ha^{-1}. In a recent study performed in Brazil (state of São Paulo), Gasparotto et al. [7] observed that the productivity limited by water (Yw) was 102.3 Mg ha^{-1}. The management Yield Gap was 34.0 Mg ha^{-1}, with 70% of the causes being related to nutritional limitation, insect damage, and sprouting failure.

Thus, a better understanding of the interaction between sugarcane development and the main agrometeorological elements can contribute to better planning, aiming to enhance productivity and minimize losses in adverse weather conditions. This understanding is important to provide the possibility of changes in climatological patterns due to climate change.

2. Sugarcane development phases

When sugarcane is cultivated in low latitude regions (0–15°), flowering occurs annually. However, in terms of agro-industrial production, aiming at the production of sugar and ethanol, flowering is not desirable. Therefore, in the management adopted for commercial crops, the crop starts to behave as a semi-perennial cycle, and after harvest, the crop begins a new cycle of development from the ratoon crop. The new implementation of the crop (reform) will be performed when agricultural productivity is low, and it is not economically viable to let the crop go through another development cycle.

2.1 Development phases

The sugarcane cultivation cycle, whether in planting or in ratoon crops, has four phases: (i) sprouting; (ii) tillering; (iii) vegetative growth; and (iv); ripening (**Figure 3**).

Sprouting is the first phase of crop development. After planting, sprouting starts from the stalks, and after cutting, sprouting starts from the ratoon. The duration of this phase ranges from 20 to 40 days. Factors such as good water availability in the soil and high temperature stimulate sprouting.

After the crop has been established, the tillering phase begins, in which sugarcane clumps are formed. Tillering is characterized by the appearance of new plants from the formation of the primary tiller. In modern cultivars, the number of tillers formed from the primary tiller ranges from 10 to 20. Factors such as high-water availability in the soil, high temperature, and high luminosity promote tillering. This phase lasts 90

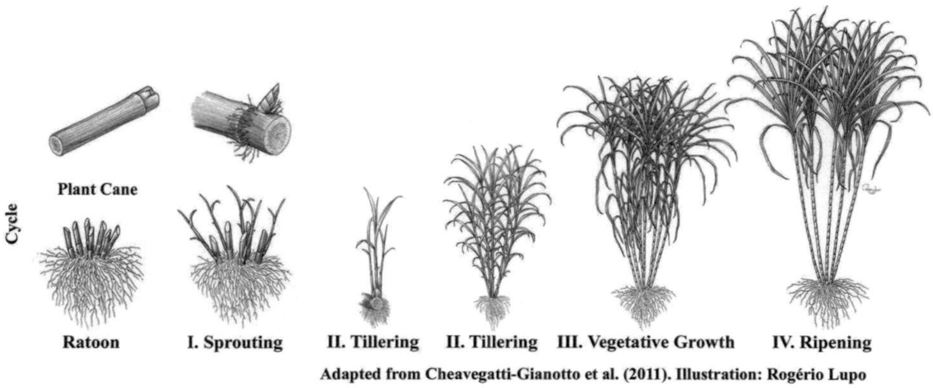

Figure 3.
Sugarcane development phases in the cane plant and ratoon cycles. I—Sprouting; II—Tillering; III—Vegetative growth; and IV—Ripening (maturation). Adapted from Cheavegatti-Gianotto et al. [8]. Illustration: Rogério Lupo.

to 120 days, being faster in ratoon sugarcane. Tillering stops when the luminosity at the base of the plant is reduced by 70% [9].

The vegetative growth is the third phase of crop development, and it is totally influenced by the agrometeorological conditions of the environment. This phase is characterized by elongation and large accumulation of stalk biomass. At the beginning of this phase, stalk biomass represents approximately 20% of the total biomass. At the end of vegetative growth, stalk biomass will represent 70–80% of the plant's total biomass. At this stage, sugarcane can form four to five internodes per month, and the accumulation of green biomass can reach 1.0–1.5 t ha^{-1} per day. Vegetative growth occurs due to the combination of good water availability in the soil, high temperatures, and high luminosity.

The sugarcane maturation phase is characterized by a reduction in biomass accumulation and an increase in sugar (sucrose) concentration, which means an increase in technological quality. The plant must undergo climatic stress to induce sugarcane maturation. Water deficit in the soil and low temperatures are the two main stress-inducing factors. Sugarcane maturation is reduced in regions that do not have a period of low air temperatures (Central America, for example).

Barbosa and Tiritan [10] developed a formula (Eq. (1)) to estimate the sucrose content in sugarcane stalks based on the analysis of soluble solids (SS, brix degree) that can be quantified in the field using a refractometer.

$$\text{Sucrose Content } (\%) = 1.0179 * \text{SS–}3.0614 \tag{1}$$

2.2 Phenological stages

Barbosa [11] created the phenological stages of sugarcane-based on biometric and technological parameters of the crop (**Table 1**), which provided a better understanding and detail of the sugarcane development phases.

	Phenological stages	Description
Sprouting to ripening		
S1	Initial sprouting	Appearance of the primary tiller in plant cane and ratoon cycle. (**Figure 4**)
S2	Established sprouting	All primary tillers emerged and with two leaves developed. (**Figure 4**)
T1	Initial tillering	Appearance of the secondary tiller. (**Figure 5**)
T2	Intense tillering	Leaves reach halfway between Interrow spacing (**Figure 5**)
T3	Final tillering	Appearance of the first visible internode; Interrow spacing closure. (**Figure 5**)
VG1	Initial vegetative Growth	Stalk with 2–3 visible internodes
VG2	Intermediate Vegetative growth	Stalk with 6–15 visible internodes
VG3	Final vegetative Growth	Stalk ≥ visible internodes. (stalk fresh mass is 70% ≥ that of the total fresh mass of the plant)
R1	Initial ripening (Maturation)	Sucrose content ≥14% / Soluble solids ≥18°
R2	Intermediate ripening (Maturation)	Sucrose content ≥16% / Soluble solids ≥20°

	Phenological stages	Description
R3	Final ripening (Maturation)	Sucrose content ≥18% / Soluble solids ≥22°
Flowering		
F1	Induction of flowering	Inflorescence initiation (**Figure 5**)
F2	Induction of flowering	Inflorescence development within the sheath (inflorescence not visible) (**Figure 5**)
F3	Intermediate flowering	Appearance of the inflorescence (**Figure 5**)
F4	Final flowering	Total opening of the inflorescence (**Figure 5**)

Table 1.
Phenological stages of sugarcane Barbosa [11].

Figure 4.
Phenological stages of sugarcane (sprouting in cane plant and ratoon cycle).

2.3 Agrometeorology x development phases

The center-south of Brazil is the region with the largest sugarcane cultivation in the world, with an area of 7.3 million hectares [12]. It is located between latitudes 13–25° South. This region is characterized by a hot, rainy, high-luminosity spring/summer period (October to March) (vegetative growth phase) and an autumn/winter period (April to September) with low water availability in the soil (water deficit) and low air temperature and low luminosity (maturation phase), with the occurrence of frost being uncommon. The maturation phase can be divided into three stages (**Figure 6**).

Initial maturation: the first phase of maturation (April) characterizes the transition from vegetative growth to the beginning of sugarcane maturation. This change occurs due to the reduction in photoperiod, reduction in air temperature, and the beginning of water deficit. During initial maturation, there is still a gain in biomass and the beginning of sugar accumulation in the stalk.

Figure 5.
Phenological stages of sugarcane (tillering and flowering).

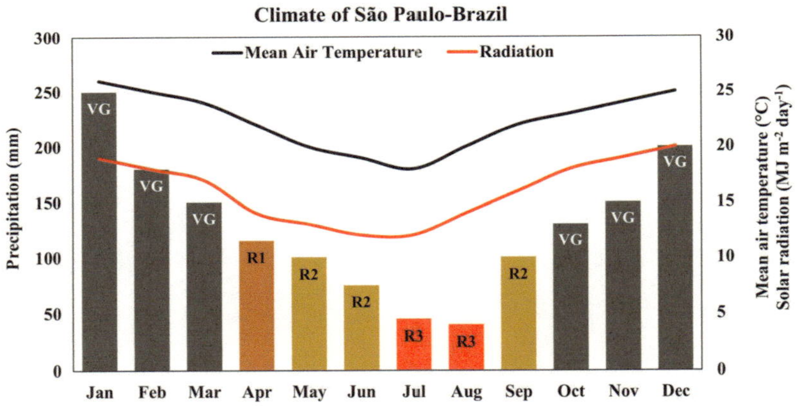

Figure 6.
Agrometeorology and sugarcane development phases. Monthly accumulated precipitation (mm), mean monthly air temperature (°C), and mean daily radiation (MJ m^{-2} dia^{-1}) in the state of São Paulo—Brazil (south hemisphere). Vegetative growth (VG); ripening phases (R1, R2, and R3).

Intermediate maturation: at this stage the biomass gain is very low and sugar accumulation intensifies due to the increase in water deficit and the reduction in air temperature.

Final maturation: this phase is characterized by the peak in sugar accumulation in sugarcane stalks due to the period of greater water deficit and lower air temperature, which means, a period of greater stress for the plant.

Long periods of drought between intermediate and final maturation impair the accumulation of sugar in the stalks. Therefore, in addition to the plant not producing sucrose, sucrose stored in the stalk may be consumed through respiration. The time period between the occurrence of the climatic factor and the plant's response ranges from 20 to 40 days.

2.4 Flowering

The flowering of sugarcane is a desirable phenomenon from the point of view of plant breeding. However, it is not desirable from an agricultural and industrial point of view, aiming at the production of sugar and ethanol. The occurrence of flowering interrupts the vegetative growth of sugarcane (formation of new internodes). Furthermore, during the panicle emission process, there is a reduction in the accumulation of sucrose in the stalk.

The inductive period for sugarcane flowering is close to 20 days. This process occurs due to the reduction of the environmental photoperiod, and the induction starts when the environmental photoperiod is close to 12.5 h. In the Southern Hemisphere, this period occurs approximately between February 20th and March 20th, while in the Northern Hemisphere, this period occurs approximately from August 20th to September 20th.

In addition to the photoperiod, which is related to the latitude of the region, other agrometeorological factors also interfere with flowering, and this is the main reason why in some years flowering occurs intensely, and in other years the plants do not flower. The best-known factors are: air temperature, water availability in the soil, and luminosity.

Regarding air temperature, sugarcane flowering is inhibited in air temperature conditions below 18°C and above 32–35°C during the inductive period [13–15]. In relation to water availability in the soil, the occurrence of water deficit during the inductive period is unfavorable to flowering [16]. Regarding luminosity, studies have shown that cloudy days increase the flowering of sugarcane, which means the greater incidence of solar radiation can inhibit flowering [17].

Barbosa [18] created a sugarcane flowering prediction index (FPI) based on soil water availability and atmospheric transmissivity, which can be seen in Eq. (2) and **Table 2**.

$$\mathrm{FPI} = (\mathrm{SW}/\mathrm{AWC} + \mathrm{RI})/2 \tag{2}$$

where SW (soil water storage, mm - lowest value in the period) can be obtained through the water balance of Thornthwaite and Mather [19]; AWC (available water capacity, mm), and RI (radiation index) are available in **Table 3**. The FPI can be calculated throughout the year using daily data or 10-days averages.

The radiation index (RI) is a function of atmospheric transmissivity (AT) (**Table 2**), which represents the fraction of solar radiation that falls on the Earth's surface in relation to the radiation that falls on the top of the atmosphere and it is calculated according to Eq. (3).

Conditions for flowering	FPI
Extremely favorable	0.90 a 1.0
Highly favorable	0.76 a 0.9
Favorable	0.51 a 0.75
Unfavorable	0.26 a 0.50
Highly unfavorable	0 a 0.25

Table 2.
Conditions for sugarcane flowering according to the flowering prediction index (FPI).

AT	RI*
≤0.5	1.00
0.501 a 0.525	0.75
0.526 a 0.55	0.50
0.551 a 0.575	0.25
≥ 0.576	0.00

Calculated according to Eq. (2).

Table 3.
Radiation index (RI) as a function of atmospheric transmissivity (AT).

$$AT = Qg/Q_o \quad (3)$$

where AT represents atmospheric transmissivity; Qg represents global solar radiation at the surface (MJ m^{-2} day^{-1}), Qo represents Global solar radiation at the top of the atmosphere (MJ m^{-2} day^{-1}).

3. Solar radiation

Solar radiation and air temperature define the productive potential of sugarcane. However, this meteorological element has been little explored in most studies on sugarcane cultivation. The efficiency of converting solar energy into biomass ranges from 2.80 to 4.54%. In addition to the characteristics of the cultivar, the variation in efficiency occurs mainly due to four factors, such as soil water balance, soil fertility, air temperature, and solar radiation [20].

Solar radiation recorded on the Earth's surface in tropical crops ranges from 10.0 to 25.0 MJ m^{-2} day^{-1} between seasons. In general, values between 15 and 25 MJ m^{-2} day^{-1} promote vegetative growth, and in this range, the air temperature also tends to be favorable for this phase of development. Regions that have solar radiation below 12 MJ m^{-2} day^{-1} are not recommended for sugarcane cultivation due to the low air temperature and the high risk of frost.

Solar radiation values ranging from 12 to 15 MJ m^{-2} day^{-1} can favor maturation for three reasons: (i) by the quantity, that even if it is low, it is enough for the sugarcane to continue producing carbohydrates (sugars) from photosynthesis; (ii) by the fact that the maturation period is associated with a water deficit in the soil, then, high values could intensify the deficit, and consequently, the stress on the plant; (iii) by the fact that normally in this value range, winter weather conditions are not yet so severe for sugarcane cultivation. The relationship between solar radiation at different latitudes and the seasons can be observed in **Figure 7**. We can observe that at latitudes above 25°C, solar radiation during winter is below 12 MJ m^{-2} day^{-1}.

The good development of the crop, the agrometeorological conditions of different regions, management, and planting times can be monitored by analyzing the interception of solar radiation by the sugarcane canopy. Radiation interception can be expressed as a percentage (%) or as an intercepted fraction.

The interception of solar radiation from the same cultivar in two regions in Brazil (Paraná and São Paulo) can be observed in **Figure 8**. The air temperature average was 2.7°C higher in the São Paulo region. This situation favored the tillering and vegetative

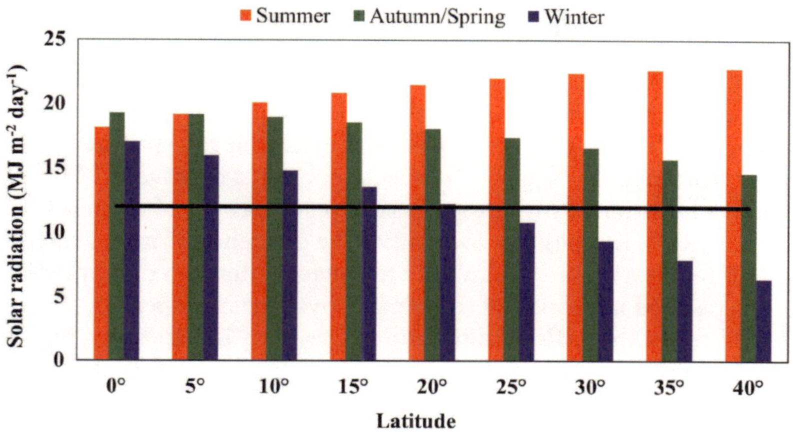

Figure 7.
Solar radiation (MJ m^{-2} day^{-1}) on the surface as a function of latitude and seasons (it was considered that 51% of the radiation from the top of the atmosphere).

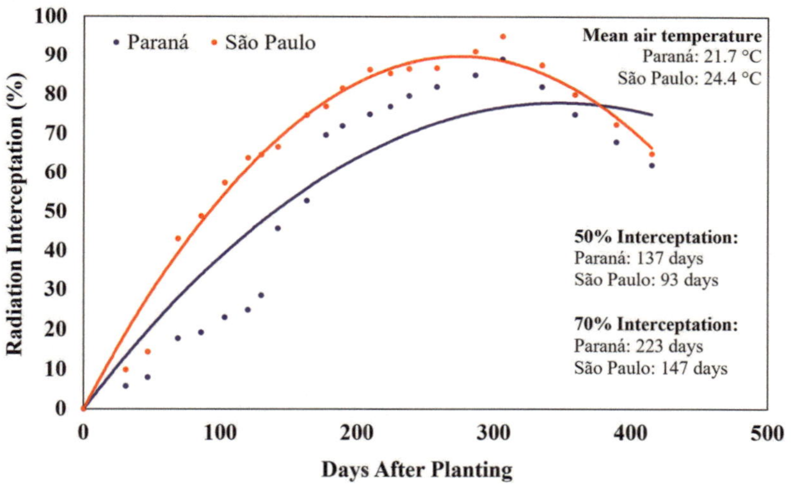

Figure 8.
Interception of solar radiation by sugarcane (cultivar RB867515) in two regions of Brazil (Paraná and São Paulo).

growth phases, with the interception of 70% of radiation being achieved after 147 days after planting. In the Paraná region, which is colder, it took 223 days for the canopy to reach 70% interception, which means 76 more days. Therefore, the choice of region and also the planting time, as well as management, which interferes with the interception of radiation (spacing, for example), have a great influence on the productive potential of sugarcane.

Radiation interception is correlated with the leaf area index (LAI). In the tillering phase, the LAI increases from 0.5–1.0 to 3.5–4.0. The highest LAI values are observed during vegetative growth, ranging from 7.0 to 8.0. From the beginning of maturation, the LAI decreases, with values of 3.0–5.0 being observed at harvest time. The radiation intercepted by the sugarcane canopy can be calculated using Beer's law [21] (Eq. (4)).

$$I = Io\ e^{-k\ LAI} \quad (4)$$

where, I is the solar radiation below the canopy; Io is the solar radiation above the canopy; k refers to the extinction coefficient.

In relation to radiation, one of the main agrometeorological parameters is the radiation use efficiency (RUE, g MJ^{-1}), which represents how many MJ of photosynthetically active radiation (PAR) are necessary to produce 1 g of dry mass per square meter. Through RUE, it is possible to evaluate the efficiency of management, cultivars, etc. RUE represents the crop's ability to convert intercepted solar radiation into biomass, being a good indication of the plant's physiological efficiency.

In irrigated crops, the RUE of sugarcane ranges from 1.63 to 2.09 g MJ^{-1}, while in non-irrigated crops, the RUE ranges from 0.71 to 1.03 g MJ^{-1} [22]. RUE values of 2.73 and 2.78 g MJ^{-1} [23] have already been observed in sugarcane. Under stress conditions, the plant's ability to convert solar radiation into biomass is reduced. This result occurs due to increased breathing and greater energy expenditure, as the plant starts to develop different physiological strategies to combat stress.

Thus, in order to obtain high sugarcane productivity, two strategies are important to be considered. The first one is to promote an increase in radiation interception by sugarcane leaves, which is mainly due to the leaf area index; the second one is to increase the efficiency of the use of radiation through management that mitigates stress in the sugarcane crop.

4. Air temperature

Air temperature can promote or inhibit sugarcane development. In general, in the vegetative growth phase, high temperatures promote stalk elongation and biomass accumulation, while low temperatures reduce vegetative growth. During the maturation phase, low temperatures promote sucrose accumulation, while high temperatures can harm sucrose accumulation.

The ideal air temperature range for the sprouting, tillering, and vegetative growth phases of sugarcane is from 25 to 35°C, with the optimum range being between 28 and 33°C (**Figure 9**). Sugarcane growth is reduced between 12 and 25°C due to decreased

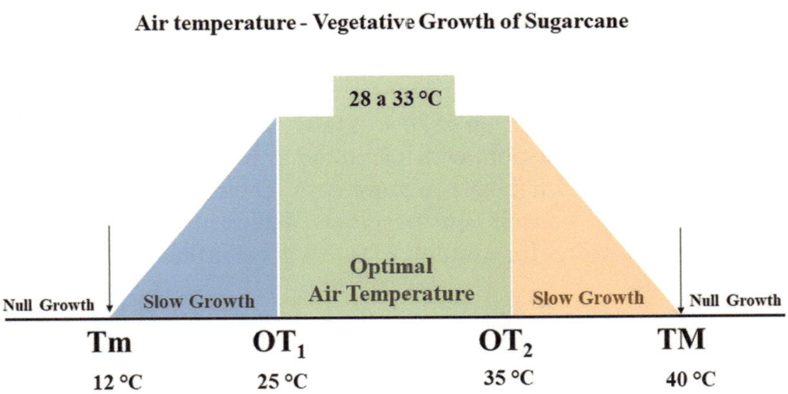

Figure 9.
Air temperature range for sugarcane vegetative growth. Minimum (Tm), maximum (TM), and optimal (TO) air temperature.

metabolism. This phenomenon occurs because the enzymes involved in C4 metabolism are less efficient in low-temperature conditions. In the range of 35–40°C, sugarcance growth is reduced due to increased respiration. In conditions below 12 and above 40°C, sugarcane growth is considered null.

Nighttime temperatures also have a great influence on the vegetative growth of sugarcane. Warm nights promote increased respiration, thus reducing the daily biomass accumulation. These conditions are observed in low-altitude regions and in regions located at latitudes between 15°N and 15°S.

During the maturation phase, low temperatures promote the accumulation of sucrose in the stalks. Maturation is favored in months with an average monthly temperature below 22°C, with the ideal range being between 18 and 22°C. Regions with an average monthly temperature below 18°C have a winter with a risk of frost.

5. Soil water balance

As sugarcane cultivation is mainly located in tropical regions, there is a high demand for evapotranspiration, and the occurrence of dry periods is also common. Therefore, water is the biggest limiting factor in the productive potential of sugarcane. It is estimated that 15–20 mm are needed to produce 1.0 ton of culm. The water demand for sugarcane cultivation ranges from 1500 to 2500 mm. However, more important than the accumulated volume is the distribution of precipitation throughout the cycle. The occurrence of water deficit during the vegetative growth phase negatively impacts stalk productivity. For every 100 mm of water deficit in the soil, there is a reduction of 10–15 ton ha^{-1} of stalks.

The greatest water demand of sugarcane occurs during the vegetative growth phase, with values reaching 8.0 mm day^{-1}. However, it is recommended that monitoring of daily crop evapotranspiration (ETc) be performed based on the reference evapotranspiration (ETo) and according to the crop coefficient (Kc) which varies according to the growth phases, i.e. ETc = ETo * Kc.

Silva et al. [24] observed the following sugarcane Kc values for each growth phase: 0.65 at sprouting, 0.85–1.0 at tillering, 1.1 at vegetative growth, and 0.85 at maturation. In a recent study of sugarcane cultivation in a tropical environment, Marin et al. [25] observed that Kc rarely exceeds the value of 1.0, suggesting a value of 0.77 for the sprouting and tillering phases and 0.87 for the vegetative growth and maturation phases in ETo conditions above 6.0 mm day^{-1}.

The occurrence of a period of water deficit is necessary to induce sugarcane maturation. Therefore, the sugarcane harvest period is performed precisely during the period of lowest precipitation. Through the agrometeorological water balance proposed by Thornthwaite and Mather [19], it is possible to verify whether the region has favorable conditions for the cultivation of sugarcane. As can be seen in **Table 4**, it is recommended to cultivate sugarcane that has an annual water deficit of 100–300 mm.

Regarding water conditions during ripening, two observations are important: (i) sugarcane cultivation cannot go through long periods without precipitation, as a minimum of soil moisture is necessary for the establishment of the crop (sprouting and tillering); (ii) the ideal is the occurrence of a water deficit of mild to moderate intensity, as the plant needs water for the sucrose production in the leaves and for the translocation and storage of sucrose in the stalks.

In crop cultivation at latitudes of 15–30°, the period of water deficit is associated with a reduction in air temperature and luminosity, factors that mitigate the effects of

Annual water deficit	Recommendation	Observation
No deficit	Not recommended	No period of deficit for ripening
< 100 mm	Recommended with observations	Low deficit for ripening
100–200	Recommended	Deficit favorable for ripening
200–300	Recommended with observations	High deficit for ripening
> 400 mm	Not recommended	High deficit in sugarcane cultivation

Regions with high water deficit may become recommended with the use of irrigation.

Table 4.
Recommendation for sugarcane cultivation depending on the annual water deficit.

low water availability in the soil and favor the sugarcane maturation. In crop cultivation at latitudes 0–15°, the occurrence of water deficit is associated with high air temperature and high luminosity. These factors intensify the effects of the water deficit and harm the sugarcane maturation, as the plant can reach severe stress condition.

The physiological condition of the plant can be checked through its water potential (**Table 5**). The plant must not undergo periods of stress in the vegetative growth. Sugarcane maturation begins when the water potential reaches −0.7 MPa and it is

Water potential	Stress intensity	Plant impacts
−0.1 to −0.6 MPa	No tress	Vegetative growth
−0.7 to −0.9 MPa	Mild stress	Reduction in VG - Ripening begins
−1.0 to −1.3 MPa	Moderate stress	Interruption of VG - Ripening
−1.4 to −1.5 MPa	Severe stress	Ripening begins is impaired
< −1.5 MPa	Permanent wilting point	Plant death (**Figure 10b**)

VG: Vegetative growth.

Table 5.
Water potential and its impacts on sugarcane.

Figure 10.
Effects of water deficit on sugarcane. (a) Water deficit reduces stalk development; (b) plant at permanent wilting point.

enhanced with water potential between −1.0 and − 1.3 MPa. Very severe stress conditions (below −1.4 MPa) are unfavorable to maturation and can lead to the death of the plant.

There is still a relationship between soil water availability and soil texture. Soils with higher clay content have greater water storage (180–200 mm). This characteristic means that sugarcane goes through fewer periods of water deficit. In sandy soils, which have a low water storage capacity (80–100 mm), water limitation is more frequent, reducing stalk yield. However, maturation occurs more quickly and intensely in sandy soils.

In the field, the effects of water deficit are visible from moderate stress, wherein there is a reduction in leaf area (reduction in the number of green leaves per plant) and a decrease in the formation of internodes [26]. During vegetative growth, many times the plants go through mild stress, and in this situation, the effects of water deficit are not visible. However, they are sufficient to reduce the productive potential of the crop, mainly due to less leaf expansion, which will result in less LAI, and consequently, a reduction in the interception of solar radiation.

Author details

Alexandrius de Moraes Barbosa
Unoeste - University of Western São Paulo, Presidente Prudente, São Paulo, Brazil

*Address all correspondence to: alexandrius@unoeste.br;
alexandriusmb@yahoo.com.br

References

[1] Van Ittersum MK, Cassman KG, Grassini P, Wolf J, Tittonell P, Hochman Z. Yield gap analysis with local to global relevance—A review. Field Crops Research. 2013;**143**:4-17. DOI: 10.1016/j.fcr.2012.09.009

[2] Sentelhas PC, Battisti R, Monteiro LM, Duarte YC, Visses F. Yield Gap—conceitos, definições e exemplos. Informações Agronômicas. IPNI (International Plant Nutrition Institute); 2016;**155**:9-12

[3] Van Ittersum MK, Rabbinge R. Concepts in production ecology for analysis and quantification of agricultural input-output combinations. Field Crops Research. 1997;**52**:197-208. DOI: 10.1016/S0378-4290(97)00037-3

[4] Lobell DB, Cassman KG, Field CB. Crop yield gaps: Their importance, magnitudes, and causes. Annual Reviews Further. 2009;**34**:179-204. DOI: 10.1146/annurev.environ.041008.093740

[5] Dias HB, Sentelhas PC. Sugarcane yield gap analysis in Brazil—A multi-model approach for determining magnitudes and causes. Science of the Total Environment. 2018;**637-638**:1127-1136. DOI: 10.1016/j.scitotenv.2018.05.017

[6] Monteiro LA, Sentelhas PC. Sugarcane yield gap: Can it be determined at national level with a simple agrometeorological model? Crop and Pasture Science. 2017;**68**:272-284. DOI: 10.1071/CP16334

[7] Gasparotto LG, Rosa JM, Grassini P, Marin FR. Developing an operational framework to diagnose yield gaps in commercial sugarcane mills. Field Crops Research. 2022;**278**:1-9. DOI: 10.1016/j.fcr.2022.108433

[8] Cheavegatti-Gianotto A, de Abreu HMC, Arruda P, et al. Sugarcane (Saccharum X officinarum): A reference study for the regulation of genetically modified cultivars in Brazil. Tropical Plant Biology. 2011;**4**:62-89. DOI: 10.1007/s12042-011-9068-3

[9] Inman-Bamber NG. Temperature and seasonal effects on canopy development and light interception of sugarcane. Field Crops Research. 1994;**1994**(36):41-51. DOI: 10.1016/0378-4290(94)90051-5

[10] Barbosa AM, Tiritan CS. Estimativa da Pol e do ATR em função dos valores de Brix na cultura da cana-de-açúcar. Boletim de Pesquisa do Programa de Pós-Graduação em Agronomia-Unoeste. 2023;**4**:51-54

[11] Barbosa AM. Estádios fenológicos da cana-de-açúcar: uma proposta baseada em parâmetros biométricos e tecnológicos. Natal-RN, Brasil: XXII CBAGRO–Congresso Brasileiro de Agrometeorologia; 2023. pp. 385-388

[12] Conab–Companhia Nacional de Abastecimento. Acompanhamento da safra brasileira. Safra 2023/24, 3° levantamento. 2023;**11**:1-56

[13] Gosnell JM. Some factors affecting flowering in sugarcane. Proceedings South African Sugar Technologists' Association. 1973;**47**:144-147

[14] Levi CA. Froracion de cana de azucar, determinacion de requerimentos indutivos. Revista Industrial de Tucuman. 1983;**60**:1-15

[15] Pereira AR, Barbieri V, Villa Nova NA. Climatic conditioning of flowering induction in sugarcane. Agricultural Meteorology. 1983;**29**: 103-110. DOI: 10.1016/0002-1571(83) 90043-2

[16] Moore PH, Berding N. Flowering. In: Moore PH, Botha FC, editors. Sugarcane: Physiology, Biochemistry & Functional Biology. 1st ed. New Jersey, USA: Wiley Blackwell; 2014. pp. 379-406

[17] Espinoza G, Quemé JL. Floración de la caña de azúcar y su manejo. In: Melgar et al., editors. El Cultivo de la Caña de Azúcar en Guatemala. Guatemala: Artemis Edinter; 2014. p. 512

[18] Barbosa AM. Índice de previsão do florescimento da cana-de-açúcar. Boletim de Pesquisa do Programa de Pós-Graduação em Agronomia-Unoeste. 2023;**4**:55-58

[19] Thornthwaite CW, Mather JR. The Water Balance. Publications in Climatology. Drexel Institute of Technology: New Jersey; 1955

[20] Barbosa AM, Zilliani RR, Tiritan CS, Souza GM, Silva MA. Energy conversion efficiency in sugarcane cultivars as a function of production environments in Brazil. Renewable and Sustainable Energy Reviews. 2021;**150**:1-11. DOI: 10.1016/j.rser.2021.111500

[21] Monsi M, Saeki T. Uber den lichtfacktor in den pflanzen-gesellschaften und seine bedeutung fur die Stoffproduktion. Japanese Journal of Botany. 1953;**14**:22-52

[22] Silva ALC, Costa WAJ. Growth and radiation use efficiency of sugarcane under irrigated and rain-fed conditions in Sri Lanka. Sugar Tech. 2012;**14**: 247-254. DOI: 10.1007/s12355-012-0148-y

[23] Ferreira Junior RA, Souza JL, Lyra GB, Escobedo JF, MVC S. Energy conversion efficiency in sugarcane under two row spacings in northeast of Brazil. Revista Brasileira de Engenharia Agrícola e Ambiental. 2015;**19**:741-747. DOI: 10.1590/1807-1929/agriambi. v19n8p741-747

[24] Silva TGF, Moura MSB, Zolnier S, Vieira JSV SJM, Gomes Júnior WF. Demanda hídrica e eficiência do uso de água da cana-de-açúcar irrigada no semiárido brasileiro. Revista Brasileira de Engenharia Agrícola e Ambiental. 2011; **2011**(15):1257-1265. DOI: 10.1590/ S1415-43662011001200007

[25] Marin FR, Inman-Bamber G, Silva TGF, Vianna MS, Nassif DSP, Carvalho KS. Sugarcane evapotranspiration and irrigation requirements in tropical climates. Theoretical and Applied Climatology. 2020;**140**:1349-1357. DOI: 10.1007/ s00704-020-03161-z

[26] Barbosa AM, Guidorizi KA, Catuchi TA, Marques TA, Ribeiro RV, Souza GM. Biomass and bioenergy partitioning of sugarcane plants under water deficit. Acta Physiologiae Plantarum. 2015;**37**:142. DOI: 10.1007/ s11738-015-1887-7

Section 4

Biological and Ecosystem Interactions

Chapter 11

Biological Pest Control and Sustainable Agroecosystems

Zill-e-Huma Aftab, Tehmina Anjum, Waheed Akram, Najat A. Bokhari, Areeba Rehman, Shama Sharif, Humaira Rizwana and Bareera Munir

Abstract

Long ago, the human population profoundly relied on agri-farming, which was principally directed toward speedy, protected, and enhanced yields. These systems have undoubtedly fostered exploding populations with high food and fiber demands. To meet all these demands, the soil microbial ecosystem has been overlooked because of its long-term effects on the health and sustainability of the planet. Sustainable agricultural practices can not only highlight the microbial health of the soil and environment but also increase yields with minimal input. These sustainable practices benefit from the following advantages: (1) soil health preservation; (2) promotion of microbial diversity; (3) future food supply security; (4) support of farmer livelihoods; and (5) combating climate change. Biological pest control is an important and environmentally friendly strategy for sustainable agriculture in which natural microbial predators, including pathogens, nematodes, insects, and weeds, are applied to control pests, thus reducing pesticide dependency. Through a cautious and comprehensive understanding of the ecological niche and complex associations among living organisms, biological control can act as an encouraging approach to increase and stabilize crop yield while promoting congruent coexistence between agriculture and the ecological biosphere.

Keywords: sustainable agriculture, biological pest control, pathogens, fungi, bacterial insects

1. Introduction

An *ecosystem* is an intricate form of organization composed of both nonliving (abiotic) components, such as the surrounding environment, and living (biotic) components, such as communities. Each ecosystem's constituents can be classified as *autotrophs* (producers), *heterotrophs* (consumers), or decomposers. Food chains and webs are beneficial resources for illustrating the interactions between different groupings. Decomposers transform dead organisms into inorganic elements, particularly mineral ions that can be used by plants, whereas producers are responsible for capturing the energy and recycling it to produce inorganic materials into organic matter. Populations comprise communities inside ecosystems. Energy moves through the ecosystem, and activities are regenerated through the interactions between these populations and the environment.

An *agroecosystem* is an ecosystem that has been altered for agricultural purposes. All of the interactions between biotic and abiotic elements in an agricultural region are included in agroecosystems. Agroecosystems can refer to a wide variety of habitats, including farms, individual fields, and entire ecozones. A key component of the food chain is the agroecosystem. Understanding how energy moves across ecosystems will help us build more efficient and thus more sustainable agroecosystems by reducing energy loss. To meet present economic and social requirements, sustainable development ensures that those of future generations can be met as well. The method by which matter can be recycled within an ecosystem is through food chains and food webs. The cycle of matter contributes to the maintenance of a sustainable ecosystem.

According to the older disease/pest triangle paradigm, a susceptible host, a virulent infection, and a conducive environment must all exist at the same time for a disease to manifest [1]. A "perfect storm" for insect infestations is created by monocultures' biologically deficient soils, a lack of plant diversity, and the use of pesticides. It is necessary to develop a new plant health model that can be applied to agroecosystems with improved biodiversity and low chemical transfer. Mutualistic above- and below-ground interactions between plants, insects, and soil microbial populations result in the overall protection of plants in such varied agricultural systems [2].

Restoring the balance and resilience of agroecosystems is the aim of treatments according to an agroecologically oriented approach to plant health if the source and appearance of a pest or disease are perceived as imbalances. This method focuses on the root causes of pest and disease outbreaks rather than only addressing symptoms, which may or may not result in healthier plants. By reorganizing and operating farm systems to restore the variety of "in-built" preventions and restructure key regulating functions, including immune (biological pest control mechanisms) and metabolic (activation of soil biotic activity, organic matter turnover, and nutritional cycling in crops) pest-resistant agroecosystems, pest-resistant agroecosystems can be created. This preventive approach is based on overlooked synergies between plant diversity and the soil microbial community, which are caused by the addition of organic matter and polyculture patterns, two essential techniques in the creation of pest-resistant and thriving agroecosystems (**Figure 1**) [2, 3].

Biological control can frequently be included in a farming system to enhance the effects of other control techniques, and it can occasionally rescue the industry when chemical pesticides have failed [4]. The creation of innovative bioinoculants that can provide affordable, rational, and enticing alternatives to expensive agrochemicals may benefit from the utilization of consortia, including multitrait microorganisms that promote plant development. The inoculant industry is confronted with several obstacles in its quest to develop higher-quality formulations that are affordable and easy to use for farmers while also providing a long shelf life and more viable and resistant cells in rhizosphere environments [5].

A sustainable approach to managing pests known as integrated pest management, or IPM, has been in practice for a long time. Previous models have focused mostly on the ecological and, to a lesser extent, evolutionary components of pest management, even though several sources describe integrated pest management (IPM) in different ways [6, 7]. IPM implementation is dependent on several factors. According to some

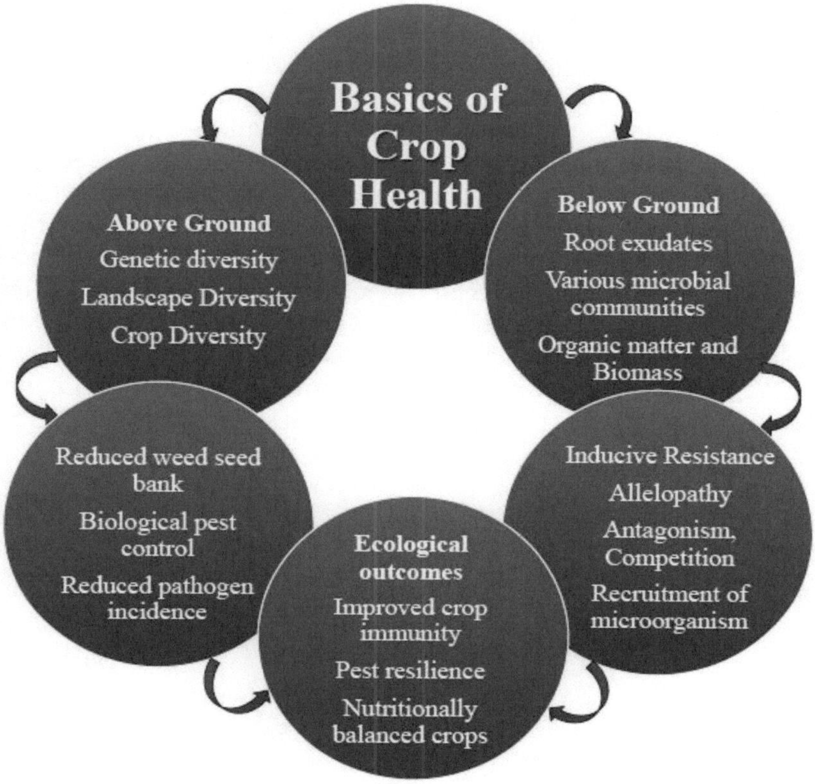

Figure 1.
The interactions between microbial community ecological activity and plant diversity were explored.

reports, some of the factors that may influence the uptake of IPM include education level, economic and social status, environmental consciousness, intellectual capacity, ethical practices, legal requirements, government policies and measures, the availability of IPM implements, extension education, consumer choice, and marketing strategies [8–11].

Biological control, often known as biocontrol, is the practice of managing pest populations through living organisms such as mites, insects, or fungal diseases. Some of the specialized natural enemies that aid in controlling the invasive species in their native habitat is returned to the playing field [12].

BCAs are naturally occurring living organisms, such as viruses, bacteria, fungi, insects, mites, nematodes, yeasts, and protozoa, and are also incorporated with the integrated pest management (IPM) technique, which distributes BCAs for the suppression of pests in a biological way [13]. The simplest level of parasitism occurs when predators kill and eat their prey, and insect parasitoids lay their eggs on their hosts, which are then consumed by offspring. Bacteria and viruses harmfully penetrate the body and destroy hazardous organisms; they are ingested by the latter, similar to entomopathogenic organisms, particularly fungi, which can breach the outer cuticle of insects and trigger systemic infection [14].

When a pest is growing at a faster rate than its natural enemies or when its populations start to rise in areas and times where natural enemies are not initially prevalent, augmentative parasitoid/predator release is an eco-friendly method of controlling pest populations [15]. Natural enemies can aid in the long-term growth of "low-prevalence" and "fly free" agricultural zones when combined with the sterile inspection technique (SIT) [16].

In a greenhouse system, if the resident population of natural enemies is insufficient to help decrease the number of pests, multiple releases of natural enemies are generally used as a biological control strategy. Both Inoculative Biological Control (IBC) and Conservation Biological Control (CBC) employ the "Predators and parasitoids-in-first" approach. There has been increasing concern about approaches that can facilitate the preventive introduction of biological control agents (standing army) [17]. As a result, two types of biological control, namely, Inoculative Biological Control (IBC), which focuses on the liberation of mass-reared natural enemies, and Conservation Biological Control (CBC), which seeks to preserve or improve naturally occurring biocontrol organisms, can be facilitated by providing natural enemies (**Figure 2**) [18].

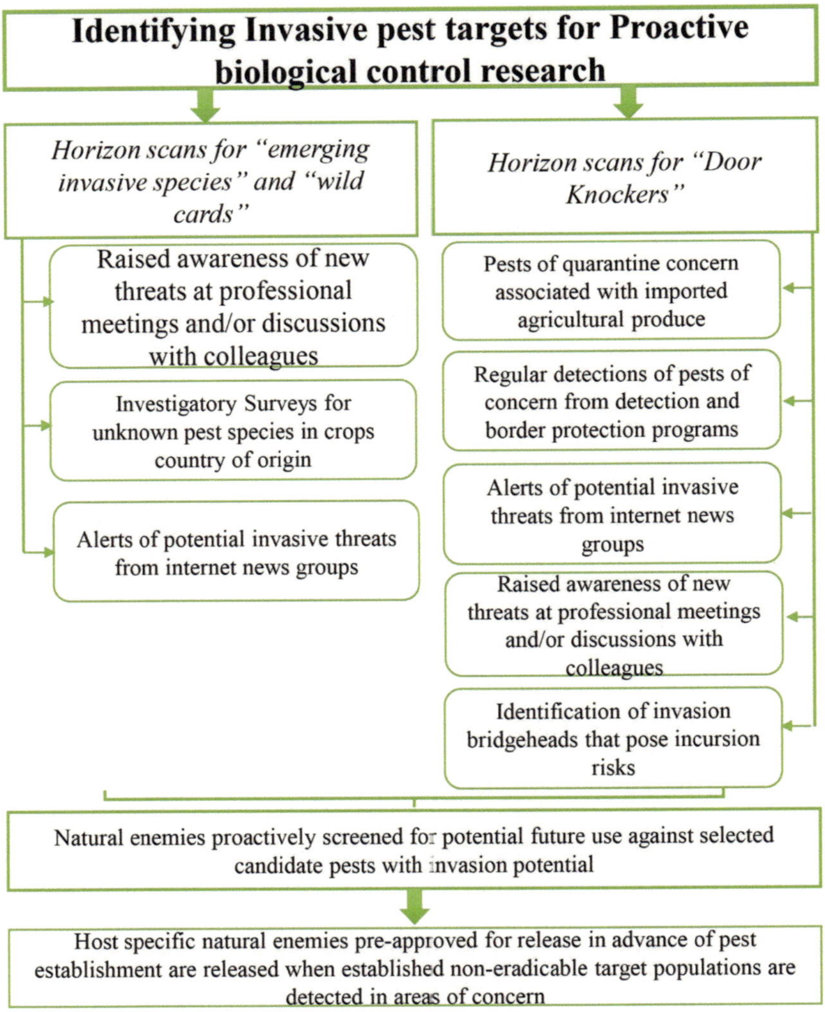

Figure 2.
Practical actions and outcomes when considering invasive pests for proactive biological control initiatives.

One of the categories of beneficial microorganisms that assist plants in disease reduction through nutrient competition is plant growth-promoting rhizobacteria (PGPR), which have siderophore molecules that inhibit infections in plants by competing for iron in the soil [19]. In this context, by enhancing the capacity of a plant's inherent immune system, mycorrhizal fungi, rhizobia, and a variety of growth-promoting rhizobacteria and fungi can also inhibit plant diseases. This process is known as induced systemic resistance (ISR) [20]. The competitive ability and potential for triggering resistance in rhizobacteria that belong to the genera *Pseudomonas* and *Bacillus* are largely due to their roles in the salicylic acid-dependent SAR pathway. To develop and evaluate more effective disease biocontrol techniques, many inoculants that cause resistance are being developed [21]. When specific pathogens are suppressed during agroecological transition, the use of such natural formulations represents an input substitution approach that should be employed, if necessary, while considering the risks to the functional biodiversity of the agroecosystem [2].

A wide variety of microorganisms, such as bacteria, fungi, protists, nematodes, and viruses, can grow and proliferate in the many niches that plants offer (the plant microbiome). Plant-associated microbiomes provide their host plants with benefits such as nutrient intake, growth promotion, stress tolerance, and disease resistance. Although microorganisms have long been used as inoculants for biocontrol or biostimulation, the effectiveness of these applications in the field varies depending on the soil type, climate, and other environmental conditions. Innovative methods that acknowledge that microorganisms in managed and wild systems are usually part of multispecies communities are needed to properly harness the microbiome. Although the explicit study of basic ecological mechanisms underlying the formation of complex microbial communities is still in the early stages, it is essential for the engineering and management of agricultural microbiomes [22].

Core IPM techniques, such as pest warning, field monitoring, record-keeping, and the selection of affordable, environmentally friendly management strategies, can help reduce the potential effects of climate change. Agricultural scientists can concentrate their research efforts on potential future solutions that can assist in minimizing and adapting to the hazards of projected climate change by providing a better understanding of how climate change affects different pests, especially crop pests [23].

2. Bacteria

2.1 Plant growth-promoting rhizobacteria (PGPR)

The primary functions of soil microorganisms (microbial ecosystem services) in agriculture are to improve soil quality and support plant health and nutrient availability [24]. Plant growth-promoting rhizobacteria are helpful bacteria that help plants grow more effectively in the context of environmental stressors by producing compounds that shield plants from infection. The indirect effects of PGPR include induced systemic resistance, antibiosis, and the generation of bioactive metabolites [25].

2.1.1 PGPR in plant growth

Atmospheric nitrogen can be fixed by PGPR and made available to plants. This nitrogen-fixing action is efficiently carried out by *Rhizobium, Enterobacter, Sinorhizobium, Gluconacetobacter, Bradyrhizobium, Mesorhizobium,* and *Frankia* [26].

A formulation of live or latent efficient strains of bacteria (PGPR) mixed with a carrier material is called a biofertilizer. It can be applied to soil, seeds, or plant surfaces. It can colonize the rhizosphere or interior of a plant, promoting growth by increasing the amount of primary nutrients available to the host plant [27]. The use of PGPR increased the levels of the soil microbial population, total nitrogen, available phosphorus, and biofertilizers in the form of organic soils [28].

Beneduzi et al. [29] suggested that the application of techniques that involve the antagonistic potential of rhizoplane bacteria and fungi can increase plant development, but identifying and describing the antagonistic activity is very difficult. The competitive nature of PGPR with other local bacteria for iron proves the capacity of PGPR to increase growth. Such mechanisms are vital for the operation of antagonistic bacteria through the principle of induced resistance. However, antagonistic and resistance-inducing PGPR can be used to produce new inoculants that protect plants against diseases and improve productivity without having adverse effects on the environment [29]. Additionally, a vast number of structural, intracellular, and extracellular polysaccharides are synthesized by specific bacteria. Biofilms are formed by another component called exopolysaccharides or EPS. As supplements, they assist in the capture of insoluble free phosphorous from insoluble phosphorous in soils. They also assist in the uptake of essential nutrients by plants for proper growth and development. Moreover, these EPSs also shield plants from pathogenic microorganisms associated with plant diseases [30]. Adedeji et al. [31] noted that the majority of PGPR synthesize antifungal compounds such as hydrogen cyanides or lysing enzymes that hamper the proliferation of fungi that provoke diseases (**Table 1**).

Bacteria	Mode of action	Phytopathogen	Crop	Disease
Fluorescent Pseudomonas	Inhibit reproductive phase of pathogens	*Phytopthora capsica*	Pepper	Foot rot disease
P. aeruginosa	Produce toxic volatile compounds	*Fusarium oxysporum*	Egg plant	Fusarium wilt
P. chlororaphis	Siderophores, extracellular antibiotics, production of volatiles	*Erwinia carotovora subsp atroseptics*	Sorghum	Charcoal rot of sorghum
P. aeruginosa	Inhibits the cell wall degrading enzyme polygalacturonase, cellulase).	*Sclerotium rolfsii*	Groundnut	Stem rot disease
P. fluorescens	Produces siderophore to inhibit the growth of phytopathogens	*Erwinia carotovora*	Potato	Bacterial stem rot
Pseudomonads sp.	Produces Pyoverdme siderophore	*Fusarium oxxsporum*	Potato	Wilt disease
B. subtilis	Produce siderophore for biocontrol of pathogens	*Fusarium oxysporum*	Pepper	Fusarium wilt of pepper
P. fluorescens	Produces Phenazine-1-carboxylic acid and cyclic lipopolypeptide	*Gaeumanomyces graminis*	Wheat	Take-all disease of wheat
P. fluorescens	Induced systematic resistance	*Rhizoctonia solani*	Rice	Sheath blight

Table 1.
PGPR as biocontrol agents.

2.1.2 Mechanism by which PGPR provides resistance against disease

i. PGPR produces phytohormones or secondary metabolites

Auxins, gibberellins, and cytokinins; these three organic substances, which are present in extremely small quantities, are called plant growth regulators, as they strongly influence the biochemical, morphological, and physiological metabolic processes in plants.

It is synthesized from gibberellin [32], abscisic acid [33], and indole acetic acid [34]. Some of the inputs indicated that the interaction between a PGPR and the plant is similar to that modeled by the interaction between rhizobia and legumes, including hormone-type signaling. Endogenous chemicals that PGPR releases undergo a sequence response mechanism known as systemic acquired resistance (SAR) [35]. For example, signal transduction, the formation of phytoalexins, lignification, and the response to oxidative stress are some of the features of SAR [36, 37]. Salicylic acid (SA), jasmonic acid (JA), and ethylene are pivotal for each of these defense systems. These hormones also increase plant growth. When plants are treated with phytopathogens, rhizobacteria help induce defense-related genes (pathogenesis-related proteins) and gene pathways that depend on SA [38]. However, with respect to *Rhizoctonia solani* and *M. phaseolina*, system resistance induced by *Pseudomonas aeruginosa* occurs through an SA-dependent pathway (**Figure 3**) [39, 40].

ii. Antimicrobial substances

Numerous antimicrobial compounds, including organic acids, lysozymes, bacteriocins, and broad-spectrum nonribosomal antibiotics, are produced by rhizobacteria

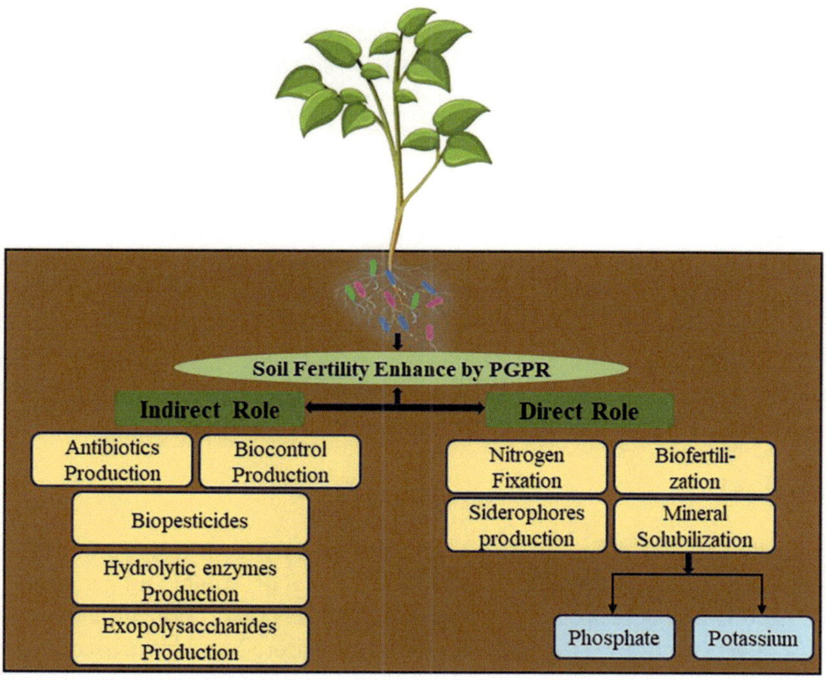

Figure 3.
Effects of PGPR on plant growth and metabolism.

that promote plant growth [41]. Bacteriocins are cationic peptides produced by ribosome synthesis that attach to the phospholipid membrane of the target strain to prevent the growth of the target organism. For example, the growth of phytopathogenic strains of *Pseudomonas* and *Xanthomonas* was inhibited by the bacteriocin generated by *P. fluorescens* SF39a, which was isolated from the rhizosphere of wheat [42].

To date, 18 distinct bacteriocins have been isolated and purified from *B. thuringiensis*. Among these include tochicin, entomocin110, morricin269, thuricin, and thuricin7 [43]. These substances stimulate the production of IAA, SA, and ABA, increase the photosynthetic rate, modulate protein activities, activate the antioxidant cascade, and alter host plant roots to increase the uptake of water and nutrients from the soil [41, 44].

iii. Antibiotics

Currently, PGPR is applied in integrated pest management. Rhizospheric PGPR are effective biocontrol inoculums of the host plant and are effective in controlling soil-borne diseases in the host plant due to the presence of fungi, bacteria, viruses, and nematodes. They also secrete antibiotics in combination with planta, which are oligopeptides that are specific in their function and actions. Antibiotics slow the formation of initiation complexes on the small ribosome subunit, thereby slowing protein synthesis in pathogen cells [45]. Antibiotics also belong to a class of compounds that inhibit several crucial enzymes, such as aldose reductase, which is the isomerase of glucose to fructose [46].

The following antibiotics are of PGPR origin: oomycin A, phenazine, pyoluteorin, pyrrolnitrin, tensin, tropolone, cyclic lipopeptide oligomycin A, kanosamine, zwittermicin A, 2-hydroxymethyl-chroman-4-one, phloroglucinols, and D-gluconic acid [47]. Compared with control, several species of Pseudomonas create 2,4-diacetylphloroglucinols that inhibit the growth of *R. solanacearum* and fungal pathogens such as *Rhizoctonia solani, Sclerotium rolfsii, Macrophomina phaseolina,* and *Fusarium oxysporum* [48].

iv. Siderophore production

Low-molecular-weight proteins called siderophores are created by PGPR to chelate iron from its insoluble form (Fe^{+3}) into its soluble form (Fe^{+2}) so that the host plant may use it. There are several types of siderophores, including mixed ligands, hydroxamate siderophores, and catecholate siderophores. The siderophore forms a strong bond with iron, limiting its availability to phytopathogens and preventing their proliferation. In soil contaminated with nickel, the inoculation of cowpeas with *P. fluorescens* and *S. acidiscabies* has dual effects [45].

2.2 Bacteria as BCA

Biological control is effective, and several microbial biocontrol agents (BCIs) have been formulated in recent decades for the control of fungal and bacterial diseases. Among all bacteria, *Bacillus* spp., *Streptomyces* spp., and *Pseudomonas* spp. are widely selected and investigated. Currently considered and identified as commercial products that have already been registered and marketed. Currently, in the EU, there are 13 bacterial-based biocontrol agents (BCIs) registered as biopesticides for the control of bacterial and fungal diseases (*Bacillus amyloliquefaciens* strains:

Bacillus amyloliquefaciens strains are approved for use in Europe by the European Food Safety Authority (EFSA) as biopesticides (QST 713, AH2, MBI 600, FZB24 and IT 45, *Bacillus amyloliquefaciens* subsp. *plantarum* strain D747, *Bacillus firmus* I-1582, *Bacillus pumilus* strain QST 2808, *Bacillus subtilis* strain IAB/BS03, *Pseudomonas* sp. strain DSMZ 13134, *Pseudomonas chlororaphis* strain MA 342, *Streptomyces* K61 and *Streptomyces lydicus* strain WYEC 108) (https://food.ec.europa.eu/plants/pesticides/eu-pesticides-database_en, accessed on 1 June 2022). However, efforts to expand the microbial biopesticides available on the market for the control of plant pathogenic diseases are still essential [49].

The efficacy of a bacterial biocontrol agent against plant diseases is also determined by the microbial agent, such as the mode of action, preparation, quantity, mode of application, plant pathogen target susceptibility, host plant cultivar type, physical properties, and environmental factors, which include chemical residues, nutrient availability, temperature, and humidity [50]. Many factors have been shown to influence biocontrol efficiency, including plant-to-plant variation, orchard and annual variation, with no efficacy or unstable field performance reported. Hence, it is important to understand the effectiveness and sustainability of biological control to conventional chemical fungicide and bactericide applications under sufficiently broad production conditions in orchards that represent different ecosystems and management systems for agriculture (**Figure 4**) [51].

Several measures are used by bacterial biocontrol agents to guard plants against invasion by pathogens. It can affect pathogens directly or utilize one or more stages to mitigate or reduce the severity of plant disease [51, 52]. The secreted antimicrobial chemicals of BCA can enter into physical contact with the pathogen and interfere with its pathogenic processes, as well as compete with the pathogen for resources and space. A large number of BCAs synthesize and emit substances that exhibit antimicrobial activity due to the suppression or even destruction of pathogens or their ability to develop. Other metabolites that are generated include bacteriocins,

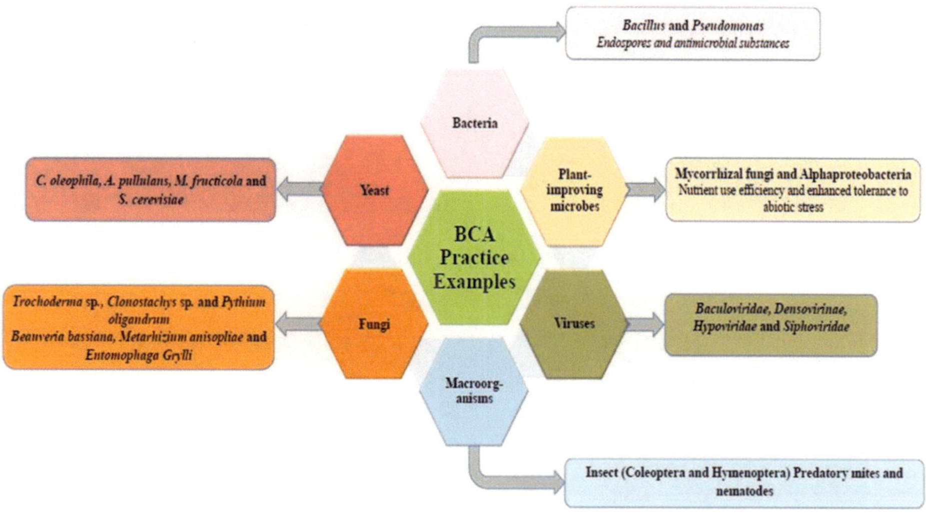

Figure 4.
BCAs are used in IPM as best practices to control a few important pests and diseases. Several mechanisms are used in these examples, such as nutritional competition, antibiosis, and mycoparasitism. By promoting tolerance or resistance to biotic or abiotic challenges, plant-improving microorganisms, such as microbial plant biostimulants, frequently increase plant health.

lipopeptides, biosurfactants, antibiotics, cell wall-degrading enzymes, and microbial volatile chemicals. In addition, BCA may interfere with the quorum sensing (QS) system of pathogens by hindering or hydrolyzing the signal molecules produced to initiate infections. The organic synthesis of QS inhibitors, such as chitinases, pectinases, and lactonases, can reduce pathogen infection and the severity of disease through the degradation of QS signaling molecules [53]. In addition, BCA limits pathogen proliferation, although it does not completely eliminate it; this acts as a competitive pressure that diminishes the prevalence of pathogen infection. In terms of nutrient intake at the infected site or tissue, highly competitive bacterial BCAs can outcompete the pathogen for essential nutrition by using low-molecular-weight siderophores that have an affinity for ferric iron. In addition, through direct antagonism, BCA can prevent or reduce plant infection indirectly by stimulating plant growth or activating defense mechanisms [51, 52, 54]. It might improve the mechanisms of the host, providing systemic resistance. This keeps the host defense shield against various infections, as it results in the formation of structural barriers and triggers several biochemical/molecular defensive responses. BCA can also enhance plant health through growth promotion *via* the production of hormones or through increased uptake of minerals and water. These complex interactions with pathogens, BCA, and plants are often achieved through multiple mechanisms simultaneously. Therefore, identifying the processes involved in biocontrol is very challenging. Since the process is natural, it very quickly becomes implausible to determine how and why biocontrol occurs. Understanding the mode of action in which a BCA provides protection is crucial for refining the strategy of biocontrol and establishing the right conditions for the BCA, the pathogen, and the host plant to occur. Therefore, it will be useful in designing proper formulations and application methods that improve plant health and contribute to sustainable agriculture (**Figure 5**).

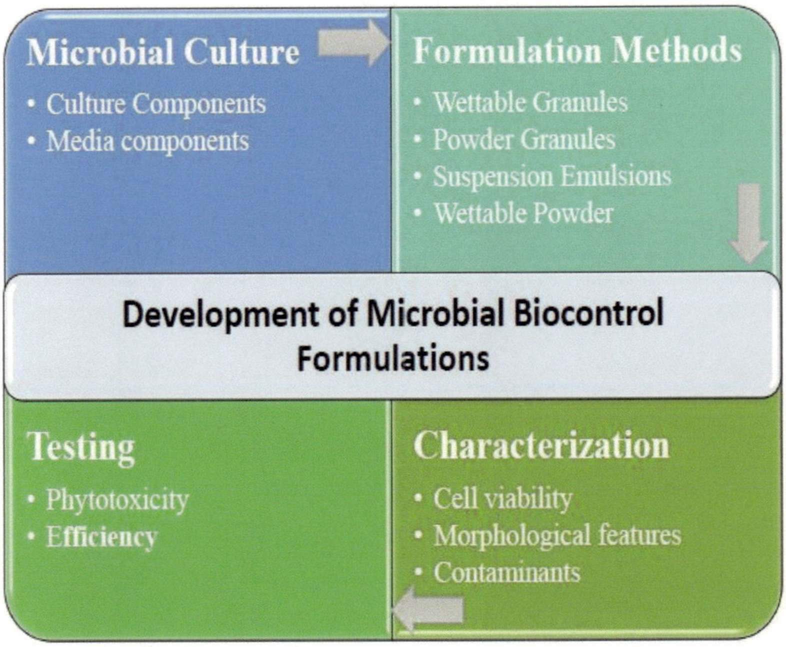

Figure 5.
General process flow for the steps required to create formulations for microbiological biocontrol.

Plant disease protection activity from a wide range of bacterial genera, such as *Agrobacterium, Alcaligenes, Arthrobacter, Bacillus, Enterobacter, Erwinia, Pseudomonas, Rhizobium, Serratia, Stenotrophomonas, Streptomyces,* and *Xanthomonas*, against fungal and bacterial pathogens has been reported. These bacteria can maximize the processes linked to the inhibition of plant pathogen growth. The induction of plant resistance, antagonistic activity based on the secretion of highly active antimicrobials such as antibiotics or cell wall lytic enzymes, colonization of infection sites, and competitive exclusion of the pathogen are some of these mechanisms of action [55].

3. Fungi

3.1 Fungi and yeasts as BCAs

Numerous fungal species have been recognized as both plant endophytic fungi and entomopathogens of insect pests [56]. A common genus of fungi, *Trichoderma*, is found in soil as saprotrophs, mycoparasites, and plant symbionts [57]. Owing to their ability to both compete with and parasitize phytopathogens (mycoparasitism), among other mechanisms of action [56–58], as well as mitigate unfavorable growth conditions, filamentous fungi in this genus have been thoroughly studied and used in agriculture as biocontrol agents against phytopathogens. Among several species that have been previously marketed as biopesticides or biofertilizers, *Trichoderma* stands out as the most studied fungal biocontrol agent [57, 58]. Over the last few years, there have been both direct and indirect conversations on the application of *Trichoderma* as a BCA for insect pests [56]. It has also been proven that *Trichoderma* directly manages insect pests through parasitism and the synthesis of insecticidal secondary metabolites, antifeedants, and repellents [56, 59]. By attracting natural enemies, plants stimulate plant responses and parasitize microorganisms associated with insects. For this reason, *Trichoderma* application in agriculture is a solution against insect pests and plant pathogens that shows great potential for increasing sustainable agriculture in the future [56]. From various *Agaricus bisporus* growth substrates from farms in Castilla-La Mancha (Spain), isolates of *Trichoderma aggressivum* were obtained. This fungus exhibited over 80% antagonistic activity against a range of phytopathogens. The most potent isolate, *T. aggressivum* f. *europaeum* TAET1, completely blocked the germination of *S. sclerotiorum sclerotia* and decreased the mycelial proliferation of fungal pathogens, including *Botrytis cinerea, Sclerotinia sclerotiorum,* and *Mycosphaerella melonis*, in a detached leaf assay. This isolate may be utilized in conjunction with different pest management techniques because of its exceptional compatibility with chemical fungicides [60]. The resistance of tomato seedlings to *A. alternata* leaf spot disease was enhanced by *Trichoderma asperellum* TaspHu1, an isolate isolated from *Juglans mandshurica* Maxim. rhizosphere soils in China [61]. Olive rhizosphere soils in northern Algeria are inhabited by *Trichoderma* spp., which have shown significant biocontrol capacity against *V. dahliae,* the wilting agent of olive plants. These isolates have shown an intriguing ability to inhibit the *in vitro* mycelial proliferation of this pathogenic fungus. The *Trichoderma* isolate T12 inhibited *V. dahliae* at a rate of 69% according to the confrontation assay [62]. It has been demonstrated that *Trichoderma* spp. efficiently suppresses Sclerospora graminicola, the pathogen that causes downy mildew disease in pearl millet. It has been clearly shown that they play a significant role in creating systemic resistance and indirectly inhibiting pathogens in the rhizosphere (**Figure 6**) [63].

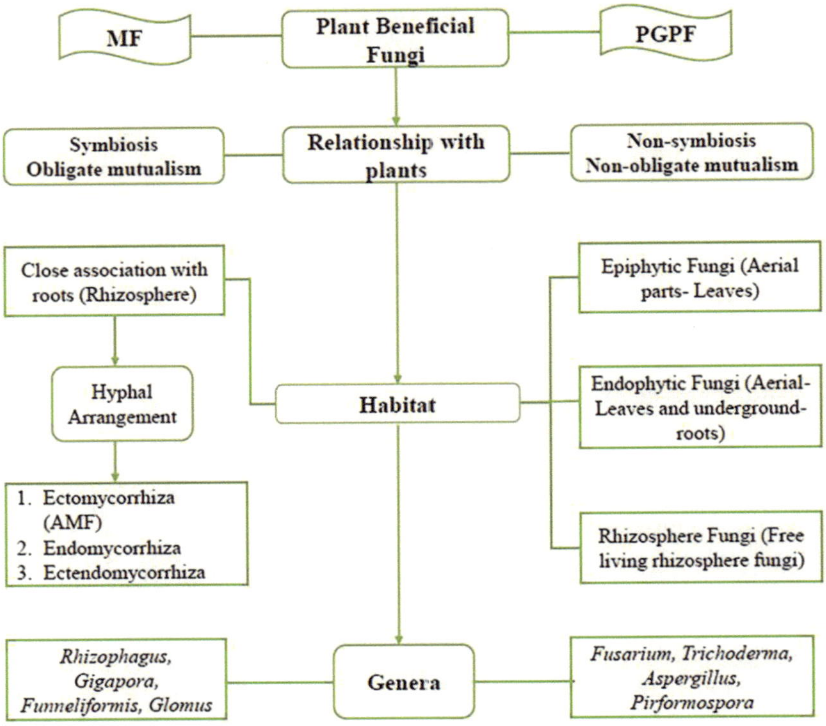

Figure 6.
Schematic illustration of the classification of beneficial fungi on the basis of their relationship with plants, habitat, nutrition, and hyphal arrangement. PGPF, plant growth-promoting fungi; MF, mycorrhizal fungi; AMF, arbuscular mycorrhizal fungi.

Furthermore, owing to their ability to regulate biocontrol, yeasts are frequently utilized in commercial formulations. They can already be used on a range of crops, such as citrus fruits and pome, after harvest. These microorganisms have been demonstrated to have considerable BCA activity due to their remarkable capacity for rapid colonization and suppression of pathogens, as well as their capability for nutritional competition, physical interaction, and the synthesis of lytic enzymes [64].

Fernandez-San Millan et al. [65] reported the effects of 69 yeast strains isolated from Spanish vineyards against soil-borne diseases such as *Fusarium oxysporum* and *V. dahliae*, as well as *Alternaria alternata*, *Penicillium expansum*, and *B. cinerea*. The goal of this study was to screen a possible yeast that has high biocontrol potential against wilt diseases and postharvest diseases. Therefore, the impact of diseases resulting from the fungi *V. dahliae* and *F. oxysporum* was reduced to 40% and 50%, respectively, through the intervention of *Wickerhamomyces anomalus* Wa-32. Furthermore, the results of postharvest tests for biocontrol effectiveness revealed high levels of biocontrol against *B. cinerea* and *P. expansum*, where the disease severity was reduced by 86% and 97%, respectively. Furthermore, two *Metschnikowia pulcherrima* strains, Mp-22 and Mp-30, exhibited good efficacy against *B. cinerea* along with *W. anomalus* Wa-32. In accordance with the results of the present study, *Candida lusitaniae* Cl-28, *Candida oleophila* Co13, *Debaryomyces hansenii* Dh-67, and *Hypopichia pseudoburtonii* Hp-54 were the most effective against *P. expansum*. This was also noted in the case of apples and grapes [66].

3.2 Mycorrhizae as BCAs

Numerous studies have demonstrated the ability of mycorrhizal associations to provide plants with resistance against a wide range of pathogenic diseases. These associations are then used to give plant roots prolonged defense against pathogens [67]. The formation of beneficial fungi in the root cortical tissue during plant growth and development, known as mycorrhizal associations, can be defined as creating an unfavorable microenvironment that prevents the spread of diseases. These mycorrhizal fungal associations serve an environmentally beneficial purpose as biocontrol agents, slowing the development of harmful diseases in plants. Numerous investigations have demonstrated that compared with nonmycorrhizal roots, tree seedlings with mycorrhizal relationships are more resistant to harmful fungi, bacteria, and nematodes when they feed on roots [68].

Ectomycorrhizae form the Hartig net, a structure similar to a net, on the surface of roots without entering inside the root cells. Through several strategies, including the production of antifungal drugs, antibiosis, and the formation of a fungal mantle that blocks the pathogen's entrance into plant roots, pathogens stop pathogenic attacks. Another crucial component of the microbial soil community that offers major advantages to plants is vesicular-arbuscular mycorrhizal fungus (VAM) [69]. In addition to enhancing the general development and growth of plants, VAM fungi provide host plant protection against pathogenic bacteria [70]. Inoculation with VAM fungi from the genus Glomus, such as *G. mosseae, G. fasciculatum, G. monosporum, G. constrictum*, and *G. macrosporum*, has been found to be relatively common to other fungi. They help reduce the risk of infection because they enhance the host plant's immune system and reduce the likelihood of pathogen invasion through the root. In addition, a previous study indicated that the occurrence of nematode attacks leading to root-knot infection is reduced [71]. Bacterial pathogens such as *Pseudomonas syringae* have been noted to drastically reduce tomato production by causing infections, which can be effectively reduced through the growth of mycorrhizal fungi on the host plant root [72]. The mycorrhizal association acts as a physical barrier and accomplishes some chemical reactions, causing some direct, in addition to indirect, effects [73]. The indirect effect enhances the potential of plants to take up nutrients, lignifies the roots, minimizes the effects of environmental stressors, and alters the microbalance of the mycorrhizospheric area for subsequent beneficial microbes [74]. As noted by Weng and Yan [75], AMF provides their host plants with systemic resistance, through which most pathogenic diseases are reduced. On the basis of the literature, the depletion of intermediate compounds synthesized in the SA-dependent defense pathway and the active depletion of components in the SA-dependent defense pathway lead to mycorrhizae-induced resistance (MIR), eliciting systemic priming of jasmonic acid-dependent defenses [76].

AMF effectively improves soil structure and fertility, plant growth and production, and regulates biotic stress [77]. AMFs are considered potentially effective biocontrol agents since they are widespread in agricultural and other terrestrial ecosystems [78]. AM fungi and plants are associated with this special type of mycorrhizal symbiosis that occurs in most terrestrial plant communities [79].

Considering that plants have mycorrhizas instead of roots and that more than 90% of flowering plants create symbiotic associations with these AMFs would not be an exaggeration. In normal cropping systems and natural ecosystems, the most common type of AMF–plant association is mycorrhizal [80]. To complete its life cycle, AMF forms an extraradical mycelium by colonizing the root cortex. This mycelium

Figure 7.
Working steps of sustainable practices in agriculture.

provides a microenvironment in the soil that aids in the plant's uptake of water and minerals. Extraradical mycelia interact with other soil-borne microorganisms and modify microbial communities to increase soil health. AM symbioses are also crucial in the ecological nutrition cycle [81]. The area of soil where roots are colonized by mycorrhizal fungi is known as the mycorrhizosphere. Unlike the rhizosphere, the mycorrhizosphere promotes microbial activity and helps with nutrient absorption.

Similar root tissues are used by AMF and soil-borne plant diseases; if colonization takes place simultaneously, there may be direct competition for these spaces. If AM fungi and plant diseases colonize the same host tissues, there may be competition for space because they both typically originate within distinct cortical cells of roots. Studies have shown that *Phytophthora* development is inhibited in AM-colonized and adjacent noncolonized root systems and that the pathogen is unable to enter precolonized root cells. The mycorrhizosphere, which is created by the interaction of plant roots and AMF, affects the soil microenvironment and promotes the growth of other beneficial microorganisms that are hostile to soil-borne diseases and their proliferation (**Figure 7**) [82, 83].

4. Botanicals such as the BCA

Plant products that are naturally occurring have considerable potential for application in modern agriculture. Certain plant products include biochemically active chemicals that are efficient against a range of pests and insects [84]. In scientific terms, the active chemicals that are removed from plants and successfully applied in integrated pest management are referred to as botanical pesticides. Owing to their significant contribution to reducing crop pathogenicity, their use is becoming widely acknowledged in modern agriculture [85].

4.1 Commonly used botanicals

Plant-based extracts: Plant extracts are intricate mixtures of substances from several chemical classes that possess bioactive characteristics. Traditionally, they are made from the entire plant as well as its constituent components, including the rhizome, root, stem, leaf, flower, tissues, and cell cultures [86]. A solvent is used to extract this complex collection of bioactive chemicals from plant or plant components. Water, acetone, acetic acid, ethanol, ethyl acetate, methanol, and hexane are among the most commonly used solvents. Importantly, the solvent selected and the experimental setup may have a significant influence on the extraction yield and composition of the constituents [87]. Many plants are rich sources of botanical extracts, including neem, garlic, eucalyptus, turmeric, ginger, and tobacco [88].

Essential oils: Essential oils (EOs) are viscous, aromatic liquids extracted from a variety of plant parts, including roots, flowers, buds, leaves, twigs, bark, herbs, wood bark and fiber, fruit, and seeds [89]. The general techniques used to obtain oils include enfleurage, expression, fermentation, and steam distillation [90]. Many different kinds of essential oils have been identified as having some proven botanical potential, including nettle oil from *Urtica* spp., thyme oil from *T. vulgaris,* eucalyptus oil from *E. globules*, rue oil from *R. graveolens*, lemongrass oil from *C. flexuosus*, and tea tree oil from *Melaleuca alternifolia* [91]. *Gel and latex*: Aloe vera is a veritable source of gel and latex.

4.2 Secondary metabolites and their mechanism of action

SMs are generally organic compounds made primarily by plants, fungi, and bacteria. These are low-molecular-weight compounds with a variety of chemical structures and biological activities [92, 93]. These substances are either nonfunctional or not necessary for growth, development, or reproduction [94]. For an organism to remain healthy, perform its primary functions, and survive, secondary metabolites must be synthesized. Owing to their wide range of applications and functions, secondary metabolites are used in the pharmaceutical, medical, and agricultural fields in addition to other industries that produce pigments and surfactants [94, 95].

Phenolics bind to CAMs, interfere with the membrane, decrease the substrate concentration, and create complexes with the cell wall that alter enzyme configuration and integration. The cell membranes and walls are destroyed by *terpenes and oils*. *Alkaloids and flavonoids* are able to cross biological membranes. *Tannins and coumarins* interact with DNA, lectins, polypeptides, and other biological components. They also generate disulfide bridges in proteins and bind to many biological components. Certain plants, such as Allium sativum (garlic), Tagetes erecta (marigold), and Chenopodium ambrosioides (epazote or goosefoot), have the remarkable ability to regulate the population of parasitic insects and exhibit antiparasitic characteristics [88].

Another type comprises bioactive compounds from plants, plant metabolites, volatiles, and enzymes from specific microorganisms. These compounds may be used as fungicides to regulate important fungal diseases that occur after harvest [96–98].

The majority of the defense of medicinal plants against antifungal activities lies in alkaloids, phenols, flavonoids, and terpenoids [96]. In parallel, some microbes can also act negatively against fungal pathogens through direct competition with them for nutrients such as iron, biofilm formation on their host or fruits, and/or the production of volatile and toxic metabolites in addition to enzymes such as glucanases, chitinases, proteases, and hydrolases that suppress pathogenic fungi [99, 100]. Jiménez-Reyes and Carrasco [101] presented comprehensive insight into the antifungal effects of plant extracts, essential oils (EOs), and secondary metabolites [102].

4.3 Plant extracts with different modes of action: Fungicides and antimicrobials

Certain plants can break down the fungal cell membrane at various phases of fungal growth, which affects the enzymatic activity and metabolic processes of fungal cells and makes them appropriate fungicides. Plant extracts can increase the levels of certain proteins that can suppress various fungal infections. These substances strengthen the defenses of plants against fungal invasions by strengthening their immune systems [88, 103, 104].

5. Nanotechnology

The connection between the physical sciences and the biological sciences is nanotechnology. It investigates the creation, production, assembly, and characterization of materials smaller than 100 nanometers, as well as the use of reduced-size, practical frameworks derived from these materials [105]. Nanotechnology is predicted to play a massive role in the agricultural sector. This sector is known as nanoagribusiness, which suggests that this development is regularly used to increase yield [106]. The manufacturing of nanoparticles from plants is a simple and environmentally beneficial way to decrease toxicity, time, and expense. Similarly, when applied to mung beans (*Vigna radiata*), Ag NPs derived from neem (*A. indica*), dark tulsi (*Ocimum tenuiflorum*), and banana (*Musa balbisiana*) significantly increase shoot and root lengths [107].

5.1 The role of nanoparticles in plant protection

NPs have been shown to increase crop productivity and shield plants from a number of diseases. Crop diseases are managed by a variety of techniques, including integrated pest management, novel insecticides, sanitation programs, and genetic breeding. Nanotechnology has offered new perspectives for enhancing and adjusting current crop management techniques. Chemicals and fertilizers for plant protection are typically applied *via* methods such as broadcasting and spraying. However, leaching, hydrolysis, and microbiological processes prevent minimal amounts of chemicals and nutrients from reaching the target site.

More than 90% of the pesticides employed for pest management are either lost in the environment or cannot reach the intended sites [108]. The use of pesticides has increased costs and has resulted in the overall degradation of the climate. Thus, the creation of nanoformulations, or the encapsulation of pesticides, was a better initiative that was required in the agricultural sector to protect plants from microbial diseases. These nanoformulations contain insecticides as the active ingredient and only a small number of microscopic particles. Plant diseases caused by different phytopathogens have been managed through the use of carbon, silver, silica, and alumina silicate nanoparticles.

Better results are obtained by the optimized nanoformulations, as they promote the regulated entry of dynamic fixes into the plant's targeted zones. According to Dwivedi et al. [109], there are fewer issues with traditional pesticides, such as low solubility, increased resistance, and nanoformulations. Therefore, the importance of nanotechnology has increased significantly to achieve greater harvest. According to a study published in 2017 by Petosa et al. [110], pesticide nanoformulations increased crop output by improving pesticide efficacy by controlling pesticide transport potential. The authors enriched polymeric nanocapsules by adding pyrethroid bifenthrin (Ncap-BIF), and the latter turned out to be a great carrier for plant protection. Since trypsin is a viral protease, its catalytic activity was inhibited by the synthesized bioactive AuNPs, resulting in excellent performance in controlling insects. These changes in catalytic activity were attributed to the fact that proteins bonded with the metallic nanoparticles [111]. For example, zinc oxide nanoparticles effectively reduce the growth of *Penicillium expansum*, *Alternaria alternata, Rhizopus stolonifer, A. flavus, Fusarium graminearum*, and pathogenic bacteria [109, 112]. Moreover, Si and TiO_2 have been shown to control crop diseases through their antifungal properties. Pesticides, herbicides, and fungicides constitute one of the most significant aspects of sustainable agricultural development, where nanomaterial encapsulation plays a significant role.

5.2 Nanoparticles improved plant yield

The use of nanoparticles has increased crop productivity and profitability. A novel approach based on the application of nanoparticles has been widely used in the agricultural industry to solve issues with crop productivity and efficiency. Nanotechnology in agriculture may increase global food production, enhance food quality, screen for plant growth, identify diseases in plants, provide protection for plants, and reduce waste [10, 113–117].

The plants hereditarily elicited by nanomaterial-based compounds can play an important role in increasing agrarian productivity [90]. Tomato seed experiments demonstrated that the germination efficiency of seeds enriched with inserted carbon nanotubes (CNTs) increased many-fold because CNTs increased the water absorption ability of seeds [118]. The emergence of spinach increased with the addition of TiO_2 nanoparticles, which increased the activity of the Rubisco enzyme and light absorbance [119]. It was also observed that nitrogen metabolism was activated by TiO_2 nanoparticles; therefore, the growth of spinach improved. DeRosa et al. [120] reported that the germination of corn and ryegrass seeds was reduced by ZnO nanoparticles. The application of ZnO-based nanomaterials affects some porous areas in the root systems of these plants and hence affects nutrient delivery systems [121].

6. Others

6.1 Biochar

Biochar is characterized by a high carbon content and is obtained through pyrolysis; this material is widely used for carbon capture and the enhancement of soil properties [122]. One of the mechanisms by which it seems to exert positive effects is by altering the composition of microorganisms in the soil. Owing to the large surface area contributed by pores, biochar serves as a conducive environment for microbial activities [123]. The use of biochar for soil amendment to reduce negative plant–soil feedback affects the bacterial and fungal communities of the soil, induces the proliferation of beneficial bacteria such as *Bacillus* and *Lysobacter*, and depletes the density of plant pathogens, including *Rhizoctonia solani, Fusarium,* and *Ilyonectria* [124–127]. Biochar can also improve the root microbiome of PGPR and fungi, which helps reduce the susceptibility of plants to soil-borne bacteria as well as fungal and airborne diseases [125, 126, 128–131]. Despite the useful application of biochar in agricultural fields, some problems related to the application of biochar exist, such as its long-term safety and impact. With respect to the impact of biochar on soil biota, the overall strategies have not been completely explained or assessed qualitatively. Additionally, the physical and chemical properties of biochar differ according to feedstock and production parameters; thus, assessing the impacts of biochar over the long term is not easy [132]. More studies are needed to identify whether the use of biochar in the provision of fertilizers has positive or negative impacts.

Biochar can also help lessen climate change by promoting the long-term storage of carbon and regulating the cycling of other powerful greenhouse gases in the soil, including nitrous oxide and methane; hence, it has an impact on the emission of such strong greenhouse gases [133, 134]. Thus, biochar has spread around the world as a potential and efficient tool for resolving food security and environmental problems in the field of agriculture.

6.2 Chitin and its derivative chitin, a polymer of N-acetylglucosamine

It is the main component of the exoskeletons of arthropods and the cell walls of fungi. Chitin and its derivatives, chitosan, and oligosaccharide derivatives are other products that have multiple uses, including in the purification of agricultural and industrial wastewaters and urban homes [135], the sequestration of phosphates and hydrosulfides from agricultural washings and the addition of additives to soils and fruits for the control of crop pathogens [136]. These substances have multiple effects on viruses, bacteria, fungi, oomycetes, and nematodes, as well as the activation of defense mechanisms in plants [137]. Research has shown that chitosan is an effective biocontrol agent against the formation and growth of soil-borne pathogens such as *V. dahliae* and K. aphicola, postharvest pathogens, and pathogens of thousands of plant species such as *C. gloeosporioides, B. cinerea*, and *R. stolonifer* [138]. These effects are achieved through processes such as disruption of the cell membrane, changes in the shape and size of the cell walls, and interference with normal cell metabolic functions. Additionally, chitosan enhances bacterial chitinase production and alters the number of members of the *Actinobacteria* and Oxalobacteraceae families that play critical roles in chitin decomposition in the soil. This leads to improved soil suppression of plant diseases [139]. However, even though chitin did not directly affect the resting spore population, the application of this crab shell (23% chitin) to infested soil resulted in a reduction in potato wart disease infestation, a disease caused by the obligatory soil-borne fungus *Synchytrium endobioticum* (**Figure 8**) [50].

6.3 RNA interference (RNAi)

Studying the interactions between plants and pathogens has made it possible to gain a better understanding of the molecular processes that define disease and/or

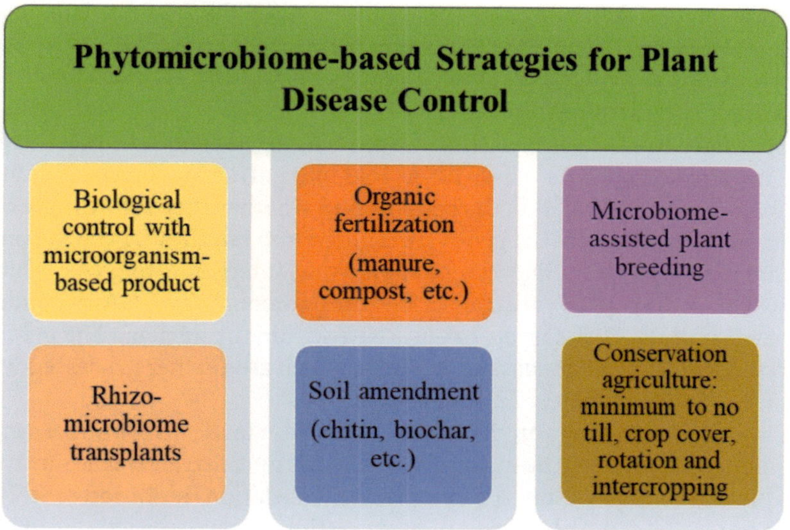

Figure 8.
Framework of phytomicrobiome-related agricultural practices that may lead to plant disease suppression and management.

resistance, as well as to create new methods for managing plant diseases. There have previously been several breakthroughs with RNA interference (RNAi) techniques, and current research is opening new prospects for the management of plant diseases.

In eukaryotes, RNA interference is a natural method of gene control. It starts with the sequence-specific identification of mRNAs and ends with the repression of their expression by translation inhibition or ribonuclease-directed destruction. The two primary RNAi pathways in plants involve two types of short single-stranded RNAs (~21–24 nucleotides) called miRNAs and siRNAs. The plant genome encodes miRNAs, which primarily regulate gene expression. The genome can also be used to derive siRNAs, but plants have evolved an RNA surveillance system that can identify and degrade particular RNAs, producing siRNAs. This RNA surveillance system in plants also serves as a strong antiviral defense system. The identification of RNAs for degradation is the first step in the plant-specific siRNA process. These RNAs, which are detected in the cytoplasm of the cell, are frequently dsRNAs or highly structured RNAs. The ribonuclease activity of Dicer-like proteins (DCLs) targets and cleaves these RNAs.

6.3.1 Mechanism behind RNA interference

According to Gasciolli et al. (2005), *Arabidopsis thaliana* has four DCLs (DCLs 1, 2, 3, and 4) that cleave dsRNAs or structured RNAs into short RNA duplexes, which are usually composed of 21, 22, 23, or 24 nucleotides. Different plants can have different DCLs [140]. The Argonaute protein (AGO), a component of the RNA-induced silencing complex (RISC), is now recruiting tiny RNA duplexes produced by the DCL, which is an exquisite feature of the RNAi response. AGO retains one strand of the short RNA duplex, which is the passenger strand, and is largely broken down; the other strand functions as a guide RNA. The attached RNA is degraded when the RISC carrying the guide RNA comes into contact with an RNA that is identical to the guide

RNA (by Watson–Crick base pairing). This occurs when the AGO initiates ribonuclease cleavage. Moreover, the RNAi response in plants can be amplified by endogenous and virus-encoded RNA-dependent RNA polymerases [140]. As a result, siRNAi action is strong and extremely specific.

Primary micro RNAs (pri-mi RNAs), or specialized noncoding RNAs, are the source of most plant miRNAs. These RNAs fold into particular structures. On the basis of where they are located in the genome and whether they are intronic or intergenic, many MIRNA (MIR) genes are divided into families. While intronic miRNAs are extracted from RNA transcript introns, intergenic miRNAs, which are found between coding genes, are transcribed autonomously by DNA-dependent RNA polymerase II (Pol II). DCL targets pri-miRNAs within the nucleus. The pri-miRNAs are converted by DCL into precursor miRNAs, which are further processed into the miRNA duplex (approximately 22 nt) in the nucleus.

There are three overhangs on the duplex, which are made up of the passenger and guide strands. Like siRNAs, the mature miRNA (the guide strand) loads into the RISC once the duplex is exported into the cytoplasm. Translation inhibition or RNA degradation can be the outcome of the complex's targeting of particular RNA sequences by the mature miRNA coupled to the RISC. Therefore, the miRNA and siRNA pathways are inherent to plants. They can play significant roles in host defense in addition to their vital roles in controlling gene expression. The specifics of these pathways are well described in several outstanding reviews [141–143].

Single-stranded RNA comprises the genetic material of most viruses that infect plants, and most viruses use dsRNA intermediates to replicate their genomes in the cytoplasm of plant cells. Preprogramming plants to have an active RNAi defense against a specific viral RNA before the plant is challenged by that virus is thus a conceptual strategy that might be utilized to confer resistance against plant viruses. The virus would then be confronted by the plant's already active RNAi, which would detect and destroy the entering viral RNA before it could replicate and become established. Fortunately, it is possible to genetically modify plants to express and carry additional genes, allowing for the production of plants with targeted antiviral RNAi-based resistance.

For numerous viruses in numerous plant species, RNA interference (RNAi) has been shown to be a successful method for managing plant virus infections [143, 144]. Plants are typically modified to produce dsRNAs or structured RNAs, including hairpin RNAs, in place of untranslatable RNAs because these are potent RNAi inducers [145]. Relatively few genetically modified virus-resistant agricultural plants have obtained the required regulatory permission for use in commercial agriculture, despite the possibility of engineering antiviral immunity in situations in which no natural resistance sources are known [143, 146]. This is most likely caused by other factors, such as the costs of the intellectual property used for plant genetic engineering, governmental deregulation/approval costs and activities, and, in general, the public perception of genetically engineered plants [147]. This is not because there is no need for them or because they are unlikely to be effective. However, the advantages of using transgenic RNAi-based antiviral resistance in agricultural systems have been striking thus far. The adoption of genetically modified papaya ringspot virus (PRSV)-resistant papaya (*Carica papaya*), which was formed and employed in Hawaii when no other means of controlling PRSV were available, is one of the biggest success stories [148]. Because of this, the Hawaiian papaya sector has been practically preserved, and the transgenic papaya that is resistant to viruses currently controls the majority of the market. The ability to simultaneously develop resistance against several viruses is another advantage of RNAi-based transgenic resistance. RNA interference (RNAi) is a promising solution for managing plant virus infections and enhancing crop security in basic foods [149].

6.3.2 Biological control of insect pests

Eco-friendly biological agents have been created to ensure biological control because the use of pesticides has been linked to problems with the environment, ecology, and human health [150]. The most significant mechanism limiting damage to plants in natural ecosystems has historically been thought to be the direct consumption of pests, phytopathogens, and weeds that cause trophic cascades. It includes a number of techniques and is frequently considered the most significant type of biological control in agriculture. For example, while insect parasitoids deposit their eggs on or into their hosts, which are then consumed by the immature offspring, predators kill and eat their prey (such as pests or weed seeds) (**Figure 9**) [13].

Certain EPs are produced commercially *in vivo* (viruses and nematodes) or *in vitro* (fungi, bacteria, and nematodes) [151]. In certain circumstances, they are also produced on a small scale for local, noncommercial use [152]. One significant component of integrated pest management (IPM) for a range of pests may be the use of EPs as biopesticides. To manage pests over the long term, exotic microorganisms are imported and released; this is a traditional microbial control method that has been and is being used with EPs [153].

- *Entomopathogenic bacteria*

The genera *Pseudomonas, Photorhabdus, Xenorhabdus, Serratia,* and *Yersinia* contain non-spore-forming entomopathogenic bacteria, whereas the genera *Bacillus, Clostridium,* and *Paenibacillus* contain spore-forming bacteria. When susceptible insect hosts consume these EPs, the hosts become infected.

In its stationary development phase, Bt, a gram-positive soil bacteria, creates δ-endotoxins or Cry pro-toxins, which are parasporal bodies composed of crystals and tightly packed proteins that kill specific insects. After the bacterial walls are lysed during the sporulation phase, these proteins are released into the environment and, once consumed by insect pests, are effective against Lepidoptera (butterflies),

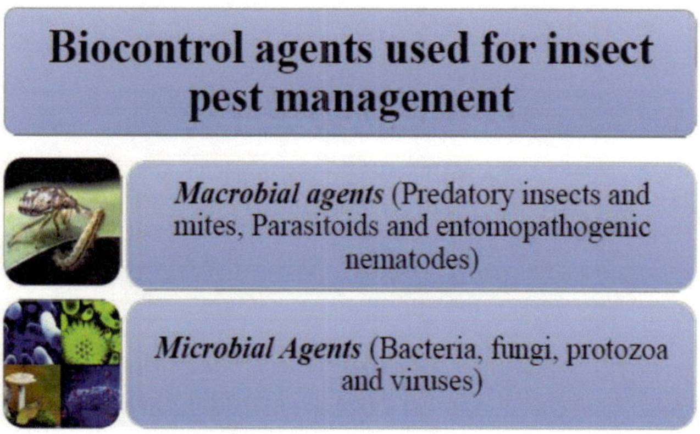

Figure 9.
An illustration of the biocontrol agents used in the management of insect pests.

Diptera (flies), and Coleoptera (beetles) larvae [154–157]. When Bt is consumed, the alkaline environment of the insect gut triggers a toxic protein (δ-endotoxin), which binds to midgut receptor sites and induces the formation of holes in midgut cells. Cell lysis, paralysis of the midgut, and loss of osmoregulation follow this process. The pH balance is interrupted when the stomach contents seep into the hemocoel and the hemolymph seeps into the gut. Bacteria that penetrate the hemocoel of the host insect result in septicemia and ultimately death. Depending on the crystal proteins, receptor locations, generation of other toxins, and spore needs, insects react to Bt toxins in different ways. With fermentation technology, Bt is manufactured commercially all around the world and sold in a range of forms, including dust, wettable powder, emulsifiable concentrate, etc. [158].

- *Entomopathogenic fungi (EPF)*

Fungi can be divided into eight major classes, four of which contain EPF, *viz*. Microsporidia, Ascomycota, Zygomycota, and Basidiomycota [159]. Zygomycota and Deuteromycota are significant EPFs with broad information [160]. To control crop insect pests, the use of entomopathogenic fungi (EPFs), such as *Beauveria, Metarhizium, Lecanicillium,* and *Isaria,* has increased recently. Because EPFs are target-specific, environmentally friendly, and safe for people, they are regarded as superior to synthetic insecticides. Long-term costs are lower, they have fewer side effects, and they can address resistance issues (**Figure 10**).

When conidia adhere to the host epicuticle, the EPF infection process begins. The conidia germinate, and the appressoria form on the insect's body after attachment.

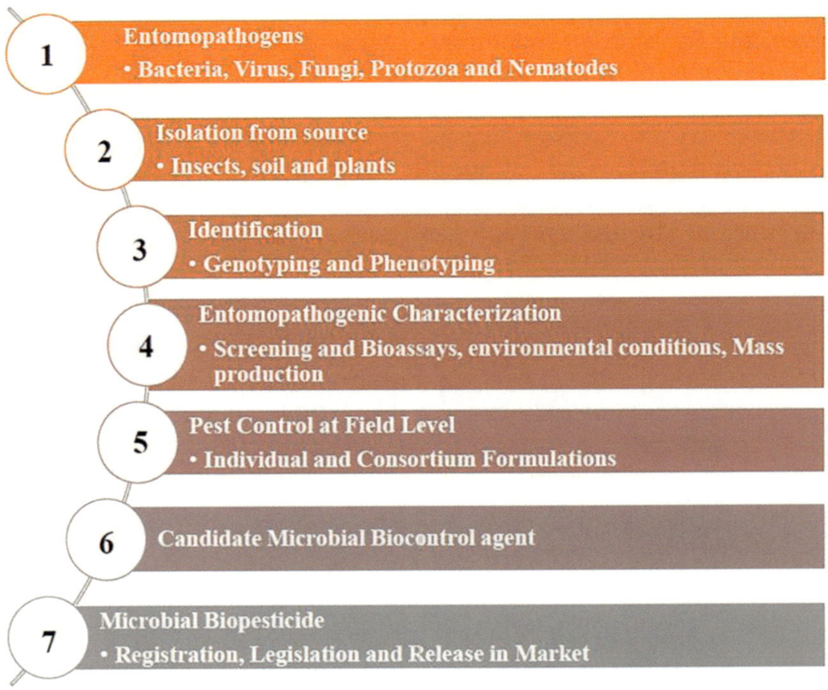

Figure 10.
Entomopathogens as microbial biocontrol agents.

After entering the procuticle, the mycelium enters the hemolymph and forms blastospores. Once the host insect has invaded and killed, the fungus reverts to mycelia and breaks through the cuticle. During cuticle disintegration, genes that were initially utilized for invasion and penetration are repurposed. Conidia spread from one host to another with the aid of abiotic elements such as wind and water [161]. EPF is mostly used as an agent for inundative biocontrol. EPF insect control mechanisms, such as those of B. bassiana, induce systemic resistance, competition, and antibiosis [162]. EPF uses a combination of mycelial growth, invasion, and toxin synthesis to either kill or incapacitate insect pests.

Author details

Zill-e-Huma Aftab[1*], Tehmina Anjum[1], Waheed Akram[1], Najat A. Bokhari[2], Areeba Rehman[3], Shama Sharif[1], Humaira Rizwana[2] and Bareera Munir[3]

1 Department of Plant Pathology, University of the Punjab, Lahore, Pakistan

2 Department of Botany and Microbiology, College of Science, King Saud University, Riyadh, Saudi Arabia

3 College of Earth and Environmental Sciences, University of the Punjab, Lahore, Pakistan

*Address all correspondence to: huma.dpp@pu.edu.pk

References

[1] Vega D et al. Agroecological management of crop diseases: A review. Agroecology and Sustainable Food Systems. 2023;**47**(7):919-949

[2] Altieri MA et al. Toward an agroecological approach to crop health: Reducing pest incidence through synergies between plant diversity and soil microbial ecology. npj Sustainable Agriculture. 2024;**2**(1):6

[3] Wyckhuys KAG, Tang FHM, Hadi BAR. Pest management science often disregards farming system complexities. Communications Earth & Environment. 2023;**4**(1):223

[4] Tanda AS. Advances in Biological Control Pest Management Technology. IK International Pvt Ltd. New Delhi, India: The Federation of Indian Publishers; 2024

[5] Ijaz M et al. Development of future bioformulations for sustainable agriculture. In: Microbiome in Plant Health and Disease: Challenges and Opportunities. New York: Springer Nature; 2019. pp. 421-446

[6] Dara SK. The new integrated pest management paradigm for the modern age. Journal of Integrated Pest Management. 2019;**10**(1):12

[7] Peterson RKD, Higley LG, Pedigo LP. Whatever happened to IPM? American Entomologist. 2018;**64**(3):146-150

[8] Jayasooriya HJC, Aheeyar MMM. Adoption and factors affecting on adoption of integrated pest management among vegetable farmers in Sri Lanka. Procedia Food Science. 2016;**6**:208-212

[9] Lefebvre M, Langrell SRH, Gomez-y-Paloma S. Incentives and policies for integrated pest management in Europe: A review. Agronomy for Sustainable Development. 2015;**35**:27-45

[10] Parsa S et al. Obstacles to integrated pest management adoption in developing countries. Proceedings of the National Academy of Sciences. 2014;**111**(10):3889-3894

[11] Rezaei R et al. Drivers of farmers' intention to use integrated pest management: Integrating theory of planned behavior and norm activation model. Journal of Environmental Management. 2019;**236**(1):328-339

[12] Hoddle MS. A new paradigm: Proactive biological control of invasive insect pests. In: BioControl. New York: Springer Nature; 2023. pp. 1-14

[13] Stenberg JA et al. When is it biological control? A framework of definitions, mechanisms, and classifications. Journal of Pest Science. 2021;**94**(3):665-676

[14] Galli M et al. Can biocontrol be the game-changer in integrated pest management? A review of definitions, methods and strategies. Journal of Plant Diseases and Protection. 2024;**131**(2):265-291

[15] Kota S et al. The role of natural enemies and biopesticides for sustainable management of major insect pests of legumes. In: Biopesticides in Organic Farming. Boca Raton, London New York: Taylor and Francis Group; 2021. pp. 207-216

[16] Cancino J et al. Application of nuclear techniques to improve the mass production and management of fruit fly parasitoids. Insects. 2012;**3**(4):1105-1125

[17] Messelink GJ et al. Approaches to conserving natural enemy populations in greenhouse crops: Current methods and future prospects. In: BioControl. Vol. 59. New York: Springer Nature; 2014. pp. 377-393

[18] Pijnakker J et al. Predators and parasitoids-in-first: From inundative releases to preventative biological control in greenhouse crops. Frontiers in Sustainable Food Systems. 2020;**4**:595630

[19] Abdelaziz AM et al. Biocontrol of soil borne diseases by plant growth promoting rhizobacteria. Tropical Plant Pathology. 2023;**48**(2):105-127

[20] Pieterse CMJ et al. Induced systemic resistance by beneficial microbes. Annual Review of Phytopathology. 2014;**52**(1):347-375

[21] Olowe OM, Akanmu AO, Asemoloye MD. Exploration of microbial stimulants for induction of systemic resistance in plant disease management. Annals of Applied Biology. 2020;**177**(3):282-293

[22] Trivedi P et al. Plant–microbiome interactions: From community assembly to plant health. Nature Reviews Microbiology. 2020;**18**(11):607-621

[23] Sen K, Banerjee A. Insect pest management under climate change scenario. ENTOMOLOGY. 2023;**133**:35

[24] Lugtenberg B. Life of microbes in the rhizosphere. In: Principles of Plant–Microbe Interactions: Microbes for Sustainable Agriculture. New York: Springer Nature; 2015. pp. 7-15

[25] Meena M et al. PGPR-mediated induction of systemic resistance and physiochemical alterations in plants against the pathogens: Current perspectives. Journal of Basic Microbiology. 2020;**60**(10):828-861

[26] Bhattacharjee R, Dey U. Biofertilizer, a way toward organic agriculture: A review. African Journal of Microbiology Research. 2014;**8**(24):2332-2343

[27] Mohammadi K, Sohrabi Y. Bacterial biofertilizers for sustainable crop production: A review. ARPN Journal of Agricultural and Biological Science. 2012;7(5):307-316

[28] Ju W et al. Impact of coinoculation with plant-growth-promoting rhizobacteria and rhizobium on the biochemical responses of alfalfa-soil system in copper contaminated soil. Ecotoxicology and Environmental Safety. 2019;**167**:218-226

[29] Beneduzi A, Ambrosini A, Passaglia LMP. Plant growth-promoting rhizobacteria (PGPR): Their potential as antagonists and biocontrol agents. Genetics and Molecular Biology. 2012;**35**:1044-1051

[30] Qurashi AW, Sabri AN. Bacterial exopolysaccharide and biofilm formation stimulate chickpea growth and soil aggregation under salt stress. Brazilian Journal of Microbiology. 2012;**43**:1183-1191

[31] Adedeji AA, Häggblom MM, Babalola OO. Sustainable agriculture in Africa: Plant growth-promoting rhizobacteria (PGPR) to the rescue. Scientific African. 2020;**9**:e00492

[32] Baghel M et al. Pleiotropic influences of brassinosteroids on fruit crops: A review. Plant Growth Regulation. 2019;**87**:375-388

[33] Kour D et al. Microbe-mediated alleviation of drought stress and acquisition of phosphorus in great

millet (*Sorghum bicolour* L.) by drought-adaptive and phosphorus-solubilizing microbes. Biocatalysis and Agricultural Biotechnology. 2020;**23**:101501

[34] Sharma S, Chandra D, Sharma AK. Rhizosphere plant–microbe interactions under abiotic stress. In: Rhizosphere Biology: Interactions between Microbes and Plants. New York: Springer Nature; 2021. pp. 195-216

[35] Kamle M et al. Systemic acquired resistance (SAR) and induced systemic resistance (ISR): Role and mechanism of action against phytopathogens. Fungal biotechnology and bioengineering. 2020;**2**:457-470

[36] Kallamadi PR et al. Host defense responses during powdery mildew (*Golovinomyces latisporus comb*. nov.) infection in sunflower (*Helianthus annuus* L.). Tropical Plant Pathology. 2022;**47**(4):495-508

[37] Yu Y et al. Induced systemic resistance for improving plant immunity by beneficial microbes. Plants. 2022;**11**(3):386

[38] Huang YT et al. Salicylic acid enhances sunflower seed germination under Zn^{2+} stress via involvement in Zn^{2+} metabolic balance and phytohormone interactions. Scientia Horticulturae. 2021;**275**:109702

[39] Rahman MU et al. Microrheology of pseudomonas aeruginosa biofilms grown in wound beds. npj Biofilms and Microbiomes. 2022;**8**(1):49

[40] Tewari S, Arora NK. Role of salicylic acid from *Pseudomonas aeruginosa* $PF23^{EPS+}$ in growth promotion of sunflower in saline soils infested with phytopathogen *Macrophomina phaseolina*. Environmental Sustainability. 2018;**1**:49-59

[41] Nazari M, Smith DL. A PGPR-produced bacteriocin for sustainable agriculture: A review of thuricin 17 characteristics and applications. Frontiers in Plant Science. 2020;**11**:916

[42] Godino A, Príncipe A, Fischer S. A ptsP deficiency in PGPR pseudomonas fluorescens SF39a affects bacteriocin production and bacterial fitness in the wheat rhizosphere. Research in Microbiology. 2016;**167**(3):178-189

[43] Nazari M. The Potential Role of a PGPR-Produced Bacteriocin for Sustainable Agriculture: Assessment of Thuricin 17 Efficacy on Canola Production. Lausanne, Switzerland: Frontiers in Plant Science; 2024

[44] Subramanian S, Smith DL. Bacteriocins from the rhizosphere microbiome–from an agriculture perspective. Frontiers in Plant Science. 2015;**6**:909

[45] Bano A et al. Phytostimulants in sustainable agriculture. Frontiers in Sustainable Food Systems. 2022;**6**:801788

[46] Barbier T et al. Erythritol availability in bovine, murine and human models highlights a potential role for the host aldose reductase during *Brucella* infection. Frontiers in Microbiology. 2017;**8**:1088

[47] Maheshwari DK, Saraf M, Aeron A. Bacteria in Agrobiology. Disease Management. Berlin: Springer Publisher; 2013

[48] Suresh P et al. Partial purification and characterization of 2,4-diacetylphloroglucinol producing *Pseudomonas fluorescens* VSMKU3054 against bacterial wilt disease of tomato. Saudi Journal of Biological Sciences. 2021;**28**(4):2155-2167

[49] Robin DC, Marchand PA. Evolution of the biocontrol active substances in the framework of the European pesticide regulation (EC) No. 1107/2009. Pest Management Science. 2019;**75**(4):950-958

[50] Chen W, Modi D, Picot A. Soil and phytomicrobiome for plant disease suppression and management under climate change: A review. Plants. 2023;**12**(14):2736

[51] Köhl J, Kolnaar R, Ravensberg WJ. Mode of action of microbial biological control agents against plant diseases: Relevance beyond efficacy. Frontiers in Plant Science. 2019;**10**:845

[52] Legein M et al. Modes of action of microbial biocontrol in the phyllosphere. Frontiers in Microbiology. 2020;**11**:1619

[53] Kalia VC et al. Quorum sensing inhibitors as antipathogens: Biotechnological applications. Biotechnology Advances. 2019;**37**(1):68-90

[54] Elnahal ASM et al. The use of microbial inoculants for biological control, plant growth promotion, and sustainable agriculture: A review. European Journal of Plant Pathology. 2022;**162**(4):759-792

[55] Berendsen RL, Pieterse CMJ, Bakker PAHM. The rhizosphere microbiome and plant health. Trends in Plant Science. 2012;**17**(8):478-486

[56] Poveda J. *Trichoderma* as biocontrol agent against pests: New uses for a mycoparasite. Biological Control. 2021;**159**:104634

[57] Alfiky A, Weisskopf L. Deciphering *Trichoderma*–plant–pathogen interactions for better development of biocontrol applications. Journal of Fungi. 2021;**7**(1):61

[58] Pfordt A et al. *Trichoderma afroharzianum* ear rot–A new disease on maize in Europe. Frontiers in Agronomy. 2020;**2**:547758

[59] Degani O, Dor S. *Trichoderma* biological control to protect sensitive maize hybrids against late wilt disease in the field. Journal of Fungi. 2021;**7**(4):315

[60] Sánchez-Montesinos B et al. Biological control of fungal diseases by *Trichoderma aggressivum* f. *europaeum* and its compatibility with fungicides. Journal of Fungi. 2021;**7**(8):598

[61] Yu Z et al. Biocontrol and growth-promoting effect of *Trichoderma asperellum* TaspHu1 isolate from *Juglans mandshurica* rhizosphere soil. Microbiological Research. 2021;**242**:126596

[62] Reghmit A et al. *Trichoderma* spp. isolates from the rhizosphere of healthy olive trees in northern Algeria and their biocontrol potentials against the olive wilt pathogen, *Verticillium dahliae*. Organic Agriculture. 2021;**11**:639-657

[63] Nandini B et al. Trichovariability in rhizosphere soil samples and their biocontrol potential against downy mildew pathogen in pearl millet. Scientific Reports. 2021;**11**(1):9517

[64] Bonaterra A et al. Bacteria as biological control agents of plant diseases. Microorganisms. 2022;**10**(9):1759

[65] Fernandez-San Millan A et al. Successful biocontrol of major postharvest and soil-borne plant pathogenic fungi by antagonistic yeasts. Biological Control. 2021;**160**:104683

[66] Lahlali R et al. Biological control of plant pathogens: A global perspective. Microorganisms. 2022;**10**(3):596

[67] Bharti L, Yadav K, Chaubey AK. *Trichoderma* spp.: Approach for bio-control agent. In: Challenges in Plant Disease Detection and Recent Advancements. London: IntechOpen Limited; 2024

[68] Schouteden N et al. Arbuscular mycorrhizal fungi for the biocontrol of plant-parasitic nematodes: A review of the mechanisms involved. Frontiers in Microbiology. 2015;**6**:1280

[69] Sukhada M, Manjula R, Rawal RD. Evaluation of arbuscular mycorrhiza and other biocontrol agents against *Phytophthora parasitica* var. *nicotianae* infecting papaya (*Carica papaya* cv. Surya) and enumeration of pathogen population using immunotechniques. Biological Control. 2011;**58**(1):22-29

[70] Ziedan ESH et al. Application of mycorrhizae for controlling root diseases of sesame. Journal of Plant Protection Research. 2011;**51**(4):355-361

[71] Adigun BA et al. Opportunistic fungi, plant, and nematode interactions in agricultural crops. In: Opportunistic Fungi, Nematode and Plant Interactions: Interplay and Mechanisms. New York: Springer; 2024. pp. 11-32

[72] Song Y et al. Enhanced tomato disease resistance primed by arbuscular mycorrhizal fungus. Frontiers in Plant Science. 2015;**6**:786

[73] Ibrahim DS et al. Unraveling the mysteries of mycorrhiza-plant interactions: Mechanisms of protection and ecological factors influencing symbioses. In: Mycorrhizal Symbiosis and Agroecosystem Restoration. New York: Springer; 2024. pp. 197-226

[74] Malobane M, Madzivhandila M. Enhancement of soil arbuscular mycorrhizal fungi: A step toward restoring marginal soils. In: The Marginal Soils of Africa: Rethinking Uses, Management and Reclamation. New York: Springer; 2024. pp. 263-277

[75] Weng W et al. Roles of arbuscular mycorrhizal fungi as a biocontrol agent in the control of plant diseases. Microorganisms. 2022;**10**(7):1266

[76] De Mandal S et al. Plant–microbe association for mutual benefits for plant growth and soil health. In: Current Trends in Microbial Biotechnology for Sustainable Agriculture. Switzerland: Springer Nature; 2021. pp. 95-121

[77] Diagne N et al. Roles of arbuscular mycorrhizal fungi on plant growth and performance: Importance in biotic and abiotic stressed regulation. Diversity. 2020;**12**(10):370

[78] Delaeter M et al. Arbuscular Mycorrhizal fungi as biostimulant and biocontrol agents: A review. Microorganisms. 2024;**12**(7):1281

[79] Berrios L et al. Positive interactions between mycorrhizal fungi and bacteria are widespread and benefit plant growth. Current Biology. 2023;**33**(14):2878-2887. e4

[80] Wahab A et al. Role of arbuscular mycorrhizal fungi in regulating growth, enhancing productivity, and potentially influencing ecosystems under abiotic and biotic stresses. Plants. 2023;**12**(17):3102

[81] Nabi M. Role of microorganisms in plant nutrition and soil health. In: Sustainable Plant Nutrition. Amsterdam, Netherlands: Elsevier; 2023. pp. 263-282

[82] Lioussanne L. The role of the arbuscular mycorrhiza-associated rhizobacteria in the biocontrol of soilborne phytopathogens. Spanish Journal of Agricultural Research. 2010;**1**:51-61

[83] Vishwakarma SK et al. Arbuscular mycorrhizal fungi (AMF) as potential biocontrol agents. In: Rhizosphere Microbes: Biotic Stress Management. New York: Springer; 2022. pp. 197-222

[84] Lengai GM, Muthomi JW, Mbega ER. Phytochemical activity and role of botanical pesticides in pest management for sustainable agricultural crop production. Scientific African. 2020;7:e00239

[85] Pandit MA et al. Major biological control strategies for plant pathogens. Pathogens. 2022;**11**(2):273

[86] Hao D-C, Gu XJ, Xiao PG. Medicinal Plants: Chemistry, Biology and Omics. Cambridge, Cambridgeshire, United Kingdom: Woodhead Publishing; 2015

[87] Sasidharan S et al. Extraction, isolation and characterization of bioactive compounds from plants' extracts. African Journal of Traditional, Complementary and Alternative Medicines. 2011;**8**(1):1-10

[88] Gulzar A, Islam T, Hamid M. The role of plant extracts in sustainable agriculture. In: Bandh SA, editor. Sustainable Agriculture: Technical Progressions and Transitions. Cham: Springer International Publishing; 2022. pp. 61-67

[89] Bhuvaneswari P et al. Potential uses of natural antimicrobial agents and their applications as biopreservatives. In: Antimicrobials in Food Science and Technology. England: Taylor and Francis Group; 2024. pp. 154-168

[90] Aswini T. et al. Essential Oils: Characteristics, Extraction and Pharmacological Activities. International Journal of Science and Research Archive, India. 2024

[91] Al-Samarrai G, Singh H, Syarhabil M. Evaluating eco-friendly botanicals (natural plant extracts) as alternatives to synthetic fungicides. Annals of Agricultural and Environmental Medicine. 2012;**19**(4):673-676

[92] Ahmed E et al. Secondary metabolites and their multidimensional prospective in plant life. Journal of Pharmacognosy and Phytochemistry. 2017;**6**(2):205-214

[93] Mosunova O, Navarro-Muñoz JC, Collemare J. The biosynthesis of fungal secondary metabolites: From fundamentals to biotechnological applications. In: Encyclopedia of Mycology. Amsterdam, Netherlands: Elsevier; 2021. pp. 458-476

[94] Barrios-González J. Secondary metabolites production: Physiological advantages in solid-state fermentation. Current Developments in Biotechnology and Bioengineering. 2018;**3**:257-283

[95] Delgoda R. Pharmacognosy: Fundamentals, Applications and Strategies. London and Washington, DC: Academic Press; 2016

[96] Bhutia DD et al. Antifungal activity of plant extracts against Colletotrichum musae, the post harvest anthracnose pathogen of banana cv. Martaman. Nutrition & Food Science. 2016;**46**(1):2-15

[97] Lopes MR et al. *Saccharomyces cerevisiae*: A novel and efficient biological control agent for *Colletotrichum acutatum* during preharvest. Microbiological Research. 2015;**175**:93-99

[98] Yazdani D et al. A review on bioactive compounds isolated from plants against plant pathogenic fungi. Journal of Medicinal Plants Research. 2011;**5**(30):6584-6589

[99] Qin X et al. *Hanseniaspora uvarum* prolongs shelf life of strawberry via volatile production. Food Microbiology. 2017;**63**:205-212

[100] Zhou Y et al. Mechanisms of action of the yeast *Debaryomyces nepalensis* for control of the pathogen *Colletotrichum gloeosporioides* in mango fruit. Biological Control. 2018;**123**:111-119

[101] Jiménez-Reyes MF et al. Natural compounds: A sustainable alternative to the phytopathogens control. Journal of the Chilean Chemical Society. 2019;**64**(2):4459-4465

[102] Matrose NA et al. Plant extracts and other natural compounds as alternatives for postharvest management of fruit fungal pathogens: A review. Food Bioscience. 2021;**41**:100840

[103] Kalkışım Ö. In Vitro Antifungal Evaluation of Various Plant Extracts against Walnut Anthracnose (Gnomonia Leptostyla (Fr.) Ces et de Not.). Journal of Food Agriculture and Environment. Finland: WFL Publisher; 2012

[104] Sahan Y. Effect of *Prunus laurocerasus* L.(cherry laurel) leaf extracts on growth of bread spoilage fungi. Bulgarian Journal of Agricultural Science. 2011;**17**(1):83-92

[105] Nadaroglu H, Alayli A, Ince S. Synthesis of nanoparticles by green synthesis method. International Journal of Innovative Research and Reviews. 2017;**1**:6-9

[106] Duhan JS et al. Nanotechnology: The new perspective in precision agriculture. Biotechnology Reports. 2017;**15**:11-23

[107] Banerjee P et al. Leaf extract mediated green synthesis of silver nanoparticles from widely available Indian plants: Synthesis, characterization, antimicrobial property and toxicity analysis. Bioresources and Bioprocessing. 2014;**1**:1-10

[108] Nuruzzaman MD et al. Nanoencapsulation, nanoguard for pesticides: A new window for safe application. Journal of Agricultural and Food Chemistry. 2016;**64**(7):1447-1483

[109] Dwivedi S et al. Understanding the role of nanomaterials in agriculture. In: Microbial Inoculants in Sustainable Agricultural Productivity: Vol. 2: Functional Applications. Switzerland: Springer Nature; 2016. pp. 271-288

[110] Petosa AR et al. Assessing the transport potential of polymeric nanocapsules developed for crop protection. Water Research. 2017;**111**:10-17

[111] Patil CD et al. Trypsin inactivation by latex fabricated gold nanoparticles: A new strategy toward insect control. Enzyme and Microbial Technology. 2016;**92**:18-25

[112] Vanathi P, Rajiv P, Sivaraj R. Synthesis and characterization of *Eichhornia*-mediated copper oxide nanoparticles and assessing their antifungal activity against plant pathogens. Bulletin of Materials Science. 2016;**39**:1165-1170

[113] Biswal SK et al. Applications of nanotechnology in agriculture and food sciences. International Journal of Science Innovation Discoveries. 2012;**2**(1):21-36

[114] Ditta A. How helpful is nanotechnology in agriculture? Advances in Natural Sciences: Nanoscience and Nanotechnology. 2012;**3**(3):033002

[115] Frewer LJ et al. Nanotechnology in the Agri-Food Sector: Implications for the Future. Hoboken, New Jersey: John Wiley & Sons; 2011

[116] Pérez-de-Luque A, Hermosín MC. Nanotechnology and its use in agriculture. In: Bio-Nanotechnology: A Revolution in Food, Biomedical and Health Sciences. Hoboken, New Jersey: Wiley; 2013. pp. 383-398

[117] Sonkaria S, Ahn S-H, Khare V. Nanotechnology and its impact on food and nutrition: A review. Recent Patents on Food, Nutrition & Agriculture. 2012;**4**(1):8-18

[118] Krumova S et al. Seed priming with single-walled carbon nanotubes grafted with pluronic p85 preserves the functional and structural characteristics of pea plants. Nanomaterials. 2023;**13**(8):1332

[119] Al-Moaikal RM, Alabdallah NM, Alsanie SI. Interaction of nanoparticles with photosynthetic machinery. In: Molecular Impacts of Nanoparticles on Plants and Algae. Amsterdam: Elsevier; 2024. pp. 159-194

[120] DeRosa MC et al. Nanotechnology in fertilizers. Nature Nanotechnology. 2010;**5**(2):91-91

[121] Nazneen H et al. The role of plant-mediated biosynthesized nanoparticles in agriculture. In: Sustainable Agriculture: Technical Progressions and Transitions. Germany: Springer Nature; 2022. pp. 97-117

[122] Beesley L, Moreno-Jiménez E, Gomez-Eyles JL. Effects of biochar and greenwaste compost amendments on mobility, bioavailability and toxicity of inorganic and organic contaminants in a multielement polluted soil. Environmental Pollution. 2010;**158**(6):2282-2287

[123] Zhu X et al. Effects and mechanisms of biochar-microbe interactions in soil improvement and pollution remediation: A review. Environmental Pollution. 2017;**227**:98-115

[124] Jaiswal AK et al. *Rhizoctonia solani* suppression and plant growth promotion in cucumber as affected by biochar pyrolysis temperature, feedstock and concentration. Soil Biology and Biochemistry. 2014;**69**:110-118

[125] Jaiswal AK et al. Linking the belowground microbial composition, diversity and activity to soilborne disease suppression and growth promotion of tomato amended with biochar. Scientific Reports. 2017;**7**(1):44382

[126] Jaiswal AK et al. Nonmonotonic influence of biochar dose on bean seedling growth and susceptibility to *Rhizoctonia solani*: The "shifted R max-effect". Plant and Soil. 2015;**395**:125-140

[127] Wang W et al. Biochar application alleviated negative plant–soil feedback by modifying soil microbiome. Frontiers in Microbiology. 2020;**11**:799

[128] Bakker PAHM et al. Induced systemic resistance and the rhizosphere microbiome. The Plant Pathology Journal. 2013;**29**(2):136

[129] Mehari ZH et al. Induced systemic resistance in tomato (*Solanum lycopersicum*) against *Botrytis cinerea* by biochar amendment involves jasmonic acid signaling. Plant and Soil. 2015;**395**:31-44

[130] Samain E, Aussenac T, Selim S. The effect of plant genotype, growth stage, and *Mycosphaerella graminicola* strains on the efficiency and durability of wheat-induced resistance by *Paenibacillus* sp. strain B2. Frontiers in Plant Science. 2019;**10**:587

[131] Samain E et al. Biological control of septoria leaf blotch and growth promotion in wheat by *Paenibacillus* sp. strain B2 and *Curtobacterium plantarum* strain EDS. Biological Control. 2017;**114**:87-96

[132] Brtnicky M et al. A critical review of the possible adverse effects of biochar in the soil environment. Science of the Total Environment. 2021;**796**:148756

[133] Brassard P, Godbout S, Raghavan V. Soil biochar amendment as a climate change mitigation tool: Key parameters and mechanisms involved. Journal of Environmental Management. 2016;**181**:484-497

[134] Rashid M et al. Prospects of biochar in alkaline soils to mitigate climate change. In: Environment, Climate, Plant and Vegetation Growth. Germany: Springer Nature; 2020. pp. 133-149

[135] Riaz Rajoka MS et al. Chitin/ chitosan derivatives and their interactions with microorganisms: A comprehensive review and future perspectives. Critical Reviews in Biotechnology. 2020;**40**(3):365-379

[136] de Carvalho JC et al. Agro-industrial wastewaters for algal biomass production, biobased products, and biofuels in a circular bioeconomy. Fermentation. 2022;**8**(12):728

[137] Sato K, Kadota Y, Shirasu K. Plant immune responses to parasitic nematodes. Frontiers in Plant Science. 2019;**10**:1165

[138] Riseh RS et al. Nano/ microencapsulation of plant biocontrol agents by chitosan, alginate, and other important biopolymers as a novel strategy for alleviating plant biotic stresses. International Journal of Biological Macromolecules. 2022;**222**:1589-1604

[139] Cretoiu MS et al. Chitin amendment increases soil suppressiveness toward plant pathogens and modulates the actinobacterial and oxalobacteraceal communities in an experimental agricultural field. Applied and Environmental Microbiology. 2013;**79**(17):5291-5301

[140] Muhammad T et al. RNA interference: A natural immune system of plants to counteract biotic stressors. Cells. 2019;**8**(1):38

[141] Dalakouras A et al. Genetically modified organism-free RNA interference: Exogenous application of RNA molecules in plants. Plant Physiology. 2020;**182**(1):38-50

[142] Guo Z, Li Y, Ding S-W. Small RNA-based antimicrobial immunity. Nature Reviews Immunology. 2019;**19**(1):31-44

[143] Rosa C et al. RNA interference mechanisms and applications in plant pathology. Annual Review of Phytopathology. 2018;**56**(1):581-610

[144] Lindbo JA, Falk BW. The impact of "coat protein-mediated virus resistance" in applied plant pathology and basic research. Phytopathology. 2017;**107**(6):624-634

[145] Avila YI et al. Cracking the code: Enhancing molecular tools for progress in nanobiotechnology. ACS Applied Bio Materials. 2024;**7**(6):3587-3604

[146] Dong OX, Ronald PC. Genetic engineering for disease resistance in plants: Recent progress and future perspectives. Plant Physiology. 2019;**180**(1):26-38

[147] Pixley KV et al. Genome editing, gene drives, and synthetic biology: Will they contribute to disease-resistant crops, and who will benefit? Annual Review of Phytopathology. 2019;**57**(1):165-188

[148] Vieira MS et al. Tropical fruit virus resistance in the era of next-generation plant breeding. SynBio. 2024;**2**(3):267-284

[149] Kuo Y-W, Falk BW. RNA interference approaches for plant disease control. BioTechniques. 2020;**69**(6):469-477

[150] Razaq M, Shah FM. Biopesticides for management of arthropod pests and weeds. In: Biopesticides. Amsterdam, Netherlands: Elsevier; 2022. pp. 7-18

[151] Deka B, Baruah C, Babu A. Entomopathogenic microorganisms: Their role in insect pest management. Egyptian Journal of Biological Pest Control. 2021;**31**:1-8

[152] Grzywacz D et al. Mass production of entomopathogens in less industrialized countries. In: Mass Production of Beneficial Organisms. Amsterdam, Netherlands: Elsevier; 2023. pp. 431-462

[153] Ufaq, Nabi A, Shajar F, Ul Rehman R. Microbe assisted remediation of xenobiotics: A sustainable solution. In: Microbes Based Approaches for the Management of Hazardous Contaminants. Hoboken, New Jersey: Wiley; 2024. pp. 20-41

[154] Ben-Dov E. *Bacillus thuringiensis* subsp. *israelensis* and its dipteran-specific toxins. Toxins. 2014;**6**(4):1222-1243

[155] Schünemann R, Knaak N, Fiuza LM. Mode of action and specificity of *Bacillus thuringiensis* toxins in the control of caterpillars and stink bugs in soybean culture. International Scholarly Research Notices. 2014;**2014**(1):135675

[156] Silva-Filha MHNL et al. Bacterial toxins active against mosquitoes: Mode of action and resistance. Toxins. 2021;**13**(8):523

[157] Sujayanand GK et al. Distribution and toxicity of *Bacillus thuringiensis* (Berliner) strains from different crop rhizosphere in indo-Gangetic plains against polyphagous lepidopteran pests. International Journal of Tropical Insect Science. 2021;**41**:1-19

[158] Sabbahi R et al. A global perspective of entomopathogens as microbial biocontrol agents of insect pests. Journal of Agriculture and Food Research. 2022;**10**:100376

[159] Bergman ME, Davis B, Phillips MA. Medically useful plant terpenoids: Biosynthesis, occurrence, and mechanism of action. Molecules. 2019;**24**(21):3961

[160] Sinha KK, Choudhary AK, Kumari P. Entomopathogenic fungi. In: Ecofriendly Pest Management for Food Security. Amsterdam, Netherlands: Elsevier; 2016. pp. 475-505

[161] Shin TY et al. Pathogenesis-related genes of entomopathogenic fungi. Archives of Insect Biochemistry and Physiology. 2020;**105**(4):e21747

[162] Sharma A, Sharma S, Yadav PK. Entomopathogenic fungi and their relevance in sustainable agriculture: A review. Cogent Food & Agriculture. 2023;**9**(1):2180857

Chapter 12

Bio-Organic Management of Root-Knot Nematodes in Sustainable Agroecosystems

Jaseem KP, Ameer Favas V, Shana Sherin and Sonia Waqar

Abstract

Root-knot nematodes (RKN) are a serious threat to sustainable agroecosystems. They attack a wide range of crops all over the world and lead to significant yield losses. This chapter explores the use of bio-organic management strategies instead of harmful chemical control measures. We evaluate the possibilities of decreasing the RKN population by effectively using cropping practices such as crop rotation, resistant cultivars, and soil solarization. Most importantly, the chapter discusses about biological control agents (BCA) such as nematode-trapping fungi, endoparasitic fungi, *Bacillus pumilus*, *Pasteuria penetrans*, and other useful microorganisms to suppress the nematode population. Finally, we conclude by discussing the use of soil amendments such as botanicals, plant extracts, oil cakes, and other organic materials like manure and vermicompost to inhibit the RKN population and enhance soil health. This chapter provides us with valuable information on how to control nematode populations effectively and promote sustainable and environmentally friendly agricultural practices.

Keywords: bio-organic nematode control, sustainable agroecosystems, root-knot nematodes, plant parasitic pests, integrated pest management

1. Introduction

Root-knot nematodes (RKNs) are a specific concern to global food security, since they present a substantial obstacle to crop production caused by plant-parasitic nematodes (PPNs) [1]. Among the class of plant-parasitic nematodes, the *Meloidogyne* genus is well known for its scientific and substantial economic significance. Due to their great plasticity and flexibility, these nematodes can be found in a wide range of environments, including tropical and subtropical regions found in North and South America, Asia, Africa, and Europe [2]. Globally, the devastation of about $100 billion in US dollars a year is attributed to the *Meloidogyne* species alone [3].

In addition, *Meloidogyne* species have the ability to build dynamic disease complexes (MDCs) with bacterial and fungal species, which can intensify disease outbreaks and severely harm a variety of important crops around the world, such as

IntechOpen

tomato and coffee [4]. The use of soil fumigants containing poisonous and expensive nematicides is a major component of the issues involved in chemical management of worms. Furthermore, some farming systems, like organic or regenerative, might not be a good fit for these techniques. Chemical nematicide is essential for agriculture, but their widespread use has created serious worries about how the soil's biodegradation systems are being enhanced [5]. Using helpful bacteria (like *Pseudomonas fluorescens* and *Bacillus firmus*), beneficial fungus (like *Trichoderma longibrachiatum* and *T. asperellum*), botanicals, and essential oils has become more popular in this context as effective biological options for nematode management [6]. Finding efficient environmentally friendly management solutions is therefore essential. There have been reports of the effectiveness of several bioorganic products against plant parasitic nematodes, including neem-based treatments, bioagent (*Trichoderma harzianum*), and organic amendments (vermicompost). But because of variations in their biochemical characteristics, the environment they are in, and the characteristics of the soil, their efficacy against parasitic nematodes could be uneven and erratic [7]. Creating cultivars with broad-based resistance would make nematode management easier and boost host plant resistance to root-knot nematodes [8]. In order to stabilize vegetable production, researchers worldwide are trying to standardize management strategies of root-knot nematode by utilizing non-chemical and environment friendly alternatives like fertilization, soil management, sanitation, organic amendments, heat-based techniques, and biological control [9].

Using bio-organic management tactics, this chapter offers a sustainable way to control root-knot nematodes. The life cycle and deceptive tactics used by these pests to take advantage of their plant hosts are examined. The strategy builds a healthy agroecosystem by utilizing resistant plant types and advantageous microorganisms. The soil microbiota is enhanced by organic amendments, which gives plants more resistance against these pests. With this knowledge, home gardeners and farmers can protect crops and build a sustainable future for food systems.

2. Biology and ecology of root-knot nematodes (RKN)

2.1 Life cycle of RKN

Nematodes are ubiquitous organisms that are found everywhere. Root-knot nematodes (RKN), which parasitize plants, are classified as sedentary (sessile) endoparasites. *Meloidogyne* spp. are characterized by four juvenile stages that undergo four molts before achieving the adult stage (male or female). The female lays 500–1500 eggs into the gelatinous matrix, which is made up of high- and low-molecular weight glycoproteins that protect the eggs and regulate humidity and temperature, to create the first-stage juvenile (J1). Prior to hatching, the molts give rise to the second-stage juvenile (J2) that emerges from the egg. Chemotaxis is the mechanism through which the root plant attracts the second-stage juvenile, the only infectious form. This phenomenon, which can occur as various stimuli and signal gradients of chemical, volatile, and non-volatile exudates, is the primary method by which RKNs locate a host plant. After the penetration, J2 establishes a feeding site inside the cortex and enters into the vascular cylinder. Using their stylet, they inject a mixture of hydrolytic enzymes such as endoglucanase, endoxylnase, pectatelase, amylase, cellulase, and pectinases which cause cell degradation. This process gives rise to large, multinucleate cells termed as giant cells. The plant cells around the nematodes ingest the cytoplasm and cause three molts. After the molts are completed, the sexual dimorphic pear-shaped sedentary endoparasitic, 0.5–1.0 mm long female and vermiform males are formed. The J3 and J4 stages can be identified by the number of previous molts and the non-functional stylet [10–14].

For almost 3 months, these infectious nematodes can survive in moist soil. They last for a few weeks at most under dry conditions. When the right conditions prevail, the eggs hatch, and within 5 to 6 weeks—depending on the temperature— the new juveniles emerge to finish the life cycle. A feeding site made up of five to seven enlarged cells known as giant cells, which are multinucleate, hypertrophied cells with dense cytoplasm, and a high metabolic activity, is induced by root-knot nematodes. They are produced by successive mitoses rather than cytokinesis, which results in an increase in the size and quantity of nuclei in a cell. Massive cells have the potential to grow up to 100 times larger than typical root vascular parenchyma cells and can have up to 100 nuclei (**Figure 1**) [15–18].

2.2 Host range and damage

The root systems of ornamental plants, fruits, and vegetables are frequently impacted by root-knot nematodes. The global distribution of the parasitizing species *Meloidogyne incognita*, *M. javanica*, and *M. chitwoodi* allows them to infect a large number of roots worldwide, and others are restricted to geographical areas or are growing on fewer hosts, such as *M. hispanica* that infects peach and tomato and *M. graminicola* on rice plants which causes serious loses and thus threatens food security in India. Soybeans, potatoes, tomatoes, lettuce, onions, and carrots are a few plants that have been known to be susceptible to the root-knot nematode [19]. One factor that affects how much root galling occurs in plants is the species of the host plant. A diverse range of plant-parasitic nematode species, primarily *Meloidogyne* spp., is found in vegetables, such as spinach (*Spinacia oleracea*), tomato (*Solanum*

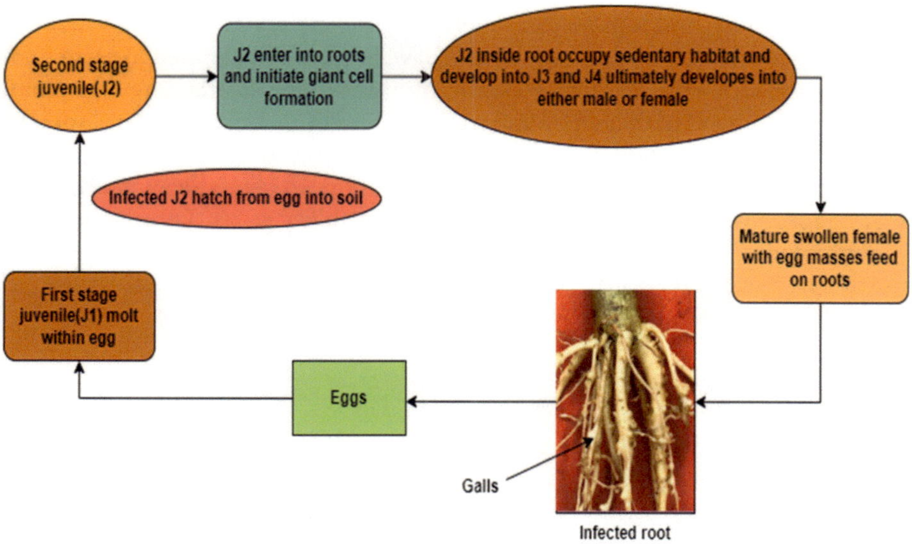

Figure 1.
Life cycle of root-knot nematodes.

lycopersicon), peppers (*Capsicum* spp.), brinjal (*Solanum melongena*), beetroot (*Beta vulgaris*), and carrot (*Daucus carota*) [20–23].

When root-knot nematodes infect good host crops severely, the yield or quality is typically significantly reduced, and in certain situations, the crop may completely collapse [20, 24]. *M. incognita* and *M. javanica* are the most common root-knot nematode species on tomatoes worldwide in tropical and subtropical areas [25]. Both species infect vegetable crops wherever they are planted, and if appropriate nematode control techniques are not used, this could lead to significant output losses [21, 22, 26]. Due to *Meloidogyne* spp. infection, tomato yield loss of 20–40% [22] and more than 50% [25] has been documented. One of the main causes of the substantial crop losses produced by root-knot nematodes is their ability to produce large numbers of offspring, especially in warmer climates where they can go through many life cycles in one season. The carrot taproot can be severely distorted by the root-knot nematodes *M. chitwoodi*, *M. hapla*, and *M. fallax*, resulting in significant losses to the root and forming the forking ends with galls on the delicate secondary root [27]. According to a study in [28], obstacles caused by *Meloidogyne* spp. resulted in losses of eggplant (*Solanum melongena* L.) yield in Egypt; this genus was shown to occur with a frequency of 62.5% [29]. Up to 96.26% of the fields in these types of areas had the species identified [30].

3. Bio-organic management strategies

3.1 Host plant resistance and breeding programs

The host plant resistance management of nematode genera with specialized host-parasite interaction during their life cycles has been particularly successful. For instance, in the control of *Aphelenchoides, Ditylenchus, Globodera, Heterodera, Meloidogyne, Pratylenchus, Radopholus, Rotylenchulus,* and *Tylenchulus*. Nematode-resistant varieties have been created in various cultivated crops such as cotton (*Gossypium hirsutum*), chickpea (*Cicer arietinum*), peach (*Prunus persica*), tobacco (*Nicotiana tabacum*), and tomato (*Solanum lycopersicum*) against *Meloidogyne*. Recent advances in genome editing and sequencing have been widely applied in plant nematology trends to study the intricate genomic/transcriptomic interactions that plants undergo in response to pathogen invasions. New developments in the development of nematode-resistant varieties are gaining attention, including spray-induced gene silencing (SIGS), RNA interference (RNAi), host-induced gene silencing (HIGS), cross-kingdom RNAi, and engineering plant susceptibility genes. Future research on these contemporary technologies should be conducted because they have the potential to provide practical and affordable methods of controlling nematodes and minimizing crop yield losses brought on by these parasites [31].

3.1.1 Cotton (Gossypium hirsutum)

Infestations of plant-parasitic nematodes, including the Root-Knot Nematode (RKN) *Meloidogyne incognita*, pose a global risk to cotton production. Host-plant resistance is a very successful RKN management tactic [32]. Over 143 nations worldwide have seen a rapid spread of the biological hazard posed by nematode damage brought on by the southern root knot nematode (RKN), *Meloidogyne incognita* Kofoid and White (Chitwood) [33]. Stunted growth and decreased yields are the result of the afflicted tissues reducing the amount of water and nutrients that are transported from the root system into the aerial area of the plant [34]. Nematode resistance has received particular attention in breeding programs for crop species such as cotton, soybean, tomato, and common bean, among others, because of the numerous detrimental effects of RKN infestation in agriculture [35, 36]. RKN resistance has been used in Gh cotton breeding programs from multiple sources of germplasm. The 'Acala NemX' (NemX) cultivar is one of these sources of resistance [37–39]. The resistance gene is also located on chromosome D02 of the Wild Mexico Jack Jones line, and this gene helps in suppressing the production of nematode eggs [40–42].

Host-plant resistance proves to be a very successful method for controlling damage caused to crops by root knot nematodes (*Meloidogyne* spp.). In the context of cotton, NemX was introduced by the California Planting Cotton Seed Distributors (CPCSD) three decades ago as a variety possessing strong resistance to RKN. The RKN resistance found in NemX is extremely effective in safeguarding against the impact of RKN infection. It has been utilized as a source of nematode resistance in breeding and genetic research studies [36, 39, 43].

3.1.2 Tomato (Solanum lycopersicum)

Plant-parasitic nematodes are small roundworms that live in the soil and affect the root systems of host plants. Among these nematodes, the root knot nematodes (RKN) such as *Meloidogyne incognita*, *Meloidogyne javanica*, and *Meloidogyne arenaria* cause extensive damage to vegetable crops, including tomatoes [44]. The most practical, cost-effective, and environmentally friendly approach to controlling RKN is through the utilization of resistant tomato varieties [45]. Resistance genes against RKN are discovered in wild tomato species and other plant species. However, in the case of tomatoes, there is a well-known single dominant resistance gene (Mi-1) against RKN, as well as against major insect pests like white flies and aphids [46]. The Mi-1 gene is utilized in numerous breeding initiatives to create high-yielding hybrid tomato varieties that are resistant to root knot nematodes [47]. Tomatoes use resistance mechanisms against root knot nematodes that impede the juvenile (J2) nematode's ability to penetrate during invasion. However, the quantity of penetrating J2 varies according to the genotype of the tomato, post-inoculation time, and nematode population [48, 49].

In another research, carried out to assess how local and commercially available tomato varieties react to *M. javanica*, it was observed that root gall formation was notably reduced in the case of commercial resistant varieties [50]. The tomato cultivar grafted onto a root stock with the Mi resistance gene showed a much smaller number of galls compared to the control or non-grafted plants [51]. The resistance mediated by Mi-1 considerably decreases the reproductive capability of root knot nematodes [52]. The presence of both different and identical alleles for the Mi gene in tomatoes impacted the reproductive capacity of *M. incognita* [53]. The involvement of plant hormones, such as salicylic acid (SA) and jasmonic acid (JA), in the transmission of signals related to tomato resistance to root knot nematodes has been demonstrated [54–56]. Many plant compounds exuded from the root have been found to have repellent, attractant, or inhibitory effects, as well as being toxic to plant parasitic nematodes (**Table 1**) [57].

Technique	Efficacy	Environmental Impact	Sustainability
Cropping Practices			
Crop Rotation	Moderate – Breaks the continuity of the host, disrupting the life cycle of the nematode.	Low – Enhances soil and increases biodiversity	High – Renewable resource that lessens the need for resistant nematodes to be selected
Resistance Cultivars	High – Nematode damage can be considerably decreased by resistant cultivars.	Moderate – May need specific cultivars that are not usually available	Moderate – In order to combat developing nematodes, new resistant cultivars must be developed
Soil Solarization	Moderate to High – Useful for eliminating nematodes in upper soil layers	Moderate – Plastic tarps are necessary, and it may disturb the microbial populations in soil	Moderate – Reusable tarps, although the heating process might need energy input
Intercropping	Moderate – Interrupts the nematode's life cycle and brings in useful microorganisms	Low – Enhances soil and increases biodiversity	High – Decreases dependency on a single crop and is renewable
Biological Control Agents			
Fungi	Moderate – Numerous processes, including toxin generation, parasitism, and predation	Low – Able to encourage plant development and disease resistance	High – Renewable resource that increases the possibilities for natural management
Bacteria	Moderate – Destroying nematodes through parasitism, competition, and toxin production	Low – Able to encourage plant development and disease resistance	High – Renewable resource that increases the possibilities for natural management
Nematode Trapping Fungi	Moderate – Traps and eats nematodes	Low – Integrates with soil ecosystem	High – Renewable resource that increases the possibilities for natural management
Endoparasitic Fungi	Moderate – It harms and eliminates nematode eggs and females	Low – Incorporates with the soil ecosystem	High – Renewable resource that increases the possibilities for natural management
Egg and Female Parasitic Fungi	Moderate – Focuses on particular nematode stages	Low – Incorporates with the soil ecosystem	High – Renewable resource that increases the possibilities for natural management
Toxin Producing Fungi	Moderate – Discharges poisons that destroy nematodes	Low – Incorporates with the soil ecosystem and some may need special treatment	High – Renewable resource that increases the possibilities for natural management
Particular bacteria			
Bacillus thuringiensis	Moderate – Generates poisons that destroy nematodes	Low – Able to encourage plant growth	Moderate – Needs particular strains, further research is required
Bacillus pumilis	Moderate – Feeds on the eggs of nematodes	Low – Able to encourage plant growth	High – Strains that are easily accessible and have the potential for economic use
Pasteuria penetrans	Moderate – Nematodes are infected and killed by spores	Low – Incorporates with the soil ecosystem	High – Renewable resource, further research is required for commercial uses

Technique	Efficacy	Environmental Impact	Sustainability
Soil Amendments			
Botanicals	Moderate – Plant extracts possessing nematicidal effects	Low – Decomposes in soil into natural products	High – Renewable resource, requires further research
Plant Extracts	Moderate – Produced from particular plants that have nematicidal qualities	Low – Decomposes in soil into natural products	High – Renewable resource that certain plants can utilize easily
Oil Cakes	Moderate – Produces fatty acids containing nematicidal effects	Low – Increases the fertility of the soil	High – Renewable resource that certain plants can utilize easily
Biochar	Moderate – Changes the soil's environment to make nematodes less favorable	Low – Increases the soil health and fertility of the soil	High – Renewable resource with possible advantages over nematode control
Fly Ash	Low to Moderate – Includes substances that possess nematicidal effects.	Moderate – Fly ash composition can vary, potential for heavy metal contamination	Low – Waste product handling needs to be done carefully
Animal Manures	Moderate – Contains beneficial microorganisms that enhance soil health	Low – Increases the fertility of the soil	High – Abundant and easily available renewable resource
Poultry Wastes	Moderate – Contains beneficial microorganisms that enhance soil health	Low – Increases the fertility of the soil	High – Abundant and easily available renewable resource
Vermicompost	Moderate – Contains beneficial microorganisms that enhance soil health	Low – Increases the fertility of the soil	High – Renewable resource that enhances biodiversity and soil health
Chemical Nematicides	High – Able to get rid of nematodes quite effectively	High – Hazardous to beneficial soil organisms and capable of contaminating water supplies	Low – Nonrenewable and may cause nematodes to develop resistance

Table 1.
Different resistant cultivars showing resistance to various nematodes.

3.2 Biological control agents

Chemical methods are more widely used and are effective in controlling plant parasitic nematodes, but they are not environment-friendly and have the potential to seriously compromise the delicate balance of the ecosystem. Therefore, it is essential to find environmentally friendly methods of controlling plant-parasitic nematodes in order to prevent the extremely harmful effects of nematicides on humans, animals, and the environment [58]. One of the greatest substitutes for pesticides in the safe management of root-knot nematodes is biological control [59]. Although synthetic nematicides have increased crop yields and shielded the soil to some extent, health risks and environmental contamination are impeding their widespread use. Because many of the most widely used pesticides are costly or are being taken off the market because of their negative effects on people, their persistence in the soil, or their contamination of groundwater, research efforts are shifting their attention to biological control agents [60].

3.2.1 Bacterial and fungal biocontrol agents

Intimate relationships are formed between plants and plant growth-promoting bacteria (PGPB), which can improve plant growth, defense against abiotic stressors, and resistance to disease [61, 62]. The biological control potential of several species of PGPB including *Agrobacterium, Arthrobacter, Azotobacter, Clostridium, Desulfovibrio, Pasteuria, Serratia, Burkholderia, Azospirillum, Bacillus, Chromobacterium, Pseudomonas*, and *Corynebacterium* against phytoparasitic nematodes, has been investigated [63, 64]. *Pseudomonas* and *Bacillus* are of particular interest because they are widely found in natural environments, are currently used and sold, and have shown the highest level of biological control efficacy over the past 20 years [64–68].

Bacillus methylotrophicus strain R2-2 and *Lysobacter antibioticus* strain 13-6 showed the highest antagonistic activity against the tomato root knot nematode *Meloidogyne incognita* in plate and greenhouse pot experiments, according to a study on the 19 bacterial strains isolated in Yunnan (China) from rhizosphere soils and plant tissues. When the two strains were used as soil drenches or seed treatments in greenhouse tests, they decreased the occurrence and severity of root knots on tomatoes when compared to no-bacteria controls. Soil drench treatment with either strain decreased root knot disease levels and enhanced yields in tomato field trials conducted in different years compared to the control [69].

The biological control of *Pseudomonas fluorescens* over *Meloidogyne javanica*, the root knot nematode, was studied in laboratory and greenhouse settings. In comparison to the control, the results demonstrated that 109 (CFU/ml) of *P. fluorescens* significantly reduced nematode infection and other parameters. *P. fluorescens* was able to both break down the nematode egg mass matrix and drastically lower the rate at which nematode eggs hatch. The resistance-associated enzymes phosphorus acid lyase (PAL) and peroxidase (POX) were markedly more active and were found in plants which are infected with *P. fluorescens*. Five days after inoculation, PAL and POX reached their peak activity. The findings demonstrated that *P. fluorescens* uses two primary defense mechanisms against nematodes: destroying eggs and producing defense mechanisms in plants that result in systemic resistance [70].

In an in vitro study by Safni et al. [71] to evaluate ability of four rhizobacteria isolated from a potato field to withstand the damaging effects of the root-knot nematode (*Meloidogyne* spp.) on potato crops, the morphology, physiology, and biochemistry of pure cultures of rhizobacteria were examined, along with their potential applications as fertilizers and biopesticides. From the in vitro experiment, four possible antagonistic bacteria that could eliminate *Meloidogyne* spp. were chosen. The four bacteria that were found *were Aeromonas* sp., *Serratia marcescens, Vibrio* sp., *and Serratia* sp. Nematode mortality ranged from 98.20 to 99.87% against second stage nematode juveniles. *Aeromonas* sp. was the nematode species with the highest percentage of mortality (99.87%). Furthermore, all rhizobacterial strains were regarded as biopesticides because they generated the enzymes gelatinase, chitinase, and protease. This experiment demonstrated the potential of all the chosen potato-associated bacterial strains as environmentally friendly biocontrol agents using the potato root knot nematode.

The *Meloidogyne* population was isolated in a study utilizing tomato samples taken from fields in the Iranian province of West Azarbaijan, from infected cucumber roots in a greenhouse. The influence of six chosen isolates on nematode infection of a sensitive tomato cultivar, as well as the impacts of the bacterial isolates on hatching and death of nematode second-stage juveniles, was assessed in a greenhouse setting. *Meloidogyne incognita* was recognized as the root-knot nematode. The outcomes demonstrated that each isolate strongly prevented nematode hatching. The kids' deaths were also brought on by the bacterial isolates. Following the selection of six isolates with a high potential for biocontrol, additional testing and genetic identification of the chosen bacterial isolates were carried out. Isolates 1, 2, 5, 7, 10, and 11 were identified as *Bacillus wiedmannii* (MW405861), *Pseudoxanthomonas mexicana* (MW405860), *Pseudomonas thivervalensis* (MW405862), *Serratia liquefaciens* (MW405864), *Pseudomonas chlororaphis* (MW405863), and *P. fluorescens* (MW405825). For the selected isolates, the experiment produced good results in terms of nematode gall and egg mass reduction [72]. For many years, the use of fungus microorganisms has been a part of the management of RKN and other agricultural pests [73].

Among the most well-known fungi for the control of RKN are *Actylellina, Arthrobotrys, Aspergillus, Catenaria, Dactylellina, Hirsutella, Pochonia, Purpureocillium,* and *Trichoderma* [74, 75]. Arthrobotrys and *Dactylellina* can impede the nematode's invasion by trapping RKN J2s in the soil through the use of their hyphal structures. However, a variety of factors may affect the extent of RKN mitigation because soil is a dynamic matrix. [76–78]. Commentable nematicidal effects on eggs and J2s of *M. javanica* in vitro conditions were shown by species of *Trichoderma viz., T. harzianum,* and *T. viride* [79].

In tomato, varying densities of *T. harzianum* and *T. viride* inoculum against *M. javanica* inhibited the growth of nematodes and the formation of root galling. Both fungi became more effective as their inoculum densities rose. Experiments were done to determine how *Arthrobotrys oligospora* stops *Meloidogyne incognita* from infecting tomato plants. *A. oligospora's* effect on the rate of in vitro trapping and the capture of juvenile *M. incognita* was calculated. Three different soil weight densities—1, 3, and 5%—were used to evaluate the effects of *A. oligospora* on nematodes and plant growth parameters of tomato plants grown in greenhouses. Information showed that *A. oligospora* formed three-dimensional sticky networks and that juvenile *M. incognita* could be held in its second stage by its trapping organs. As part of an in vitro experiment, the juvenile nematodes were exposed to an *A. oligospora* culture for 24, 48, and 72 hours. The higher juvenile trapping rate by the trapping organs was caused by the longer exposure period (72 hours). Compared to the untreated control group, the in vitro test results showed a significant drop in the nematode criteria.

The efficacy of fungal biocontrol agents, such as *Trichoderma asperellum* and *Trichoderma viride*, against the tomato root knot nematode, *M. incognita*, was examined in [80]. The effectiveness of two fungi known to be biocontrol agents—*T. viride* and *T. asperellum*—in suppressing the root-knot nematode, *M. incognita*, has been assessed. The best treatment for increasing tomato plant growth and significantly reducing root-knot nematode reproduction on tomatoes was found to be *T. viride* applied to the soil at a rate of 15 ml/kg. This was closely followed by *T. viride* applied at a rate of 10 ml/kg. *T. asperellum* at 15 ml/kg soil and *T. viride* at 10 ml/kg soil both had nearly identical effects on nematode reproduction parameters and plant growth. Affokpan et al. [81] examined the effects of various isolates of *Trichoderma* sp. on tomato against *Meloidogyne* spp. According to their findings, tomato yield increased, root galling was suppressed, and nematode reproduction was significantly inhibited by *Trichoderma* isolates. The potential of *Trichoderma harzianum,* a fungal biocontrol agent, to suppress the root knot nematode *M. javanica* was assessed. In greenhouse experiments, after applying *Trichoderma* to the soil prior to nematode infection, there was a decrease in root galling and an increase in top fresh weight. In sterile in vitro tests, every *Trichoderma* strain demonstrated the capacity to colonize eggs separated from *M. javanica* and second-stage juveniles (J2) [82]. The capacity of *Trichoderma viride* and *harzianum* to lower the prevalence of the root knot nematode *M. incognita* was examined. Comparing the tested isolates to the control, in vitro investigations showed that each one was successful in producing J2 mortality [83].

Another fungus studied for its anti-RKN impacts was *Metarhizium guizhouense* PSUM02. The fungus spores and crude extracts of the fungus grown on a protein-enriched medium combined to produce a protease that totally prevented the *M. incognita* eggs from hatching in vitro within 24 hours. About 48 hours after treatment, J2's death rate rose to 100%. The juveniles within the eggs and the hatched juveniles resisted the mycelia's invasion of the immature eggs. Observations using electron microscopy revealed that both individual and combined treatments severely damaged the eggshell and cuticle [84]. Many studies were also conducted which used combinations of both bacterial and fungal BCAs.

Pasteuria penetrans and *Verticillium chlamydosporium* were applied singly or in combination during two crop rotations in a microplot field experiment against *Meloidogyne javanica*. The application of microbial antagonists in combination, as opposed to separately, resulted in a significant reduction in the densities of root knot nematodes. When both biocontrol agents were applied in combination, crop yield considerably increased, and plant growth was enhanced. When comparing *P. penetrans* applied plots to both microbial agents, the number of bacterial spores burdened with root knot juveniles was significantly higher. In plots where fungal antagonists were applied individually as opposed to other treatments, the densities of *V. chlamydosporium* in the rhizosphere and rhizoplane increased over time in both crop sequences. At the final harvest, there was a variable increase in *V. chlamydosporium* egg parasitism in both crop rotations [85].

In pot and greenhouse experiments, Huang et al. [86] investigated the in vitro ovicidal and larvicidal activities of *Paecilomyces lilacinus, Syncephalastrum racemosum,* and *Bacillus cereus* on cucumber and also their combined effect on *M. incognita*. A comparison was made with the chemical nematicide fosthiazate. The ovicidal and larvicidal actions of all the bioagents, with the exception of the *B. cereus* bioagent at a concentration of 2%, demonstrated nematicidal potential. At a 50% concentration, the combination of *P. lilacinus* and *S. racemosum* showed the highest ovicidal activity, which led to a roughly 70% decrease in egg hatching when compared to the control. The treatment containing 50% of *S. racemosum* exhibited the highest larvicidal activity, resulting in a mortality rate of 96.7%. The application of *P. lilacinus* along with *S. racemosum* under pot conditions resulted in significantly fewer galls and nematodes than the untreated control. In comparison to the untreated control, the combination of *S. racemosum* and *P. lilacinus* enhanced the cucumber yield under greenhouse conditions while also reducing the severity of nematodes. The application of *P. lilacinus and S. racemosum* together was the most effective treatment among all of them for controlling *M. incognita*.

Predatory nematodes can be employed as biological control agents against phytoparasitic nematodes, but this is challenging to accomplish because a number of ecological factors need to be considered in order to assess their potential, including their biology, behavior, feeding preferences, relationships with prey, and other ecological parameters.

The orders Mononchida, Rhabditida, Diplogasterida and Aphelenchida; the superfamilies Nygolaimoidea, Dorylaimoidea and Actinolaimoidea; and the families Monohysteridae, Ironidae, Thalassogeneridae, and Oncholaimidae are home to predatory nematodes. They can be divided into three categories based on the various feeding apparatuses, methods of finding and capturing prey, and feeding mechanisms they possess [87, 88]. (i) Some feed by slicing the prey's body and sucking its contents because they are unable to consume the prey as whole (Diplogasterida); (ii) others feed by combining cutting and sucking and occasionally consuming the prey as whole (Mononchida); and, (iii) those that only puncture the cuticle of the prey and suck out the contents of the body (suborders Aphelenchida, Dorylaimina and Nygolaimina).

By evaluating the galling index and number of egg masses of *M. javanica*, the second stage juvenile antagonism in soil, and the J2s reproduction in the root, [89] assessed the antagonistic activity of five EPN strains isolated from different fields in Morocco. In greenhouse experiments, RKN J2s were added to the soil along with strains of *Steinernema* sp. (EL30), *Steinernema feltiae* (EL45 and SF-MOR9), and *Heterorhabditis bacterioph*ora (HB-MOR7 and EL27). When comparing the EPNs treatments to the positive control, a significant decrease in the densities of *M. javanica* in the soil and roots was found.

3.2.2 Botanicals

The potential of essential oils (EOs) as biopesticides for the control of RKNs has been assessed by researchers and industry [73]. In order to control *M. incognita* on tomatoes in greenhouse, EOs were assessed as soil biofumigants [90]. Five distinct plant species' EO concentrations (50, 100, and 200 µl/kg soil) were examined. At all concentrations, essential oils of *Pelargonium asperum and Eucalyptus globulus* markedly decreased gall formation and nematode multiplication on roots while simultaneously increasing biomass in both the top and roots. About 29 plant-based EOs were tested against *M. incognita* on tomatoes in vivo and in vitro in a different study [91]. At 1000 µg/ml, 14 EOs exhibited nematicidal activities ranging from 8 to 100%. The Mexican tea (*Dysphania ambrosioides*) EO showed the most notable activity, reducing the number of galls and eggs by 99.5 and 100%, respectively.

Above 100 species of plants have been evaluated for the nematicidal characteristics; a few have demonstrated prominent outcomes in terms of eliminating root-knot nematodes. The root-knot juveniles have been shown to be inhibited by *Euphorbia caducifolia, Calotropis procera*, *Azadirachta indica,* and *Nerium oleander* [71, 92, 93].

Similar results have been observed when crude extracts from the roots and stem of various weed species, like *Fumaria parviflora*, have suppressed *M. incognita* second stage juveniles and its eggs on tomato. Alkaloids, sterol, and alcohol were among the nematicidal phytochemicals found in *F. parviflora* root extracts. In tomato roots, these compounds have significantly decreased nematode reproduction and root galling. Compounds like nonacosane-10-ol and 23α-Homostigmasten-3β-ol, for instance, demonstrated synergistic effects on *M. incognita*'s J2s in tomato [94].

3.2.3 Nematicidal effect of plant essential oils and their compounds

Essential oils, sometimes referred to as volatile or ethereal oils, are complex blends of naturally occurring volatile compounds that give over 3000 plants their unique flavor or scent [95]. Secondary metabolites called EOs play a significant part in a plant's defense mechanisms against pathogens. Different plant parts, such as seeds, stems, leaves, wood, flowers, buds, roots, and fruits, may yield them [96–100]. A lot of research has been done on the impact of essential oils from various plant species on *Meloidogyne* spp. The studies were carried out in vivo as well as in vitro [97, 101].

Apiaceae, Lamiaceae, Rutaceae, and Myrtaceae are examples of studied plant families, containing species belonging to the genera Mentha, Rosmarinus, Artemisia, and Thymus. For instance, a 1000 ppm in vitro test of *Eupatorium viscidum* resulted in 100% of the J2 mortality of *Meloidogyne javanica* [102]. Also, in an in vitro investigation, *Rosmarinus ofcinalis* and *Artemisia herba-alba* inhibited over 98 and 94% of *Meloidogyne incognita,* respectively [103].

Many nematicidal compounds, such as phenols, sesquiterpenes, terpenoids, and terpenes, are found in essential oils (EOs) [95]. The most active ingredients are those that contain hydroxyl and carbonyl groups, such as thymol, carvacrol, and geraniol [103–107]. For *Meloidogyne javanica and incognita*, thymol is the most effective nematicidal compound [103, 105, 107]. For instance, at 500 ppm concentration, thymol resulted in 100% mortality of *Meloidogyne javanica* second-stage juveniles [107].

According to other research, EOs and their compounds penetrate the gelatinous matrix to inhibit *Meloidogyne* spp. eggs from hatching [107–109]. Bakkali et al. [97] state that the primary components of essential oils' chemical composition are primarily responsible for their nematotoxic activity [90, 105, 110–112].

Avato et al. [103] state that the kind and concentration of active ingredients are the primary determinants of EO bioactivity. According to other authors, EOs' bioactivity involves both major and minor compounds, which can function antagonistically or cooperatively. To attain the entire toxic effect of the EOs, additional minor compounds must be added [105, 107, 109].

Wen et al. [113] examined the nematicidal activity of extracts from *Camellia oleifera and Paenoia rockii* against *M. incognita* in vitro. The hatching of eggs was inhibited by extracts from *C. oleifera* cake and *P. rockii* stems. On J2, these extracts also exhibited nematotoxic activity, with *P. rockii* showing the greatest effect. After 7 days, *M. incognita* J2 was 100% inactivated by the *P. rockii* extract at a dose of 5 mg/ml.

In a different study, vetiver (*Vetiveria zizanioides*), a known nonhost grass for some nematodes, was found to have toxic plant extracts that caused 40 and 70% J2 mortality in *M. incognita*, respectively [114].

The plant extracts, which included an acid and oil mixture, repelled J2 as well. Researchers successfully controlled *M. incognita* and *M. javanica* in a greenhouse experiment by using water extracts from Chinaberry (*Melia azedarach*) fruits, and they also observed an increase in soil biological activity [115]. In 1 gram of tomato roots, the number of *M. incognita* females dropped to only 16 for the plants treated with extract. *M. javanica* showed a similar trend, with numbers falling from 79 to 4.2. In the field studies, the number of J2 was roughly 1056 for the water-extract treated plants and 3564 for the untreated control [73].

3.2.4 Soil amendments

A study was done to find out how well two botanicals—tobacco waste dust and ground neem seed—work against root knot nematodes (RKN) in tobacco. Root knot and parasitic nematode counts were estimated in order to investigate the suppression of RKN in field conditions.

Measurements were taken of the tobacco's shoot length, fresh weight, dry weight, root length, and seedling density in order to assess the growth performance.

The most effective level of RKN suppression, as measured by the number of parasitic nematodes and root knots, was achieved by treating ground neem seeds, according to the results, out of all the bio-preparations. Tobacco waste treatment was found to be the second best. It is necessary to carry out more research to ascertain the combined impact of these bio-preparations [116].

Numerous studies have reported that neem oil cake applications reduced the *Meloidogyne* spp. population to half the density [117], and this was associated with an increase in predator and free-living nematodes [9]. Akhtar and Malik [118] repeatedly tested neem oil cake and found its particular efficacy against RKNs even at low dosages (1–2 t/ha).

The effectiveness of organic amendments, such as Farm Yard Manure (FYM), neem cake, poultry manure, mustard oil cake, and vermicompost, against the root knot nematode (*Meloidogyne incognita*) in cucumbers was tested in a field experiment. The organic amendments were applied immediately, 15 days prior to the seeding of the corresponding treatment plots. The application of neem cake at a rate of 1 t/ha was found to be effective in reducing nematode multiplication in soil, improving plant growth, and increasing cucumber yield. In comparison to the untreated control, the treatment decreased the number of galls per root system, the number of eggs per egg mass, the number of galls per root system, and the final nematode population by 76.96, 64.20, 18.87, and 55.89%, respectively [119].

Field tests were carried out in the planting seasons of 2013–2014, 2014–2015, and 2015–2016 in order to assess the efficacy of various organic material sources and the nematicide Furadan 5G in controlling *Meloidogyne incognita* infection on *Allium cepa* L., the main spice onion. Three weeks prior to transplanting onion seedlings, the soil was supplemented with organic materials such as sawdust, rice bran, mustard oil cake, and poultry waste. On the other hand, tricho-composts were added to the soil 5 days prior to seedling transplantation. The findings demonstrated that the effectiveness of various organic materials against the nematode infection varied. Every treatment showed a notable reduction in the formation of galls on the roots and an increase in plant growth and onion yield. Tricho-composts and chicken waste seemed to be the most effective treatments for reducing the growth of root knot nematodes and had a major impact on the development of the onion with the highest yield among the treatments. In addition, it was discovered that sawdust, rice bran, and the nematicide Furadan 5G were superior amendments for reducing the root knot nematode, which improved plant growth and raised onion yield [120]. Due to the presence of predatory nematodes, Kerkeni et al. [121] reported that the population of *Meloidogyne incognita* and galling index on tomato roots were significantly reduced by various composts made from sheep, cattle, horse manures, and poultry (**Figures 2–4**, **Table 2**) [122].

Figure 2.
Flow chart showing different soil amendments.

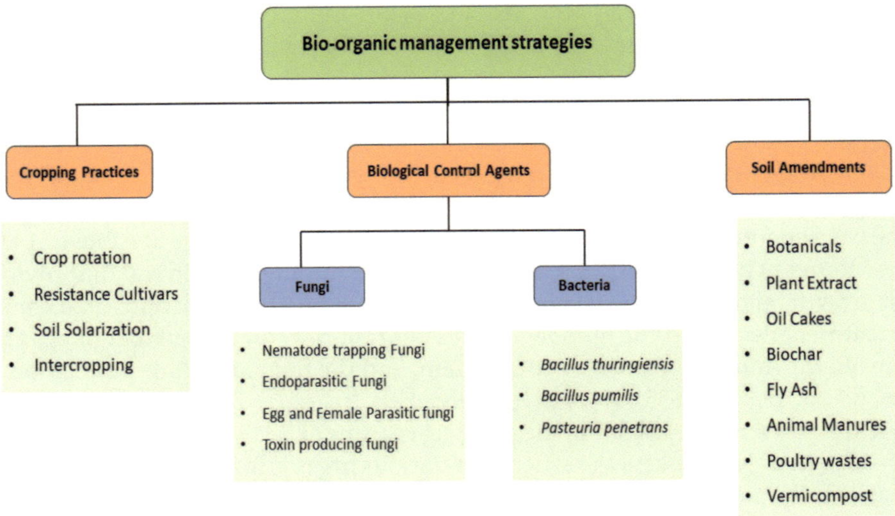

Figure 3.
Flow chart showing different bio-organic management strategies.

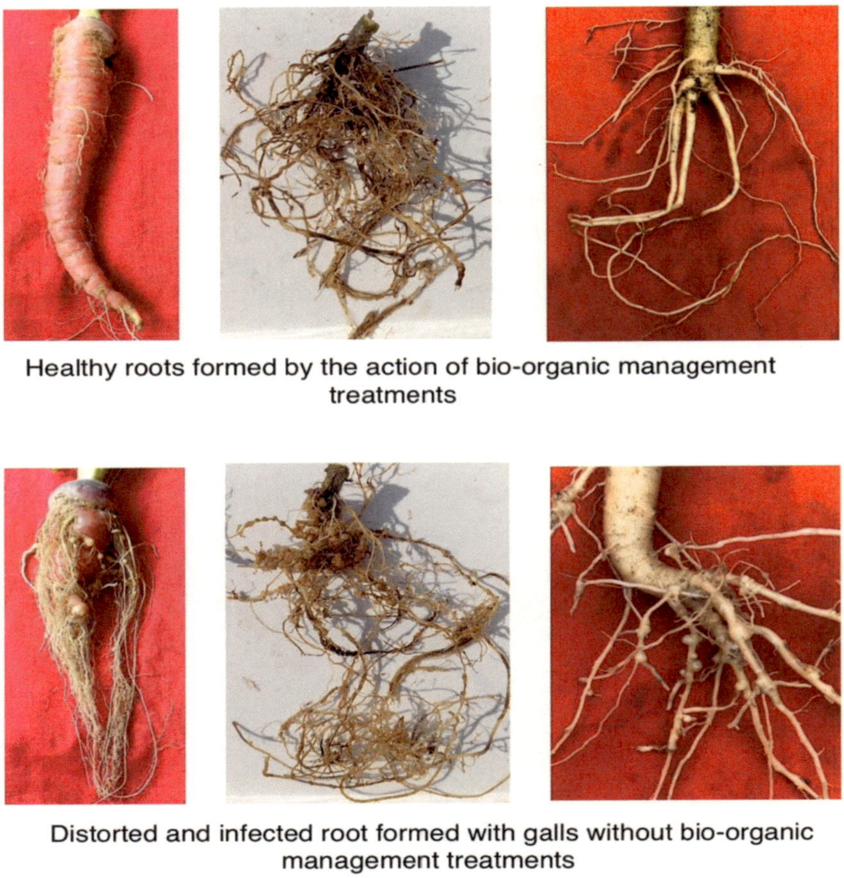

Figure 4.
Effect of bio-organic management treatments against root-knot nematode in carrot, tomato, and papaya roots.

Crop	Resistant Cultivars/Varieties	RKN Species Targeted
Tomato (*Solanum lycopersicum*)	NemaRed, Rutgers, Millennium	*Meloidogyne incognita, M. arenaria*
Cotton (*Gossypium hirsutum*)	NemX	*Meloidogyne incognita*
Bean (*Phaseolus vulgaris*)	Large Lima bean varieties, cultivars with resistance genes from wild relatives	*Meloidogyne javanica*
Cowpea (*Vigna unguiculata*)	Iron and Clay cowpea cultivars	*Meloidogyne* spp.
Sweet potato (*Ipomoea batatas*)	Beauregard, Hernandez, Covington	*Meloidogyne incognita*

Table 2.
Comparative table summarizing the effectiveness of various bio-organic management techniques against root-knot nematodes.

4. Conclusions

In conclusion, bio-organic management offers a promising and sustainable approach for controlling root knot nematodes (RKN) in agroecosystems. By harnessing the power of beneficial organisms and organic amendments, we can effectively suppress nematode populations while promoting plant health and soil fertility. This approach minimizes the reliance on harmful chemical nematicides, contributing to a healthier environment and safer food production.

However, further research is needed to optimize these bio-organic techniques for different crops, soil conditions, and RKN species. Additionally, integrating these strategies with other sustainable practices, such as crop rotation and cover cropping, will further enhance their efficacy and promote long-term RKN suppression in agroecosystems. By embracing bio-organic management, we can move toward a future of sustainable agriculture that ensures healthy crops, thriving soil ecosystems, and a safe food supply.

Conflict of interest

The authors declare no conflict of interest.

Author details

Jaseem KP*, Ameer Favas V, Shana Sherin and Sonia Waqar
Aligarh Muslim University, Aligarh, India

*Address all correspondence to: jaseemkp1196@gmail.com

References

[1] Bernard GC, Egnin M, Bonsi C. The impact of plant-parasitic nematodes on agriculture and methods of control. In: Nematology-Concepts, Diagnosis and Control. London, UK: IntechOpen; 2017

[2] Sikandar A, Jia L, Wu H, Yang S. *Meloidogyne enterolobii* risk to agriculture, its present status and future prospective for management. Frontiers in Plant Science. 2023;**13**:1093657

[3] Brand D, Soccol CR, Sabu A, Roussos S. Production of fungal biological control agents through solid state fermentation: A case study on *Paecilomyces lilacinus* against root-knot nematodes. Micología Aplicada International. 2010;**22**(01):31-48. Return to ref 2010 in article

[4] Lamelas A, Desgarennes D, López-LimaD,VillainL,Alonso-SánchezA, Artacho A, et al. The bacterial microbiome of Meloidogyne-based disease complex in coffee and tomato. Frontiers in Plant Science. 2020;**11**:136

[5] Karpouzas DG, Karanasios E, Giannakou IO, Georgiadou A, Menkissoglu-Spiroudi U. The effect of soil fumigants methyl bromide and Metham sodium on the microbial degradation of the Nematicide Cadusafos. Soil Biology and Biochemistry. 2005;**37**:541-550

[6] Subedi S, Thapa B, Shrestha J. Root-knot nematode (*Meloidogyne incognita*) and its management: A review. Journal of Agriculture and Natural Resources. 2020;**3**:21-31

[7] Singh B, Devindrappa HKK, Singh U, Gupta S. Eco–friendly management of *Meloidogyne javanica* in chickpea (*Cicer arietinum* L.) using organic amendments and biocontrol agent. Journal of Cleaner Production. 2020;**257**:1-8

[8] Ehlers JD, Matthews WC, Hall AE, Roberts PA. Breeding and evaluation of cowpeas with high levels of broad-based resistance to root-knot nematodes. In: Proceedings of World Cowpea Conference III: 2002. Ibadan, Nigeria: International Institute of Tropical Agriculture (IITA); 2002. pp. 41-51

[9] Collange B, Navarrete M, Peyre G, Mateille T, Tchamitchian M. Root-knot nematode (*Meloidogyne*) management in vegetable crop production: The challenge of an agronomic system analysis. Crop Protection. 2011;**30**(10):1251-1262

[10] Verdejo-Lucas S, Talavera M. Root-knot nematodes on zucchini (*Cucurbita pepo* subsp. pepo): Pathogenicity and management. Crop Protection. 2019;**126**:104943

[11] Singh UB, Sahu A, Singh RK, Singh DP, Meena KK, Srivastava JS, et al. Evaluation of biocontrol potential of *Arthrobotrys oligospora* against *Meloidogyne graminicola* and *Rhizoctonia solani* in Rice (*Oryza sativa* L.). Biological Control. 2012;**60**(3):262-270

[12] Bais HP, Weir TL, Perry LG, Gilroy S, Vivanco JM. The role of root exudates in rhizosphere interactions with plants and other organisms. Annual Review of Plant Biology. 2006;**57**:233-266

[13] Hamada N, Yimer HZ, Williamson VM, Siddique S. Chemical hide and seek: nematode's journey to its plant host. Molecular Plant. 2020;**13**(4):541-543

[14] Oota M, Tsai AYL, Aoki D, Matsushita Y, Toyoda S, Fukushima K,

et al. Identification of naturally occurring polyamines as root-knot nematode attractants. Molecular Plant. 2020;**13**(4):658-665

[15] Abad P, Favery B, Rosso MN, Castagnone-Sereno P. Root-knot nematode parasitism and host response: Molecular basis of a sophisticated interaction. Molecular Plant Pathology. 2003;**4**(4):217-224

[16] Das N, Mahalik J. Brinjal-root-knot and reniform nematodes. In: Nematode Problems in Crops and Their Management in South Asia. UK: Cambridge Scholars Publishing; 2024. pp. 322-331

[17] Rodiuc N, Vieira P, Banora MY, de Almeida Engler J. On the track of transfer cell formation by specialized plant-parasitic nematodes. Frontiers in Plant Science. 2014;**5**:76813

[18] Elling AA. Major emerging problems with minor *Meloidogyne* species. Phytopathology. 2013;**103**(11):1092-1102

[19] Mitkowski N, Abawi G. Reproductive fitness on lettuce of populations of *Meloidogyne hapla* from New York state vegetable fields. Nematology. 2003;**5**(1):77-83

[20] Mtshali M, Fourie H, Mc Donald AH. Incidence of root-knot nematodes in resource-poor agriculture in South Africa. African Plant Protection. 2002;**8**:14-15

[21] Sikora RA, Fernandez E. Nematode parasites of vegetables. In: Luc M, Sikora RA, Bridge J, editors. Plant Parasitic Nematodes in Subtropical and Tropical Agriculture. Wallingford: CAB International; 2005. pp. 319-392

[22] Bridge J, Starr JL. Plant nematodes of agricultural importance. In: A Color Handbook. USA: Academic Press (an imprint of Elsevier); 2007

[23] Ntidi NK, Fourie H, Mc Donald AH, et al. Plant-parasitic nematodes associated with weeds in developing agriculture in South Africa. Journal of Nematology. 2012;**14**:875-887

[24] Tefu G, Daneel MS, Steyn WP, et al. The effect of organic amendments in integrated crop production for vegetables in rural farming. African Plant Protection. 2005;**12**:11

[25] Nono-Womdim R, Swai IS, Mrosso LK, et al. Identification of root-knot nematode species occurring on tomatoes in Tanzania and resistant lines for their control. Plant Disease. 2002;**86**:127-130

[26] Mcsorley R. Overview of organic amendments for management of plant-parasitic nematodes, with case studies from Florida. Journal of Nematology. 2011;**43**(2):69

[27] Wesemael WM, Moens M. Quality damage on carrots (*Daucus carota* L.) caused by the root-knot nematode *Meloidogyne chitwoodi*. Nematology. 2008;**10**(2):261-270

[28] Abd-Elgawad MMM. Plant-parasitic nematodes of strawberry in Egypt: A review. Bull NRC. 2019a;**43**:7

[29] Ibrahim IKA. Diseases and Pests of Vegetable Crops and Control Methods. Alexandria: Monshaat Al-Maarf Publisher; 2006

[30] Bakr RA, Mahdy ME, Mousa EM. A survey of root-knot and citrus nematodes in some new reclaimed lands in Egypt. Pakistan Journal of Nematology. 2011;**29**(2):165-170

[31] Mashavakure N, Bandaru G. Host resistance, current status, and

emerging advances. In: Khan MR, editor. Novel Biological and Biotechnological Applications in Plant Nematode Management. Singapore: Springer; 2023. DOI: 10.1007/978-981-99-2893-4_4

[32] Ojeda-Rivera JO, Ulloa M, Roberts PA, Kottapalli P, Wang C, Nájera-González HR, et al. Root-knot nematode resistance in *Gossypium hirsutum* determined by a constitutive defense-response transcriptional program avoiding a fitness penalty. Frontiers in Plant Science. 2022;**13**:858313

[33] Bebber DP, Holmes T, Gurr SJ. The global spread of crop pests and pathogens. Global Ecology and Biogeography. 2014;**23**(12):1398-1407

[34] Williamson VM, Hussey RS. Nematode pathogenesis and resistance in plants. Plant Cell. 1996;**8**:1735-1745. DOI: 10.1105/TPC.8.10.1735

[35] Roberts PA. Current status of the availability, development, and use of host plant resistance to nematodes. Journal of Nematology. 1992;**24**:213-227

[36] Ulloa M, Wang C, Roberts PA. Gene action analysis by inheritance and quantitative trait loci mapping of resistance to root-knot nematodes in cotton. Plant Breeding. 2010;**129**:541-550. DOI: 10.1111/J.1439-0523.2009.01717.X

[37] Ogallo JL, Goodell PB, Eckert J, Roberts PA. Evaluation of NemX, a new cultivar of cotton with high resistance to *Meloidogyne incognita*. Journal of Nematology. 1997;**29**:531-537

[38] Wang C, Ulloa M, Roberts PA. Identification and mapping of microsatellite markers linked to a root-knot nematode resistance gene (rkn1) in Acala NemX cotton (*Gossypium hirsutum* L.). Theoretical and Applied Genetics. 2006a;**112**:770-777. DOI: 10.1007/s00122-005-0183-0

[39] Roberts PA, Ulloa M. Introgression of root-knot nematode resistance into tetraploid cottons. Crop Science. 2010;**50**:940-951. DOI: 10.2135/CROPSCI2009.05.0281

[40] Gutiérrez OA, Jenkins JN, McCarty JC, Wubben MJ, Hayes RW, Callahan FE. SSR markers closely associated with genes for resistance to root-knot nematode on chromosomes 11 and 14 of upland cotton. Theoretical and Applied Genetics. 2010;**121**:1323-1337. DOI: 10.1007/S00122-010-1391-9/TABLES/6

[41] He Y, Kumar P, Shen X, Davis RF, Van Becelaere G, May OL, et al. Re-evaluation of the inheritance for root-knot nematode resistance in the upland cotton germplasm line M-120 RNR revealed two epistatic QTLs conferring resistance. Theoretical and Applied Genetics. 2014;**127**:1343-1351. DOI: 10.1007/S00122-014-2302-2

[42] Kumar P, He Y, Singh R, Davis RF, Guo H, Paterson AH, et al. Fine mapping and identification of candidate genes for a QTL affecting *Meloidogyne incognita* reproduction in upland cotton. BMC Genomics. 2016;**17**:1-12. DOI: 10.1186/S12864-016-2954-1/TABLES/2

[43] Wang C, Ulloa M, Nichols RL, Roberts PA. Sequence composition of bacterial chromosome clones in a transgressive root-knot nematode resistance chromosome region in tetraploid cotton. Frontiers in Plant Science. 2020;**11**:1730. DOI: 10.3389/fpls.2020.574486

[44] Sikora RA, Fernandez E. Nematode parasites of vegetables. In: Plant parasitic nematodes in subtropical and tropical agriculture. Wallingford UK: CABI Publishing; 2005. pp. 319-392

[45] Sorribas FJ, Ornat C, Verdejo-Lucas S, Galeano M, Valero J. Effectiveness and profitability of the mi-resistant tomatoes to control root-knot nematodes. European Journal of Plant Pathology. 2005;**111**:29-38

[46] Casteel CL, Walling LL, Paine TD. Behavior and biology of the tomato psyllid, *Bactericerca cockerelli*, in response to the Mi-1.2 gene. Entomologia Experimentalis et Applicata. 2006;**121**:67-72

[47] Shrestha S, Kumar R, Behera TK, Sharma HK. Inheritance of resistance to root-knot nematode (*Meloidogyne incognita* race 1) in tomato (*Solanum lycopersicum* L.) 'Pusa 120′. Journal of Horticultural Science and Biotechnology. 2012;**87**:211-216

[48] MelilloMT, LeonettiP, BongiovanniM, Castagnone-Sereno P, Bleve-Zacheo T. Modulation of reactive oxygen species activities and H_2O_2 accumulation during compatible and incompatible tomato-root-knot nematode interactions. New Phytologist. 2006;**170**:501-512

[49] Verdejo-Lucas S, Sorribas FJ. Resistance response of the tomato rootstock SC 6301 to *Meloidogyne javanica* in a plastic house. European Journal of Plant Pathology. 2008;**121**:103-107

[50] Rumbos CI, Khah EM, Sabir N. Response of local and commercial tomato cultivars and rootstocks to *Meloidogyne javanica* infestation. Australian Journal of Crop Science. 2011;**5**:1388

[51] Wubie M, Temesgen Z. Resistance mechanisms of tomato (*Solanum lycopersicum*) to root-knot nematodes (*Meloidogyne* species). Journal of Plant Breeding and Crop Science. 2019;**11**(2):33-40

[52] Corbett BP, Jia L, Sayler RJ, Arevalo-Soliz LM, Goggin F. The effects of root-knot nematode infection and Mi-mediated nematode resistance in tomato on plant fitness. Journal of Nematology. 2011;**43**:82

[53] Jacquet M, Bongiovanni M, Martinez M, Verschave P, Wajnberg E, Castagnone-Sereno P. Variation in resistance to the root-knot nematode *Meloidogyne incognita* in tomato genotypes bearing the Mi gene. Plant Pathology. 2005;**54**:93-99

[54] Cooper W, Jia L, Goggin L. Effects of jasmonate-induced defenses on root-knot nematode infection of resistant and susceptible tomato cultivars. Journal of Chemical Ecology. 2005;**31**:1953-1967

[55] Zhang L, Lei H, Yang R, Chen H, Cheng J, Wang S. Effects of Jasmonatc on proteinase inhibitor II and variation in tomato response to root-knot nematodes. Acta Horticulturae. 2011;**917**:249-252

[56] Zinovieva SV, Vasyukova NI, Udalova ZV, Gerasimova NG. The participation of salicylic and jasmonic acids in genetic and induced resistance of tomato to *Meloidogyne incognita* (Kofoid and White, 1919). Biology Bulletin. 2013;**40**:297-303

[57] Curtis RH, Robinson AF, Perry RN. Hatch and host location. In: Root-Knot Nematodes. Wallingford UK: CABI; 2009. pp. 139-162

[58] Abawi GS, Widmer TL. Impact of soil health management practices on soil-borne pathogens, nematodes and root diseases of vegetable crops. Applied Soil Ecology. 2000;**15**:37-47

[59] Khan MR. Prospects of microbial control of root-knot nematodes infecting vegetable crops. In: Biotechnology: Plant Health Management. Lucknow, India: International Book Distributing Co.; 2007. pp. 643-665

[60] Jatala P. Biological control of nematodes. In: Sasser JN, Carter CC, editors. Advanced Treatise on *Meloidogyne*, Biology and Control. United States Agency for International Development. Issue 1. North Carolina State University: CABI Publishing; 1985. pp. 303-308

[61] de Souza R, Ambrosini A, Passaglia LMP. Plant growth-promoting bacteria as inoculants in agricultural soils. Genetics and Molecular Biology. 2015;**38**:401-419. DOI: 10.1590/S1415-475738420150053

[62] Gupta G, Parihar SS, Ahirwar NK, Snehi SK, Singh V. Plant growth promoting rhizobacteria (PGPR): Current and future prospects for development of sustainable agriculture. Journal of Microbial & Biochemical Technology. 2015;7:96-102. DOI: 10.4172/1948-5948.1000188

[63] Jatala P. Biological control of plant-parasitic nematodes. Annual Review of Microbiology. 1986;**24**:453-489

[64] Migunova VD, Sasanelli N. Bacteria as biocontrol tool against phytoparasitic nematodes. Plants. 2021;**10**:1-15. DOI: 10.3390/plants10020389

[65] Berlitz DL, Knaak N, Cassal MC, Fiuza LM. Bacillus and biopesticides in control of phytonematodes. In: Sahayaraj K, editor. Basic and Applied Aspects of Biopesticides. New Delhi: Springer; 2014. pp. 3-16. DOI: 10.1007/978-81-322-1877-7_1

[66] Meena B. Biological control of pest and diseases using fuorescent pseudomonads. In: Sahayaraj K, editor. Basic and Applied Aspects of Biopesticides. New Delhi: Springer; 2014. pp. 17-29. DOI: 10.1007/978-81-322-1877-7_2

[67] Dehghanian SZ, Abdollahi M, Charehgani H, Niazi A. Combined of salicylic acid and *pseudomonas fuorescens* CHA0 on the expression of PR1 gene and control of *Meloidogyne javanica* in tomato. Biological Control. 2020;**141**:104134. DOI: 10.1016/j.biocontrol.2019.104134

[68] Tapia-Vázquez I, Montoya-Martínez AC, De los Santos-Villalobos S, Ek-Ramos MJ, Montesinos-Matías R, Martínez-Anaya C. Root-knot nematodes (*Meloidogyne* spp.) a threat to agriculture in Mexico: Biology, current control strategies, and perspectives. World Journal of Microbiology and Biotechnology. 2022;**38**(2):26

[69] Zhou L, Yuen G, Wang Y, Wei L, Ji G. Evaluation of bacterial biological control agents for control of root-knot nematode disease on tomato. Crop Protection. 2016;**84**:8-13

[70] Norabadi MT, Sahebani N, Etebarian HR. Biological control of root-knot nematode (*Meloidogyne javanica*) disease by *Pseudomonas fluorescens* (Chao). Archives of Phytopathology and Plant Protection. 2014;**47**(5):615-621

[71] Safni I, Lisnawita L, Lubis K, Tantawi AR, Murthi S. Isolation and Characterization of Rhizobacteria for Biological Control of Root-Knot Nematodes in Indonesia. In: Journal of the International Society for Southeast Asian Agricultural Sciences. Philippines; 2018

[72] Moslehi S, Pourmehr S, Shirzad A, Khakvar R. Potential of some endophytic bacteria in biological control of root-knot nematode *Meloidogyne incognita*. Egyptian Journal of Biological Pest Control. 2021;**31**:1-11

[73] Forghani F, Hajihassani A. Recent advances in the development of environmentally benign treatments to control root-knot nematodes. Frontiers in Plant Science. 2020;**11**:1125

[74] Timper P. Conserving and enhancing biological control of nematodes. Journal of Nematology. 2014;**46**:75-89

[75] Abd-Elgawad MMM, Askary TH. Fungal and bacterial nematicides in integrated nematode management strategies. Egyptian Journal of Biological Pest Control. 2018;**28**:74. DOI: 10.1186/s41938-018-0080-x

[76] Saxena G, Dayal R, Mukerji KG. Interaction of nematodes with nematophagus fungi: Induction of trap formation, attraction and detection of attractants. FEMS Microbiology Letters. 1987;**45**:319-327. DOI: 10.1016/0378-1097(87)90018-8

[77] Hsueh Y-P, Mahanti P, Schroeder FC, Sternberg PW. Nematode-trapping fungi eavesdrop on nematode pheromones. Current Biology. 2013;**23**:83-86. DOI: 10.1016/j.cub.2012.11.035

[78] Wang X, Li GH, Zou CG, Ji XL, Liu T, Zhao PJ, et al. Bacteria can mobilize nematode-trapping fungi to kill nematodes. Nature Communications. 2014;**5**:1-9. DOI: 10.1038/ncomms6776

[79] Khattak B. Biological management of root knot nematode *Meloidogyne javanica* (Treub) with *Trichoderma harzianum* Rifai in tomato at Peshawar [PhD thesis]. Khyber Pakhtunkhwa Agricultural (KPK) University: Pakistan; 2008

[80] Kumari M, Sharma MK, Baheti BL. Biological control of root-knot nematode, *Meloidogyne incognita* infecting tomato (*Solanum lycopersicum*). Journal of Entomology and Zoology Studies. 2020;**8**(1):35-39

[81] Affokpon A, Coyne DL, Htay CC, Agbèdè RD, Lawouin L, Coosemans J. Biocontrol potential of native *Trichoderma* isolates against root-knot nematodes in West African vegetable production systems. Soil Biology & Biochemistry. 2011;**43**:600-608

[82] Sharon E, Bar-Eyal M, Chet I, HerreraEstrella A, Kleifeld O, Spiegel Y. Biological control of the root-knot nematode *Meloidogyne javanica* by *Trichoderma harzianum*. Phytopathology. 2001;**91**:687-693

[83] Dababat AA, Sikora RA. Use of *Trichoderma harzianum* and *Trichoderma viride* for the biological control of *Meloidogyne incognita* on tomato. Jordan Journal of Agricultural Sciences. 2007;**3**(3):297-308

[84] Thongkaewyuan A, Chairin T. Biocontrol of *Meloidogyne incognita* by Metarhizium guizhouense and its protease. Biological Control. 2018;**126**:142-146. DOI: 10.1016/j.biocontrol.2018.08.005

[85] Amer-Zareen ZMJ, Abid M, Gowen SR, Kerry BR. Management of root knot nematode (*Meloidogyne javanica*) by biocontrol agents in two crop rotations. International Journal of Biology and Biotechnology. 2004;**1**:67-73

[86] Huang WK, Cui JK, Liu SM, Kong LA, Wu QS, Peng H, et al. Testing various biocontrol agents against the root-knot nematode (*Meloidogyne incognita*) in cucumber plants identifies a combination of *Syncephalastrum racemosum* and *Paecilomyces lilacinus* as being most effective. Biological Control. 2016;**92**:31-37

[87] Bilgrami AL. In: Ciancio A, Mukerji KG, editors. Biological Control Potentials of Predatory Nematodes. Dordrecht: Springer; 2008. DOI: 10.1007/978-81-322-1877-7_110.1007/978-1-4020-6063-2

[88] Roy P, Borah A. Role of Predeceous nematodes in plant disease management. The Pharma Innovation Journal. 2020;**9**:463-467

[89] El Aimani A, Houari A, Laasli SE, Mentag R, Iraqi D, Diria G, et al. Antagonistic potential of Moroccan entomopathogenic nematodes against root-knot nematodes, *Meloidogyne javanica* on tomato under greenhouse conditions. Scientific Reports. 2022;**12**(1):2915

[90] Laquale S, Candido V, Avato P, Argentieri MP, D'Addabbo T. Essential oils as soil biofumigants for the control of the root-knot nematode *Meloidogyne incognita* on tomato. The Annals of Applied Biology. 2015;**167**:217-224. DOI: 10.1111/aab.12221

[91] Barros AF, Campos VP, De Oliveira DF, De Jesus Silva F, Jardim IN, Costa VA, et al. Activities of essential oils from three Brazilian plants and benzaldehyde analogues against *Meloidogyne incognita*. Nematology. 2019a;**21**:1081-1089. DOI: 10.1163/15685411-00003276

[92] Qamar E, Saeed M, Khan A, Seema N. Control of anguinosis of wheat using plant extracts. Pakistan Journal of Science. 1989;**32**:260-261

[93] Zureen S, Khan MI. Nematicidal activity in some plant lattices. Pakistan Journal of Nematology. 1984;**2**:69-77

[94] Yuhui B, Gao C, Yu Z. Rhabdopeptides from Xenorhabdus budapestensis SN84 and their nematicidal activities against *Meloidogyne incognita*. Journal of Agricultural and Food Chemistry. 2018;**66**(15):3833-3839

[95] Azlay L, El Boukhari MEM, Mayad EH, Barakate M. Biological management of root-knot nematodes (*Meloidogyne* spp.): A review. Organic Agriculture. 2023;**13**(1):99-117

[96] Van de Braak SAA, Leijten GCJ. Essential oils and oleoresins: A survey in the Netherlands and other major markets in the European Union. In: CBI, Centre for the Promotion of Imports from Developing Countries. Netherlands; 1994

[97] Bakkali F, Averbeck S, Averbeck D, Idaomar M. Biological effects of essential oils - a review. Food and Chemical Toxicology. 2008;**46**:446-475. DOI: 10.1016/j.fct.2007.09.106

[98] Figueiredo AC, Barroso JG, Pedro LG, Schefer JJ. Factors affecting secondary metabolite production in plants: Volatile components and essential oils. Flavour and Fragrance Journal. 2008;**23**(4):213-226. DOI: 10.1002/ffj.1875

[99] Christaki E, Bonos E, Giannenas I, Florou-Paneri P. Aromatic plants as a source of bioactive compounds. Agriculture. 2012;**2**:228-243. DOI: 10.3390/agriculture203022

[100] Negi PS. Plant extracts for the control of bacterial growth: Efcacy, stability and safety issues for food application. International Journal of Food Microbiology. 2012;**156**:7-17. DOI: 10.1016/j.ijfoodmicro.2012.03.006

[101] Isman MB. Plant essential oils for pest and disease management. Crop Protection. 2000;**19**:603-608. DOI: 10.1016/S0261-2194(00)00079-X

[102] Sosa ME, Lancelle HG, Tonn CE, et al. Insecticidal and nematicidal essential oils from Argentinean eupatorium and *Baccharis* spp. Biochemical Systematics and Ecology. 2012;**43**:132-138. DOI: 10.1016/j.bse.2012.03.007

[103] Avato P, Laquale S, Argentieri MP, et al. Nematicidal activity of essential oils from aromatic plants of Morocco. Journal of Pest Science. 2017;**90**:711-722. DOI: 10.1007/s10340-016-0805-0

[104] Echeverrigaray S, Zacaria J, Beltrão R. Nematicidal activity of monoterpenoids against the root-knot nematode Meloidogyne incognita. Phytopathology. 2010;**100**:199-203. DOI: 10.1094/PHYTO-100-2-0199

[105] Ntalli NG, Ferrari F, Giannakou I, Menkissoglu-Spiroudi U. Phytochemistry and nematicidal activity of the essential oils from 8 greek lamiaceae aromatic plants and 13 terpene components. Journal of Agricultural and Food Chemistry. 2010;**58**:7856-7863. DOI: 10.1021/jf100797m

[106] Andrés MF, González-Coloma A, Sanz J, et al. Nematicidal activity of essential oils: A review. Phytochemistry Reviews. 2012;**11**:371-390. DOI: 10.1007/s11101-012-9263-3

[107] Santana O, Andrés MF, Sanz J, et al. Valorization of essential oils from Moroccan aromatic plants. Natural Product Communications. 2014;**9**:1109-1114. DOI: 10.1177/19345 78x1400900812

[108] Oka Y. Nematicidal activity of essential oil components against the root-knot nematode *Meloidogyne javanica*. Nematology. 2001;**3**:159-164. DOI: 10.1163/156854101750236286

[109] Laquale S, Avato P, Argentieri MP, et al. Nematotoxic activity of essential oils from Monarda species. Journal of Pest Science. 2018;**91**:1115-1125. DOI: 10.1007/s10340-018-0957-1

[110] Oka Y, Nacar S, Putievsky E, et al. Nematicidal activity of essential oils and their components against the root nematicidal activity of essential oils and their components against the root-knot nematode. Phytopathology. 2000;**90**(7):676-782. DOI: 10.1094/PHYTO.2000.90.7.710

[111] Sacchetti G, Maietti S, Muzzoli M, et al. Comparative evaluation of 11 essential oils of diferent origin as functional antioxidants, antiradicals and antimicrobials in foods. Food Chemistry. 2005;**91**:621-632. DOI: 10.1016/j.foodchem.2004.06.031

[112] Ntalli NG, Nasiou E, Menkissoglu-Spiroudi U. Evaluation of essential oils from rosemary, Orange, Lavandula and false Yellowhead on hatching and motility of RootKnot nematode. Journal of Agricultural Science and Technology. 2013;**3**:603-616

[113] Wen Y, Meyer SLF, MacDonald MH, Zheng L, Jing C, Chitwood DJ. Nematotoxicity of Paeonia spp. extracts and Camellia oleifera tea seed cake and extracts to Heterodera glycines and *Meloidogyne incognita*. Plant Disease. 2019;**103**:2191-2198. DOI: 10.1094/PDIS-09-18-1663-RE

[114] Jindapunnapat K, Reetz ND, MacDonald MH, Bhagavathy G, Chinnasri B, Soonthornchareonnon N, et al. Activity of vetiver extracts and essential oil against Meloidogyne incognita. Journal of Nematology. 2018;**50**:147-162. DOI: 10.21307/jofnem-2018-008

[115] Ntalli N, Monokrousos N, Rumbos C, Kontea D, Zioga D, Argyropoulou MD, et al. Greenhouse biofumigation with Melia azedarach controls *Meloidogyne* spp. and enhances soil biological activity. Journal of Pest Science. 2018;**91**:29-40. DOI: 10.1007/s10340-017-0909-1

[116] Motha KF, Abeysekara R, Kottearachchi NS. Effect of biological agents and botanicals in controlling

root-knot nematodes, *Meloidogyne* Spp., in Nicotiana tabacum. Tropical Agricultural Research and Extension. 2011;**13**(1):1-5

[117] Akhtar M. Biological control of plant-parasitic nematodes by neem products in agricultural soil. Applied Soil Ecology. 1998;**7**(3):219-223

[118] Akhtar M, Malik A. Roles of organic soil amendments and soil organisms in the biological control of plant-parasitic nematodes: A review. Bioresource Technology. 2000;**74**(1):35-47

[119] Devi TH, Das D. Effect of organic amendments on root-knot nematode (*Meloidogyne incognita*) in cucumber. Pest Management in Horticultural Ecosystems. 2016;**22**(2):176-181

[120] Faruk I. Efficacy of organic soil amendments and a nematicide for management of root-knot nematode *Meloidogyne* Spp. of onion. American Journal of Bioscience and Bioengineering. 2019;**7**(4):57-64

[121] Kerkeni A, Horrigue-Raouani N, Ben Khedher M. Effet suppressif de cinq extraits de compost vis-à-vis du nématode à galles *Meloidogyne* incognita. Nematologia Mediterranea. 2007;**35**:15-21

[122] Mokrini F, Janati S, Houari A, Essarioui A, Bouharroud R, Mimouni A. Management of plant parasitic nematodes by means of organic amendment. Revue Marocaine des Sciences Agronomiques et Vétérinaires. 2018;**6**(3):337-344

Chapter 13

Soil Microbiome and Vegetable Crop Interaction in Climate Change Scenario

Sripriya Das, Vandana Kumari, Kiran Kumar Mohapatra, Sayon Mukherjee, Plabani Roy, Dibyajyoti Panda, Abhik Patra, Ashish Rai and Vijay Singh Meena

Abstract

Climate change, like abrupt rise in temperature, changes in precipitation patterns, and extreme weather events, are of major threat to global food security exerting negative impact on crop yields. In addition, it can affect soil microbiome, a complex community of microorganisms plays vital role in plant health. Climate change disrupts various important functions of microbiome leading to decreased crop yields and increased susceptibility to pests and diseases. Vegetable crops are particularly vulnerable to the effects of climate change as they are often grown in marginal environments, such as arid and semi-arid regions, where the soil microbiome is already under stress. In addition, vegetable crops are often harvested multiple times per year, may further deplete the soil microbiome. Thus, study of the interaction of soil microbiome with vegetable crops in this changing ecological scenario is important and discussed detail in this chapter. There are several strategies to mitigate the negative effects of climate change on the soil microbiome and vegetable crop production including use of cover crops to protect the soil from erosion and improve nutrient cycling, planting vegetable crops that are adapted to the local climate, and using sustainable agricultural practices, such as no-till farming and managing pests and diseases using integrated pest management (IPM). By taking these measures to protect and enhance the soil microbiome, we can help to improve the resilience of vegetable crops to climate change. This will help to ensure that vegetable crops can continue to provide a reliable source of food in the face of climate change.

Keywords: climate change, cover crops, integrated pest management, soil microbes, vegetable crops

1. Introduction

Soil is a complex ecosystem that hosts bacteria, fungi, and other, even though it simply considered as a nutrients source for plants [1]. It is of great concern that soil's biological activity is being negatively affected by the continuous increase in surface temperature both outside and inside of the soil. A diverse population of soil microbes has a strong impact on plants and their growth and development [2]. In order to enhance plant resilience, nutrient use efficiency, and productivity under changing climatic conditions, need to understand how climate change alters soil microbial communities' composition, diversity, and activity, as well as their functional implications for vegetable crop performance. Use of chemical fertilizers indiscriminately leads to soil pollution gradually deteriorating the soil fertility [3]. Although productivity increased appreciably to feed the starving population with the onset of green revolution, the use of chemical fertilizers and pesticides glorified then had an obvious long-term adverse effect. It is urgently needed to develop alternative strategies to solve the problem of synthetic chemicals due to their harmful effects. Hence, several studies are taken up to find the sustainable alternatives for chemical fertilizers.

In this regard, plant growth-promoting rhizobacteria (PGPR) came into light as a better plant growth promoter. It is imperative to use plant growth-promoting rhizobacteria (PGPR) to enhance soil fertility, promote plant growth, and inhibit phytopathogens in the rhizosphere through a variety of mechanisms [3]. Plant growth-promoting rhizobacteria (PGPR) and their role in improving vegetable production are still being studied in a variety of ways. There is a high level of biological activity around the root surfaces of a plant called the rhizosphere. A natural soil bacteria known as PGPR colonizes plant roots aggressively and provides benefits to the plant. In soil, PPGPR participate in a variety of beneficial processes such as decomposition of crop residues, synthesis of soil organic matter, mineralization of soil organic matter, and immobilization of mineral nutrients that improve soil fertility. There is still much research to be done in order to better understand soil PGPR communities and their roles in vegetable production. Many PGPR genera have been associated with solanaceous vegetable crops, including *Pseudomonas, Azospirillum*, *Klebsiella, Alcaligenes, Azotobacter, Enterobacter*, and *Serratia*, as well as *Pseudomonas, Bacillus*, etc.

The changing climate has led to numerous changes in the life cycle of crops and their ecological characteristics. It is of great concern that soil's biological activity is being negatively affected by the continuous increase in surface temperature both outside and inside of the soil. The loss of biodiversity, soil fertility, and degradation of the environment have caused substantial reductions in agricultural production, including vegetable production. This has resulted in inadequate food supplies for human populations. Moreover, cultivation of some cereal crops could increase the emissions of greenhouse gases. Hence, instead of cereal crops, production of vegetable crops can reduce the cost of cultivation as well GH gas emissions. Among the many benefits of vegetables is that they provide essential minerals, vitamins, and fiber, much of which cannot be found in starchy foods, and also offer significant protein and carbohydrate supply. In addition to cereals, vegetable consumption provides more comprehensive nutrition to humans. In this context, researchers are exploring for alternatives to cultivate vegetable production in climate change scenario.

Plant growth is enhanced either directly or indirectly by PGPR through multiple mechanisms. As a result of direct mechanisms, phosphates are solubilized, nitrogen is fixated, siderophore is produced, HCN is produced, ammonia is produced, vitamins are produced, and phytohormones are produced (such as auxins, cytokinins, and gibberellins) [4]. Conversely, indirect mechanisms involve that mechanism which does not directly promote growth but plays a role in synthesis [5]. The PGPR participates in all beneficial soil functions such as the decomposition of crop residues, the synthesis and mineralization of soil organic matter, the immobilization of mineral nutrients, phosphate solubilization, nitrogenation, and phytohormone synthesis, which improve soil fertility and productivity in the long run [6]. A considerable number of biomolecules is produced by rhizobacteria associated with roots, which are then mixed with the soil and improve soil quality. The use of plant growth-promoting rhizobacteria (PGPR) may contribute to decreasing the negative environmental impact caused by chemical fertilizers, herbicides, and pesticides. PGPR is primarily used for rhizobia manipulation in legumes for studies on nodulation and nitrogen fixation, which increase legume growth and development. The use of rhizobacteria that promote plant growth and pest control has been regarded as a safer and more environmentally-friendly alternative to chemical fertilizers and pesticides [6].

Agricultural systems continue to face many challenges due to rapid changes in the Earth's climate, which include shifting temperature and precipitation patterns, increased frequency of extreme weather events, and altered nutrient availability. Plant productivity, soil health, and nutrient cycling are all directly affected by these changes in soil communities. Bacteria, fungi, archaea, and viruses are among the diverse soil microbes that play a crucial role in nutrient acquisition, disease suppression, and overall plant health. These microbes form complex communities that interact with each other and with one another. The way they interact with plants, especially vegetable crops, profoundly affects plant growth, health, and nutrient uptake.

2. Impact of biotic factors on vegetable production

As a protective food, vegetables are essential sources of nutrients, vitamins, and minerals and can help overcome micronutrient deficiencies. Over the last quarter century, vegetable production has doubled, and the value of global vegetable trade has surpassed cereal trade. A temperate or sub-temperate climate in the east of Asia produces the highest yields. In terms of production, India ranks second behind China (22.5 t/ha) with 17.3t/ha. In two decades, India has increased vegetable production 2.5 times [7] from 58.5 metric tons in 1991–1992 to 146.5 metric tons in 2010–2011. In general, vegetables are sensitive to extremes in the environment, so high temperatures, limited soil moisture, and biotic factors have a significant impact on low yields. Climate change will further magnify these problems. In changing climates, crop failure is common, yields are low, quality is reduced, and pests and diseases become more prevalent, making vegetable cultivation unprofitable. Ultimately, this calls into question the availability of nutrient sources in the human diet [8]. Water shortages caused by the failure of the monsoons cause crop yields to be below average among the major drought-prone regions.

Despite direct climate change impacts vegetable cultivation, it also affects water storage and irrigation availability. Drought will cause considerable stress to vegetable production due to its limited water availability [9]. As a result of climate change, species distributions, phenologies, and ecological interactions will change over the next 100 years. The pollination of many vegetable crops is affected by changes in insect species distribution, such as cabbage, onions, and root crops. Ref. [10] mention another problem associated with weed invasion in agricultural systems. In addition to changing post-harvest quality of vegetables, changing climates can adversely affect food safety and cause serious losses during storage, such as changing population of aflatoxin secreting fungi [11]. Additionally, it affects vegetable storage and distribution, and it may worsen in extreme weather events [12]. Physical and biochemical changes caused by climate change, pests, and diseases are also affected by host-pathogen interactions, insect distributions and ecology, migration patterns, and overwintering abilities. Host susceptibility and host-parasite relationships can be affected by changes in the environment surrounding a host and pathogen [13]. Physiological changes in the host and pathogen development can be affected by climate change [14]. According to [15], fungi and bacteria grow and spread depending on rainfall, temperature, radiation, humidity, and dew. The availability of nutrients and ozone are also factors affecting plant diseases.

2.1 Temperature

The primary effect of climate change adversely affecting vegetable production is the fluctuation in daily temperatures because many physiological, biochemical, and metabolic activities of crops are temperature dependent. Potatoes require specific temperatures and day lengths for tuber formation and flowering, making them the most susceptible crops to change in climatic conditions, and there can be a decline in productivity of the states where potato are grown in India (**Table 1**) by 2050 [16]. To minimize yield losses of up to 8%, researchers suggest planting potatoes in mid-November. The growing season of potatoes is prolonged by an increase in temperature in high altitude and temperate regions such as Europe, Russia, and India. The conversion of starch into sugar occurs at low temperatures, resulting in browning of chips due to the charred sugar, and the processing industry's preference for them decreases as a result. Ultimately, this results in higher post-harvest losses over the current level of 40–50%. In winter seasons, areas with low nighttime temperatures are most likely to face this problem [17].

When it comes to assessing tomato marketability, fruit color plays a significant role. Lycopene pigment develops best at temperatures between 25 and 30°C in tomatoes. A temperature of 27°C will begin to degrade lycopene, and a temperature of 40°C will destroy it. A high temperature of 25°C affects tomato pollination and fruit set, female reproductive tissues development, pollen production, imbalances in hormone, and low carbohydrate levels are responsible for tomatoes' poor reproduction at high temperatures [18, 19].

Climate change effects	Insects/Pests of vegetables
Invasion to temperate zones due to increase in temperature	Sweet potato weevil (*Cylas prmicarius*)
Due to a lower developmental zero point and low thermal totals needed for one generation, more infestation took place due to an increase in the number of generations per year.	Aphids (*Aphis gossypi; Aulacorthumsolani*)
The number of generations per year increased, resulting in more infestations	Fruit and Pod borer (*Helicoverpaarmigera; Spodoptera litura*)
Increase in its potential northern limit by 400 km Potato growing areas in 99% of the country are at risk	Colorado potato beetle (*Leptinotarsa decemlineata*)
An increase in pest status and more frequent overwintering	Diamond back moth (*Plutellaxylostella*)
An increase in range, abundance, and diversity	Cabbage butter fly (*Pierisbrassicae*)
With a 3°C temperature increase in the UK, populations become active a month earlier	Cabbage root fly (*Delia radicum*)

Table 1.
Effect of climate change on distribution, multiplication, population, diversity, range, abundance, and expansion of some vegetables' insects and pests.

2.2 Drought and salinity

Global warming is most responsible for droughts and salinities. Drought conditions affect the seeds' germination adversely in vegetable crops like okra, onion, and sprouting of potatoes [20]. Tuber yield can also be reduced by moderate levels of water stress. In leafy vegetables such as palak, amaranthus, and spinach, succulent leaves are commercial products whose water content is reduced due to drought conditions [21]. Plant cells lose water through reverse osmosis when there is drought. As a result, several biochemical and physiological processes like respiration, photosynthesis, etc., are inhibited, reducing the productivity of vegetables [22]. When plants are exposed to salt stress, they lose turgor, reduce growth, wilt, reduce photosynthesis, experience tissue necrosis, and eventually die [23]. A saline soil is not suitable for onions, but moderately suitable for cucumbers, eggplants, peppers, and tomatoes [22]. Cheeseman [24] found that salinity greatly reduced germination percentages, germination rates, root and shoot lengths, and fresh root and shoot weights in cabbage. Potato tuber yield tends to decrease with salinity. The various effect of climatic changes on abiotic and biotic factors has been listed in **Table 2**.

Climatic elements	Effects	Vegetables	References
Temperature	Potato production in high altitudes and temperate regions will face a decline in productivity and tuber yield by 2050	Potato	[16, 17, 25]
	Browning in chips occurs when starch is converted into sugar at low temperatures		
High temperatures	A lack of pollination and low carbohydrate levels are two factors contributing to tomatoes' poor reproductive performance. High temperatures significantly inhibit the production of ripening hormones in tomatoes through various mechanisms, including the suppression of mRNA expression. This prevents continuous protein synthesis, which includes the production of ethylene and the accumulation of lycopene, as well as the dissolution of cell wall	Tomato	[19, 26]
	Fruit set is inhibited by high temperatures after pollination	Pepper	[27]
	Falling temperatures also cause flower drop, ovule abortion, poor fruit set, and fruit drop in ripe chili fruits, all of which affect seed germination	Chili	[20]
	Fruits take longer to ripen due to fluctuations.	Okra	[28]
	Climates with high humidity increase vegetative growth, resulting in poor flower production in females.	Cucumber, watermelon, winter and summer squash, pumpkin seeds Cucurbitaceous vegetables	[29, 30]
Moisture stress	Reduces the productivity bolting	French bean and cole crops	[17]
Drought and salinity	Reduces germination of seeds sprouting of tubers	Onion and okra	
Drought	Reduces quality of leafy vegetables	Amaranthus, palak and spinach	

Climatic elements	Effects	Vegetables	References
Salinity	Turgor loss, growth reduction, wilting, leaf abscission, loss of cellular integrity, tissue necrosis, and finally death.	Onion, cucumber, eggplant, pepper, and tomato	[22]
CO_2 and relative humidity	Concentration the photosynthetic rates will come down inhibits fruit ripening.	Potato	[31]
	Affects pest and disease occurrence	Tomato	[32]
	Insects feed more on leaves with lower nitrogen content in order to get more nitrogen for their metabolism		[33, 34]
	Increases caterpillar food consumption, aphid reproduction, and nitrogen-based plant defense. Benefits include increased predation by predators, Bt foliar application, carbon-based plant defense, and decreased insect development	Soyabean	[32, 35]
	By promoting plant diseases such as rusts, powdery mildews, leaf spots, and blights, warmer winter temperatures increase pathogen survival and overwintering.		[14, 35, 36]
Precipitation	Removes white flies and thrips from field	Tomato	[13]
	High humidity due to high precipitation favors insects, *e.g.*, pod borer, leaf hopper, increase in fungal pathogens favor the development and survival of pathogens; helps in increasing disease severity	Okra, Brinjal	
Atmospheric pollutants and UV radiation	Increased foliar, root attacking pathogens such as plant nematodes stimulate spore germination and invasion of fungi and egg hatching, penetration and reproduction of plant parasitic nematodes reduces dry weight, leaf area and plant height, inhibits growth	Tomato	[37]
	Glazing and bronzing and increases susceptibility to virus infection	French bean	[38]

Table 2.
Climate change effects on abiotic factors of vegetable production.

3. Effect of climate change on pest incidence in vegetable production

Insect pest populations are likely to suffer from climate change resulting in increased temperatures in a number of complex ways. Despite this, sometimes climate change affects possibly depressing insect population, and the warmer temperatures in temperate climates will result in a greater variety of insects and a higher population of them. It has also been shown that climate change affects the ecology and biology of insect pests [39]. The increased temperature increases fecundity in some insects with short life cycles, such as aphids and diamond back moths. As a result, they are capable of producing increase in numbers of generations per year than they normally do [40]. In contrast, some insects such as thrips may alter gender ratios as a result of temperature changes [41]. As soil buffers its temperature change is better than atmospheric temperature, so it provides an insulating medium, some insect species living in soil may suffer more than insects living on the surface of the soil [42]. Higher temperatures may adversely affect certain pest species in the tropics as they migrate toward higher latitudes. Das et al. [35] found that high atmospheric temperatures enhance insect development, oviposition rates, outbreaks, and the introduction of invasive species, but also reduce insect biocontrol effectiveness by fungi, economic threshold levels, insect diversity, and parasitism in ecosystems. Due to global warming, *Nezaraviridula*, a sub-tropical and tropical pest, is slowly moving toward north in Japan, replacing the more temperate *Nezaraantennata*. From 1969 to 2004, *Helicoverpa armigera* migratory numbers increased dramatically as well, according to [43]. As temperatures rise, sucking pests, *e.g.*, whiteflies, aphids, and thrips become more active and more numerous [44]. According to Fleming and Tatchell [45], aphids will arrive 8 days earlier over the next 50 years if in suction traps are used to measure aphid presence. Many viral diseases are caused by pests that damage tomatoes, chillies, potatoes, eggs, cucurbits, okra, and legumes. Soil-borne pathogens do not survive in temperate climate because of their high temperatures and frost sensitivity mechanisms, e.g., *Sclerotium rolfsii*, *Macrophomina phaseolina,* etc. [46], so their distribution is affected by temperature and frost sensitivity. Increased temperatures cause airborne pathogens to cycle faster due to the demise of frost [46, 47]. Thomas [48] and Harrington and Stork [49] reported that rising winter temperatures resulted in an earlier onset of viral diseases and an increase in insect vectors [48, 49]. Effect of climate change on pest distribution has been tabulated in **Table 2**.

4. Impact of climate change on soil microbes

Terrestrial ecosystems will be impacted by climate change both directly and indirectly, both above and below the surface. Climate change effects, such as rise in CO_2 concentrations, rising temperature, alternation in rainfall patterns, and drought, may have an impact on soil microbial ecology both directly and indirectly. Research on climate change and microbiology must prioritize understanding the processes through which microbes are impacted by the changing climate. Soil microorganisms' responses to climate change may be better understood, which can ultimately enhance through different climate models. However, the soil microbiome may be affected differently by a variety of discrete perturbations or even compounded disruptions brought on by climate change. Recent research has started to include numerous elements since it is unclear how different climate change factors interact [50]. Here, we concentrate on the responses of the soil microbes to climatic stresses and how these abiotic stresses affect their functional and community profiles in the soil.

4.1 Soil temperature

Current climate models project that global temperatures will rise by 4°C by 2100 [51]. Significant emphasis has been paid to how increased temperature affects microbial metabolism [51, 52]. In climate change models, Q10 is frequently employed to account for the temperature sensitivity of microbial activities like microbial respiration, metabolism, and decomposition. Numerous field research has investigated the soil microbiome's short- and/or long-term response to temperature increases (**Table 3**). First, the rates of short-term organic carbon breakdown are accelerated, leading to an increase in microbial biomass. According to one study, soil warming increased the number of the soil microbial community by 40–150% [62]. In addition to enriching for members of the phyla *Acidobacteria* and Actinobacteria as well as the class *Alphaproteobacteria*, it has been demonstrated that raising soil temperature increases the variety and richness of microbiome communities [63, 64].

Climatic factor/ condition	Location	Microbial activity	Remarks/Comments	References
Elevated temperature	Achenkirch, Austria (11°38′21″ E; 47°34′50″ N)	Microbial community	Elevated soil temperature showed no impact on soil bacterial, actinomycetes, or fungal population or biomass, but it did significantly increase microbial metabolic activity.	[53]
Elevated temperature	Harvard Forest in Massachusetts (42.54°N, 72.18°W)	Microbial biomass	A 26-year long-term soil warming experiment revealed that rising soil temperatures reduce microbial biomass and their metabolic activity	[54]
Elevated temperature	Nyingchi County, southeastern Tibet	Soil microbial activity (bacteria and fungi)	Fungal and bacterial diversity increased at significantly increased at subsurface layer but structural changes occurred all layer.	[55]
Cooling Temperature	Nyingchi County, southeastern Tibet	Soil bacterial and fungal communities	No significant impact on fungal and bacterial diversity but structural changes occurred in fungal communities	[55]
Elevated CO_2	Soil of 6 ecosystem in USA (Wisconsin, North Carolina, Tennessee, Florida, Maryland, and Nevada)	Soil bacterial communities	Because of the rise in CO_2 the relative abundance of Group 1 acidobacterial populations decreased by 3.5 times.	[56]
Elevated CO_2	Champaign, Illinois	Bacterial community	Increased in CO_2 concentration significantly increases biological N_2 fixer and NH_4 oxidizers activity and population	[57]
Elevated CO_2	California grassland	soil microbial communities	Decrease in copiotrophic (r-selected) bacteroidetes,	[58]
Drought	Sessile oak forest, XaverovskyHáj Natural Reserve	Soil bacterial and fungal activities	Decrease in bacterial and fungal biomass	[59]
Drought	Light texture soil of Pulawy, Poland	Microbial communities and their structural activity	Relative abundance of corticotrophic Proteobacteria and Verrucomicrobia declined drastically	[60]
Drought	Calcareous grassland in Derbyshire, UK	Bacterial and fungal activity	Gram positive: Gram-negative bacteria ratio increased, and saprophytic fungal and arbuscular mycorrhiza (AM) population increased under drought condition	[61]

Table 3.
Effect of climatic factors on soil microbial activity.

These taxonomic changes and functional profiles coexist: *Copiotrophic taxa* (*i.e.*, Bacteroidetes) are favored over oligotrophic taxa, perhaps in response to shifting soil carbon composition [65]. Warming can have an effect on microbial function directly (e.g., by accelerating enzymatic activities) or indirectly (by rhizodeposition and enhancing nutrient availability for plant growth). For instance, it has been demonstrated that warming stimulates the P and S cycle in soil ecosystem [66], although it is more challenging to draw conclusions about the cycling of carbon and nitrogen. Studies have found that under long-term warming conditions, the number of genes engaged in the breakdown of labile carbon decreases, while the number of genes involved in the metabolism of recalcitrant carbon grows [65, 66] and the variety of the functional guild that is responsible increases [67]. Long-term soil-warming experimentation conducted over 26 years reported a negative impact on soil sustainability in a mid-latitude hardwood forest, evidenced by declines in soil organic matter and microbial biomass content [54].

4.2 Impact of elevated CO_2

Elevated CO_2 (eCO_2) affects the soil microbiota both directly and indirectly (**Table 3**), by enhancing genetic signals, respiration rate, and microbial biomass on a short-term scale [68]. It also promotes plant growth and rhizodeposition, which enrich oligotrophs (K-selected) and decrease copiotrophs or r-selected (like *Acidobacteria* Groups 1 and 2) in the rhizosphere to degrade labile and resistant carbon owing to decreasing soil moisture over long timescales [56, 58, 69]. Interestingly, eCO_2 enhances both microbes and plant growth in soil but reduces soil N concentration. Nitrogen cycling has been shown to be enhanced in the presence of eCO_2, despite the fact that real enzymatic rates are frequently stable or drop [58]. This disparity might be explained by increased abundances of increases biological N_2 fixer and NH_4 oxidizers activity and population [57], albeit this is not a uniform tendency [63, 70]. All things taken into account, these results suggest that eCO_2 may interact with the plant-fungal symbiosis, increasing the incidence of endophyte infection while decreasing toxin production [52]. The majority of research to far have found that increasing CO_2 levels promote the colonization of plant growth-promoting fungus (PGPF).

4.3 Impact of drought

One possible consequence of climate change is increased drought stress. Drought or soil moisture stress is caused by high temperatures and deletion of soil moisture content. Soil moisture and its relationship to the microbial inducing biological activities are more variable and complicated than temperature and therefore has received less attention. Drought reinforces the temperature sensitivity differences between fungal and bacterial species [52]. Even with a 30% reduction in the soil WHC, soil fungal communities may transition from one dominating member to another, while bacterial communities do not change much over time. Drought alters the structure, size, and activity of soil microbial communities (**Table 3**). Their impact on microbial structure, however, is greater in soils with less in soil organic matter concentration [60]. Controlled circumstances of drought stress, for example, caused alterations in the relative abundances of certain phyla prevalent in light texture soils. Actinobacteria (particularly genera *Gaiella* and *Nocarioides*) were the most frequent in studied samples, accounting for more than 95% of overall bacterial abundance. Corticotrophic proteobacteria and Verrucomicrobia population drastically decline in compared to original population under moisture stress condition [60]. Based on phospholipid fatty acid (PLFA) tests [59] discovered that when soil moisture content fell, so did microbial (bacterial and fungal) biomass decrease. In contrast, [71] discovered that dryness had an effect on the fungal PLFA biomarker but not on the bacterial PLFA biomarkers. It is interesting to note that in previous research on the effects of drought, soil that had experienced extreme moisture stress conditions showed higher levels of Firmicutes population over ambient soil condition [72]. This suggests that soils vulnerable to drought stress have lower stress mortality. This might be due to the development of microbial adaptability. Drought generally reduces AMF colonization. ABA, ACC, and JA act as stimulants to enhance gram-negative, saprophytic fungal, and arbuscular mycorrhiza biomarkers, under drought stress conditions over control soil [61].

5. Climate change impact on plant–microbe interaction

The greatest threat to humanity is climate change. The major greenhouse gases like nitrous oxide, methane, and carbon dioxide are produced and consumed by microorganisms. Some microbes can cause diseases in plants and that may be made worse by climate change [73]. Terrestrial ecosystems are made up by interaction of biological elements like plants and microorganisms. Plant–microbe interactions typically take a variety of forms, including commensalism, mutualism, parasitism, etc., and are necessary for carrying out ecological functions like mineralization, nutrient cycling, and decomposition. Most of them are plant growth-promoting microorganisms, such as plant growth-promoting fungi (PGPF) viz. ectomycorrhizae fungi (ECM), arbuscular mycorrhizae fungi (AMF) and plant growth-promoting rhizobacteria (PGPR), whereas others prefer to enter directly in to the host's root systems and have positive impacts due to their endophytic lifestyle [74]. In agriculture, mycorrhizal fungi, endophytes, and PGPB are used as phytostimulants, biofertilizers, and biocontrol agents [73, 75] or as microorganisms that help in phytoremediation of pollutant [76].

5.1 Impact of elevated CO_2 on plant–microbe interaction

The group of bacteria includes the genera such as *Azospirillum*, *Serratia*, *Rhizobium*, *Bacillus*, *Streptomyces*, *Stenotrophomonas*, and *Pseudomonas* and is commonly referred to as PGPR. The abundance of growth-promoting chemicals (phytohormones, antimicrobials, and siderophores) secreted by PGPR has an indirect impact on the general morphology of plants [74]. In elevated CO_2 conditions, plants often encourage microbes—which are crucial for fundamental ecological processes—through accelerated carbon flow into rhizosphere. Typically, the quantity and variety of rhizosphere-dwelling bacteria mostly depend on exudates of roots or metabolites secreted by plants and are greatly influenced by environmental factors brought on by physiological changes in the plant [77]. Ma et al. [78] discovered that CO_2 concentration of >10,000 becomes lethal for some group of bacteria. The decline in the bacterial population may be due to their poor adaptation to higher CO_2. The main methane-oxidizing bacterium in rice roots, Methylocystaceae, was shown to be less abundant when elevated CO_2 was present [79]. Similar effects may occur in endophytic bacterial communities that inhabit the interior tissues of plants, including roots, stems, shoots, leaves, as well as seeds, flowers, and fruits [80].

Ninety percentage of plants form mycorrhizal association and 60% of these plants develop symbiosis with obligatory symbionts like AMF that belong to the category of PGPF [81]. In exchange for rhizosphere carbohydrate molecules, AMF are known to improve plant nutrient (mostly phosphate) uptake [82]. Climate change has the ability to considerably modify environmental conditions, which may have a negative impact on function of mycorrhizae as well as on its structure also [83]. Additionally, it has been discovered that mycorrhizal hyphal development is decreased by elevated CO_2 concentrations. The physiological functions performed by the microorganisms associated with plant roots are also expected to be considerably hampered by elevated CO_2 concentrations. According to reports, certain species' ectomycorrhizal colonization has increased [84], and some leguminous crops nitrogen fixation capacity has increased [85] as a result of altered environmental conditions. In times of water stress [86], limited nutrition availability [87], and heat stress [88], fungi endophytes can enhance plant performance. The results of Brosi et al. [89] suggested that higher CO_2 concentration has more significant effects than changes in precipitation and temperature, which may promote the symbiotic relationship between grass and fungi (*Lolium arundinaceum* and *Neotyphodium coenophialum*) and increase the incidence of endophyte infections.

The ectomycorrhizal communities are likewise affected by alteration in atmospheric CO_2 levels. Under elevated CO_2 circumstances (700 ppm), Godbold and Berntson [90] noted increased ECM colonization in *Pinus strobus* and *Boswellia papyrifera*. Similar to this Fransson et al. [91] showed that *Pinus sylvestris* biomass of ECM mycelia increased by three times under high CO_2 circumstances.

5.2 Impact of elevated temperature on plant–microbe interaction

The heating of soil brought on by rise in the ambient temperature may change the composition of the rhizosphere microbiomes formed through interactions with plants. According to Okubo et al., the methylosinus, methylocystis, and methanotrophs have significant increase in their number as a result of the hot climate [92]. However, studies suggest that increased temperatures under a scenario of climate change have been demonstrated to enhance plant growth and enable balance distribution of nutrients to below-ground ecosystems. When PGPR populations were examined for their response to high temperatures in the rhizosphere of B-95 variety of groundnut, dramatic changes were found. According to a study reported by Meena et al., PGPR is a successful, environmentally-friendly, and long-lasting method for reducing heat stress in wheat (*Triticum aestivum*) caused by *Pseudomonas aeruginosa* strain 2CpS1 [93]. In two genotypes of *Leptospirillum*, suppression was seen when Mosier et al. [94] examined the effects expression of proteins due to high temperatures, involved in carbon utilization.

According to Alexandre and Oliveira [95], it is known that rhizobia depend on heat shock proteins (HSPs) being activated in order to give thermotolerance. HSPs under high temperatures often inhibit the misfolding of proteins under abiotic stresses like heat stress, similar to arbuscular mycorrhizal fungus (AMF) (*Rhizophagus irregularis*) [96]. By intensifying mineral feeding and boosting tolerance to biotic and abiotic stressors, these fungi are well known for enhancing plant development and health [97]. The buildup of trehalose, or reserved carbohydrate, is another method used by beneficial microorganisms, such as AMF, to overcome thermotolerance [98]. It was interestingly demonstrated that heat and other stimuli stimulate enzymes from the trehalose synthase complex as well as neutral trehalose, which is in charge of breaking down internal trehalose. The yeast *Saccharomyces cerevisiae* was the first organism to exhibit this trehalose cycling phenomena under heat stress [99].

5.3 Effect of moisture stress on plant–microbe interaction

There is significant reduction in fungal mycelial growth in plant root due to high temperature which in turn evaporate the soil moisture quickly. The PGPR and PGPF give different response as per their abundance. As an illustration, citrus plants inoculated with *Glomus versiforme* (mycorrhiza) are known to increase osmotic adjustment during drought stress by having higher levels of critical minerals (K^{+}, Ca^{2+}, and Mg^{2+}) in their tissues [100]. Rhizobacterial inoculants have been utilized successfully in agriculture as a means of granting resistance to drought stress [101]. Several studies have shown that under drought conditions, microbial biomass carbon (MBC) undergoes significant changes, signifying that plants rise rhizodeposition to aid in soil water and nutrient transport and the rhizodeposition is result of root cell sloughing and root exudation [102]. The common pattern reported during moisture stress was increase in the concentration of β-cellobiosidase, chitinase, and leucine aminopeptidase [103]. When there is a drought, microorganisms often build up suitable solutes like proline to combat dehydration. The first line of defense in the strategy used by microbes is switching from the growth to the survival state, which involves the making of antioxidants (SOD, CAT, and APOX) and protective molecules like osmolytes to maintain hydration and reduce water potential and the production of chaperones to stabilize proteins. But, these processes are energy expensive and reduce the microbial growth [104].

5.4 Influence of other factors on plant–microbe interaction

Increased UV exposure alters cyanobacteria growth, photosynthetic system, and cell oxygen output [105]. However, according to Bjerke et al. [106], UV-B radiation

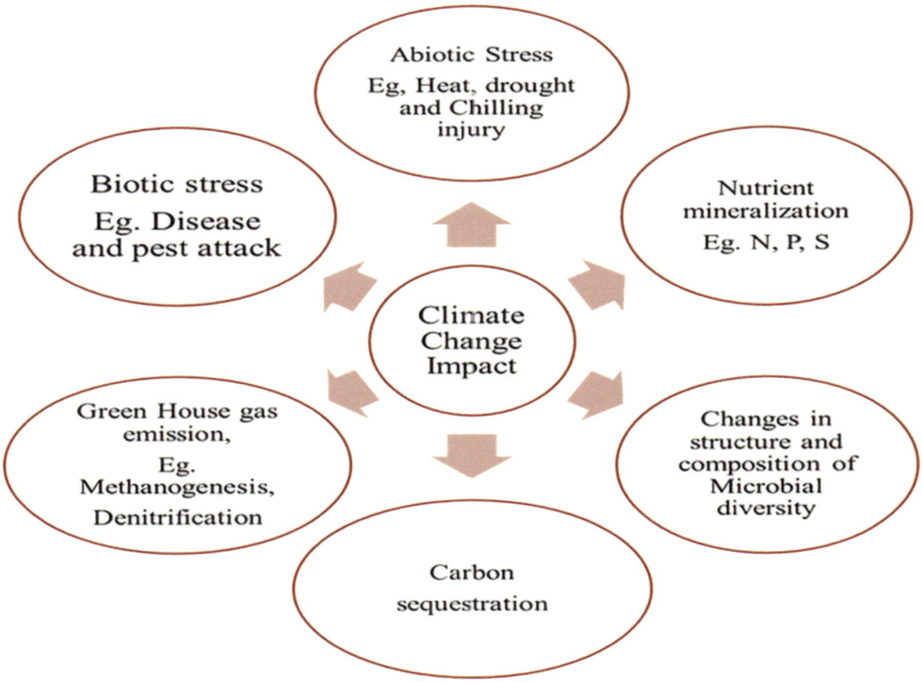

Figure 1.
Impact of climate change on plant–microbe interaction.

does not have a significant impact on lichens. Fungi that are radiation sensitive have slower development because infrared radiation damages cell proliferation. The enhanced radiations that affect phyllospheric communities more than rhizospheric communities are linked to their frequent and abrupt alterations as a result of climate changes (**Figure 1**). The climatic change causes an increase in the frequency of environmental stresses. As a result of climate change, pathogens and heterotrophic bacteria are more active, while helpful microbes are moving from one ecological niche to another [107].

6. Climate change adaptation and soil microbiome

Soil microbiome refers to the diverse community of microorganisms, including fungi, bacteria, archaea, and other microscopic organisms, living in the soil. These microorganisms perform various essential functions in soil ecosystems such as organic matter breakdown and nutrient cycling, disease suppression, and plant growth promotion. They contribute to soil fertility, structure, and overall ecosystem health.

The resilience of ecosystems and agricultural productivity are directly impacted by soil microbiome and climate change adaption. Impacts of climate change on soil microbiome stability and function, and consequently ecosystem sustainability, are a serious issue [108].

Numerous processes, such as biogeochemical cycling, bioremediation, plant development, and primary productivity, the soil microbiome contribute to the health of ecosystems [109, 110]. Given the projected changes in climate, its involvement in greenhouse gas emissions and mediating soil organic carbon (SOC) is of considerable concern. Climate change and changes in land management practices can have a favorable impact on the soil microbiome and its overall impact on soil carbon sequestration, which can also have a positive impact on soil fertility and SOC [111, 112].

The microbial community and functional characteristics of soil can be considerably changed by perturbations linked to climate change [113]. If soil carbon and/or nitrogen cycling is impacted, this may have an impact on climate change through either positive or negative feedbacks to the atmosphere, such as greenhouse gas emissions or the immobilization of carbon in microbial or plant biomass [114].

Climate change poses significant challenges to the soil microbiome and the services it provides. The structure and function of soil microbial communities can be affected by rising temperatures, changed precipitation patterns, and severe weather events. These changes can lead to shifts in microbial diversity, changes in nutrient cycling processes, and increased vulnerability to soil-borne diseases.

The soil microbiome plays a crucial role in establishing agricultural resilience under the era of climate change [115]. It discusses how soil microorganisms contribute to nutrient cycling, plant–microbe interactions, disease suppression, and soil carbon sequestration, highlighting their potential as tools for climate change adaptation. They also examine the impact of climate change on soil microbial communities and propose strategies to harness the soil microbiome's potential for mitigating the impacts of climate change and ensuring future food security.

6.1 Functions

However, understanding and harnessing the potential of the soil microbiome can contribute to climate change adaptation strategies. Here are some key aspects:

- Carbon sequestration: Soil microorganisms are involved in carbon cycling and play a vital role in storing carbon in the soil. Certain microbial communities, such as mycorrhizal fungi, can enhance plant carbon sequestration and mitigate greenhouse gas emissions. Promoting practices like cover cropping, crop rotation, and organic amendments can enhance microbial activity and carbon sequestration.

- Nutrient management: Soil microorganisms are critical for nutrient availability to plants. Climate change affects nutrient cycling processes, such as nitrogen fixation, mineralization, and nutrient uptake by plants. Managing soil microbiomes through practices like optimizing organic matter inputs and using microbial inoculants and precision nutrient management can enhance nutrient availability and reduce fertilizer requirements.

- Disease suppression: Climate change can influence the prevalence and severity of soil-borne diseases. However, certain beneficial microorganisms can suppress plant pathogens and enhance disease resistance in plants. Implementing practices that support a diverse and healthy soil microbiome, such as crop rotation, balanced fertilization, and reduced tillage, can help mitigate disease impacts.

- Ecosystem resilience: A diverse and robust soil microbiome contributes to overall ecosystem resilience in the face of climate change. It enhances soil structure, water holding capacity, and nutrient cycling efficiency. Practices like agroforestry, conservation tillage, and organic farming can support a diverse soil microbiome and improve ecosystem resilience.

- Research and innovation: Further research is needed to better understand the intricate interactions between climate change, soil microbiomes, and plant-soil interactions. Studying microbial communities, their functional roles, and responses to changing climatic conditions can guide the development of novel approaches and technologies for climate change adaptation in agriculture.

7. Function of microbes in climate resilience vegetable production

7.1 Consequences of climate change on vegetables

The current projections on climate change show a 0.9–3.5°C rise in mean temperatures worldwide and a twofold rise of carbon-dioxide (CO_2) level to more than 700 μmol mol^{-1} around the century 2100 [116]. Environmental indicators including temperature, moisture levels, and air composition might exert an immediate effect on agricultural production and quality, or they would pose a secondary impact by means of pathogen and pest infestation occurrence (**Table 4**). Vegetable survival under opposing and shifting circumstances caused by climate change is intrinsically tied to its ecological relationships with other creatures in the natural environment [133]. Although a few interactions, simultaneously hostile and favorable for plant growth, are unique to particular vegetable species, others serve broader environmental purposes [134].

- Enhanced superficial wetting of foliage and canopies, for instance, along with higher temperatures, might exacerbate the intensity of fungal infections like anthracnose triggered by Colletotrichum spp.

Microbial stimulants	Stress	Crop	References
Paraburkholderiaphytofirmans	Heat	Tomato	[117]
Pseudomonas syringae	Cold	Soybean	[118]
Pseudomonasfragi	Frost	Bean	[119]
Exophiala sp. LHL08	Drought	Cucumber	[120]
Glomus intraradices	Waterlogging	Common bean	[121]
A. piechaudii ARV8	Salinity	Tomato	[122]
Pseudomonas syringae	Cold	Tomato	[118]
P. chlororaphis TSAU13	Salinity	Cucumber	[123]
P. putida H-2-3	Drought	Soybean	[124]
Pseudomonas sp.	Hypoxia	Cucumber	[125]
P. fluorescens TDK1	Salinity	Peanut	[126]
Glomus intraradices	Drought	Carrot	[127]
B. polymyxa	Waterlogging	Tomato	[128]
P. chlororaphis TSAU13	Drought	Tomato	[123]
G. intraradices	Salinity	Lettuce	[129]
Enterobacter spp.	Anoxia	Tomato	[130]
Streptomyces sp. GMKU 336 strain	Waterlogging	*Vigna radiata*	[131]
Bradyrhizobium japonicum	Reduction of N_2O emission	Soybean	[132]

Table 4.
Execution of different microbes to mitigate stress related to climate change on different vegetable crops.

- Increased temperatures appear to promote the brutality associated with the pepper late blight fungus (*Phytophthora capsici*).
- Due to higher sugar content, greater amount of CO_2 circumstances appears to facilitate the common army worm's (*Spodoptera litura*) infestation onto mung bean seedlings.
- Under high CO_2, the green peach aphid (*Myzuspersicae*) multiplied significantly more quickly on *Brassica oleracea* crops.
- Temperature rises encourage the expeditious reproduction of whitefly (*Bemisia tabaci*) biotypes, which are serious threats to solanaceous crops because they serve as vectors for the spread of viral infections.
- Gossypii growing in cucumber developed and produced more offspring at higher temperatures not exceeding 30°C, but temperatures above this point extended development, boosted juvenile stage mortality, decreased adult life expectancy, and decreased fecundity among the species.

7.2 Use of bio-stimulants and mechanism of combating climate change

In order to promote plant development and yield improvement pursuant to extreme stress and regulatory circumstances, the usage of bio-stimulants is being regarded as an undoubtedly unique strategy [135]. Bio-stimulants are employed frequently to solve ecological problems as well as meet the demand for the development of sustainable farming practices since they pose a prospective as well as environmentally benign strategy [136]. The fundamental principles of bio-stimulant-plant interactions operate on the cellular, metabolic, and molecular scales that combat stress difficulties, which remain unattainable [137], due to intriguing potential, processes, and possible obstacles of bio-stimulants to mitigate abiotic stresses while boosting yield and quality [138]. Plants with beneficial bacteria like *Pseudomonas* and *Bacillus* and mycorrhizal fungi like *Septoglomus deserticola* and *Septoglomus constrictum* can increase the production of ROS-decaying enzymes comprising superoxide, catalase, peroxidases, and dismutase that exacerbate endurance to heat stress [139]. Microbial symbionts which generate growth-related hormones (such as auxins or gibberellins) additionally which lower ethylene content can combat the low temperature-induced reduction of plant growth. By generating ACC deaminase, microbiological bio-stimulants can reduce stress imposed through water stagnation, which minimizes the concentration of endogenous ethylene [140]. Biochar is one of the most promising agricultural techniques for mitigating the adverse impacts of climate change. By enhancing crop establishment, biochar aids in sequestering soil organic carbon. Furthermore, it reduces greenhouse gas emissions by decreasing microbial mineralization and limiting the release of organic compounds from mineral sources [141].

8. Conclusion

The soil microbiome is a complex and diverse community of microorganisms that play a vital role in plant health and productivity. Climate change is expected to have a significant impact on the soil microbiome, which could in turn affect vegetable crop production. One of the main ways in which climate change is expected to impact the soil microbiome is through changes in temperature. Increased temperatures can lead to changes in the composition and abundance of microbial communities, as well as increased microbial activity. This can have both positive and negative consequences for plant growth. For example, increased microbial activity can lead to increased nutrient availability for plants. However, it can also lead to increased competition for resources between plants and microbes and increased production of toxins by harmful microbes. Another way in which climate change is expected to impact the soil microbiome is through changes in precipitation patterns. Increased drought can lead to a decrease in soil moisture, which can reduce microbial activity and lead to changes in the composition of microbial communities. Increased flooding can also have a negative impact on the soil microbiome by displacing microbes and disrupting their interactions with plants. The impacts of climate change on the soil microbiome are expected to have a significant impact on vegetable crop production. Vegetable crops are particularly vulnerable to changes in the soil microbiome because they are typically grown in intensive agricultural systems with high nutrient inputs and limited crop rotation. This makes them more susceptible to pests, diseases, and other abiotic stresses. One way to mitigate the negative impacts of climate change on the soil microbiome and vegetable crop production is to adopt sustainable agricultural practices. These practices can help to improve soil health and resilience and promote beneficial plant–microbe interactions. By adopting sustainable agricultural practices and developing new crop varieties, it is possible to mitigate the negative impacts of climate change on the soil microbiome and vegetable crop production. This will help to ensure a sustainable food supply for future generations. By understanding the complex interactions between the soil microbiome and vegetable crops, scientists and farmers can develop new strategies to manage vegetable crop production in a changing climate.

Author details

Sripriya Das[1], Vandana Kumari[2], Kiran Kumar Mohapatra[3], Sayon Mukherjee[4], Plabani Roy[5], Dibyajyoti Panda[4], Abhik Patra[6], Ashish Rai[7]* and Vijay Singh Meena[8]

1 Krishi Vigyan Kendra, Vaishali, Dr. Rajendra Prasad Central Agricultural University, Bihar, India

2 Dr. Kalam Agricultural College, Kishanganj, Arrabari, Kishanganj, Bihar, India

3 Odisha University of Agriculture and Technology, Bhubaneswar, Odisha, India

4 Department of Soil Science and Agricultural Chemistry, Institute of Agricultural Sciences, Banaras Hindu University, Varanasi, Uttar Pradesh, India

5 Division of Soil Science and Agricultural Chemistry, ICAR-Indian Agricultural Research Institute, New Delhi, India

6 Krishi Vigyan Kendra, Parsauni, East Champaran, Dr. Rajendra Prasad Central Agricultural University, Bihar, India

7 Krishi Vigyan Kendra, Narkatiyaganj, West Champaran, Dr. Rajendra Prasad Central Agricultural University, Bihar, India

8 ICAR-Indian Agricultural Research Institute, Regional Station, Samastipur, Bihar, India

*Address all correspondence to: ashishraishiats@gmail.com

References

[1] Jacoby R, Peukert M, Succurro A, Koprivova A, Kopriva S. The role of soil microorganisms in plant mineral nutrition—Current knowledge and future directions. Frontier in Plant Science. 2017;**8**:1-19

[2] Pramanick B, Bera A, Saha P, Barman A, Paramanick B, Maitra S, et al. Soil microflora and their interaction with plants under changing climatic scenarios. In: Microbial Symbionts and Plant Health: Trends and Applications for Changing Climate. Singapore: Springer Nature Singapore; 2023. pp. 19-40

[3] Mekonnen H, Kibret M. The roles of plant growth promoting rhizobacteria in sustainable vegetable production in Ethiopia. Chemical and Biological Technologies in Agriculture. 2021;**8**:1-11

[4] Pérez-Montaño F, Alías-Villegas C, Bellogín RA, Del Cerro P, Espuny MR, Jiménez-Guerrero I, et al. Plant growth promotion in cereal and leguminous agricultural important plants: From microorganism capacities to crop production. Microbiological Research. 2014;**169**:325-336

[5] Zaidi A, Khan MS. Microbial strategies for vegetable production. In: Microbial Strategies for Vegetable Production. Springer: Berlin, Germany; 2017. pp. 1-226

[6] Rehan M, Al-Turki A, Abdelmageed AHA, Abdelhameid NM, Omar AF. Performance of plant-growth-promoting rhizobacteria (PGPR) isolated from sandy soil on growth of tomato (*Solanum lycopersicum* L.). Plants. 2023;**12**:1588

[7] Kumari M, Verma SC, Shweta S. Climate change and vegetable crops cultivation: A review. The Indian Journal of Agricultural Sciences. 2018;**88**(2):167-174

[8] Lal M, Nozawa T, Emori S, Harasawa H, Takahashi K, Kimoto M, et al. Future climate change: Implications for Indian summer monsoon and its variability. Current Science. 2001;**81**:1196-1207

[9] Verchot LV, Noordwijk MV, Kandji S, Tomich T, Ong C, Albrecht A, et al. Climate change: Linking adaptation and mitigation through agroforestry. Mitigation and Adaptation Strategies of Global Change. 2007;**12**:901-918

[10] Vermeulen SJ, Aggarwal PK, Ainslie A, Angelone C, Campbell BM, Challinor AJ, et al. Agriculture, Food Security and Climate Change: Outlook for Knowledge, Tools and Action. CCAFS Report 3. Copenhagen, Denmark: CGIARESSP Program on Climate Change, Agriculture and Food Security; 2010

[11] Cotty PJ, Jamie-Garcia R. Influences of climate on aflatoxin producing fungi and aflatoxin contamination. International Journal of Food Microbiology. 2007;**119**:109-115

[12] Costello A, Abbas M, Allen A, Ball S, Bell S, Bellamy R, et al. Managing the health effects of climate change. The Lancet. 2009;**373**:1693-1733

[13] Bhatia A, Mina U, Kumar V, Tomer R, Kumar A, Chakrabarti B, et al. Effect of elevated ozone and carbon dioxide interaction on growth, yield, nutrient content and wilt disease severity in chickpea grown in Northern India. Heliyon. 2021;7(1)

[14] Coakley SM, Scherm H, Chakraborty S. Climate change and plant

disease management. Annual Review of Phytopathology. 1999;**37**:399-426

[15] Neumeister L. Climate Change and Crop Protection-Anything Can Happen. Pesticide Action Network Asia and the Pacific: Penang, Malaysia; 2010. pp. 4-41

[16] Lim JA, Yaacob JS, Mohd Rasli SR, Eyahmalay JE, El Enshasy HA, Zakaria MR. Mitigating the repercussions of climate change on diseases affecting important crop commodities in Southeast Asia, for food security and environmental sustainability: A review. Frontiers in Sustainable Food Systems. 2023;**6**:1030540

[17] Singh AK. Climate Change Sensitivity of Indian Horticulture - Role of Technological Interventions, Souvenir of Fourth Indian Horticultural Congress. New Delhi: HSI; 2010. pp. 85-95

[18] Kalloo G, Benarjee MK, Tiwari RN. Tomato. In: Thumbraj S, Singh N, editors. Vegetables Tuber Crops and Spices. New Delhi: Directorate of Information and Publication in Agriculture, Indian Council of Agricultural Research; 2001. pp. 10-28

[19] Peet MM, Willits DH, Gardner R. Response of ovule development and postpollen production processes in malesterile tomatoes to chronic, sub-acute high temperature stress. Journal of Experimental Botany. 1997;**48**:101-111

[20] Arora SK, Partap PS, Pandita ML, Jalal I. Production problems and their possible remedies in vegetable crops. Indian Horticulture. 1987;**32**:2-8

[21] AVRDC. Vegetable Production Training Manual. Asian Vegetable Research and Training Center. Tainan: Shanhua; 1990. p. 447

[22] Pena R, Hughes J. Improving vegetable productivity in a variable and changing climate. SAT e Journal. 2007;**4**:1-22

[23] Cheeseman JM. Mechanisms of salinity tolerance in plants. Plant Physiology. 1988;**87**:547-550

[24] Jamil M, Rha ES. The effect of salinity (NaCl) on the germination and seedling of sugar beet (*Beta vulgaris* L.) and cabbage (*Brassica oleracea capitata* L.). Korean Journal of Plant Research. 2004;**7**:226-232

[25] Singh JP, Lal SS, Pandy SK. Effect of climate change on potato production in India. Central Potato Research Institute. Shimla Newsletter. 2009;**40**:17-18

[26] Lurie S, Handros A, Fallik E, Shapira R. Reversible inhibition of tomato fruit gene expression at high temperature. Plant Physiology. 1996;**110**:1207-1214

[27] Erickson AN, Markhart AH. Flower developmental stage and organ sensitivity of bell pepper (*Capsicum annuum* L.) to elevated temperature. Plant Cell Environment. 2002;**25**: 123-130

[28] Patil D, Thomas M, Kakade S, Khichi A, Raut V, Anjana G, et al. Evaluation of cultivation of tubers and spices in polypropylene bags: An option. Current Horticulture. 2022;**10**(1):29-31

[29] Manzoor N, Masoodi UH, Mufti S, Anayat R. Effect of changing climatic scenario on vegetable production and mitigation strategies: A review. SKUAST Journal of Research. 2023;**25**(2):159-172

[30] Kurtar ES. Modelling the effect of temperature on seed germination in some cucurbits. African Journal of Biotechnology. 2010;**9**:1343-1353

[31] Burke JI, Finnan JM, Donnelly A, Jones MB. The Effects of Elevated Concentrations of Carbon Dioxide and Ozone on Potato (*Solanum tuberosum* L.) Yield. Carlow, Ireland: Agriculture and Food Development authority; 2001. pp. 1-19

[32] Hamilton JG, Dermody O, Aldea M, Zangerl AR, Rogers A, Berenbaum MR, et al. Anthropogenic changes in tropospheric composition increase susceptibility of soybean to insect herbivory. Environmental Entomology. 2005;**34**:479-485

[33] Coviella C, Trumble JT. Effects of elevated atmospheric carbon dioxide on insect plant interactions. Conservation Biology. 1999;**13**:700-712

[34] Hunter MD. Effects of elevated atmospheric carbon dioxide on insect-plant interactions. Agriculture and Forest Entomology. 2001;**3**:153-159

[35] Das DK, Singh J, Vennila S. Emerging crop pest scenario under the impact of climate change – A brief review. Journal of Agricultural Physics. 2011;**11**:13-20

[36] Manning WJ, Von Tiedemann A. Climate change: Potential effects of increased atmospheric atmospheric carbon dioxide (CO_2) and ozone and ultraviolet-B (UV-B). Environmental Pollution. 1995;**88**:219-245

[37] Hao X, Hale BA, Ormrod DP. The effects of ultraviolet- B radiation and carbon dioxide on growth and photosynthesis of tomato. Canadian Journal of Botany. 1997;**75**:213-219

[38] Benda GTA. Some effects of ultra-violet radiation on leaves of french bean (*Phaseolus vulgaris. L.*). Annals of Applied Biology. 1955;**43**:71-85

[39] Jat MK, Tetarwal AS. Effect of Changing Climate on the Insect Pest Population in National Seminar on Sustainable Agriculture and Food Security: Challenges in Changing Climate. Hisar, India: Indian Horticulture; 2012. pp. 200-220

[40] FAO. Climate-Related Transboundary Pests and Diseases, Technical Background Document from the Expert Consultation. Rome: FAO; 2008. Available from ftp://ftp.fao.org/docrep/fao/meeting/013/ai785e.pdf

[41] Lewis T. Thrips as Crop Pests. CAB International, Cambridge: University Press; 1997. p. 740

[42] Bale JS, Masters GJ, Hodkinson ID, Awmack C, Bezemer TM, Brown VK, et al. Herbivory in global climate change research: Direct effects of rising temperatures on insect herbivores. Global Change Biology. 2002;**8**:1-16

[43] Cannon RJC. Annexure-1. In: Climate-Related Transboundary Pests and Diseases, Technical Background Document from the Expert Consultation. Rome: FAO; 2008

[44] Yamamura K, Kiritani K. A simple method to estimate the potential increase in the number of generations under global warming in temperate zones. Applied Entomology and Zoology. 1998;**33**:289-298

[45] Akhtar S, Sangam S, Chattopadhyay T, Naik A, Solankey SS. Emerging obstacles of vegetable production due to climate change and mitigation strategies. In: Advances in Research on Vegetable Production Under a Changing Climate. Vol. 2. Cham: Springer International Publishing; 2023. pp. 31-69

[46] Lahlali R, Taoussi M, Laasli SE, Gachara G, Ezzouggari R,

Belabess Z, et al. Effects of climate change on plant pathogens and host-pathogen interactions. Crop and Environment. 2024;**3**(3):159-170

[47] Boonekamp PM. Are plant diseases too much ignored in the climate change debate? European Journal of Plant Pathology. 2012;**133**:291-294

[48] Thomas T. Sugar beet in the greenhouse—A global warming warning. Brown Sugar. 1989;**57**:24-26

[49] Harrington R, Stork NE. Insects in a Changing Environment. London: Academic Press; 1995

[50] Weiskopf SR, Rubenstein MA, Crozier LG, Gaichas S, Griffis R, Halofsky JE, et al. Climate change effects on biodiversity, ecosystems, ecosystem services, and natural resource management in the United States. Science of the Total Environment. 2020;**733**:137782

[51] Naylor D, Sadler N, Bhattacharjee A, Graham EB, Anderton CR, McClure R, et al. Soil microbiomes under climate change and implications for carbon cycling. Annual Review of Environment and Resources. 2020;**45**:29-59

[52] Shree B, Jayakrishnan U, Bhushan S. Impact of key parameters involved with plant-microbe interaction in context to global climate change. Frontiers in Microbiology. 2022;**13**:1008451

[53] Schindlbacher AA, Rodler M, Kuffner B, Kitzler A, Sessitsch S, Zechmeister-Boltenstern. Experimental warming effects on the microbial community of a temperate mountain forest soil. Soil Biology and Biochemistry. 2011;**7**:1417-1425

[54] Melillo JM, Frey SD, DeAngelis KM, Werner WJ, Bernard MJ, Bowles FP, et al. Long-term pattern and magnitude of soil carbon feedback to the climate system in a warming world. Science. 2017;**358**:101-105

[55] Fangwei F, Li J, Li Y, Chen W, Ding H, Xiao S. Simulating the effect of climate change on soil microbial community in an *Abies georgei* var. *smithii* forest. Frontiers in Microbiology. 2023;**14**:1189859

[56] Dunbar J, Eichorst SA, Gallegos-Graves LV, Silva S, Xie G. Common bacterial responses in six ecosystems exposed to 10 years of elevated atmospheric carbon dioxide: Soil bacterial response in six ecosystems. Environmental Microbiology. 2012;**14**:1145-1158

[57] Wang P, Marsh EL, Ainsworth EA, Leakey ADB, Sheflin AM, Schachtman DP. Shifts in microbial communities in soil, rhizosphere and roots of two major crop systems under elevated CO_2 and O_3. Scientific Reports. 2017;**7**:15019

[58] Yang S, Zheng Q, Yuan M, Shi Z, Chiariello NR. Long-term elevated CO_2 shifts composition of soil microbial communities in a Californian annual grassland, reducing growth and N utilization potentials. Science of the Total Environment. 2019;**652**:1474-1481

[59] Baldrian P, Merhautová V, Petránková M, Cajthaml T, Šnajdr J. Distribution of microbial biomass and activity of extracellular enzymes in a hardwood forest soil reflect soil moisture content. Applied Soil Ecology. 2010;**46**:177-182

[60] Siebielec S, Siebielec G, Klimkowicz-Pawlas A, Gałazka A, Grzadziel J, Stuczy'nski. Short-term responses and resistance of soil microbial community structure to elevated CO_2 and N addition in grassland

mesocosms. FEMS Microbiology Letter. 2020;**364**:fnx077

[61] Sayer EJ, Crawford JA, Edgerley J, Askew AP, Hahn CZ, Whitlock R, et al. Adaptation to chronic drought modifies soil microbial community responses to phytohormones. Communications Biology. 2021;**4**:516

[62] Pold G, Billings AF, Blanchard JL, Burkhardt DB, Frey SD, Melillo JM, et al. Long-term warming alters carbohydrate degradation potential in temperate forest soils. Applied and Environmental Microbiology. 2016;**82**:6518-6530

[63] Rocca JD, Simonin M, Blaszczak JR, Ernakovich JG, Gibbons SM. The microbiome stress project: Toward a global meta-analysis of environmental stressors and their effects on microbial communities. Frontiers in Microbiology. 2019;**9**:3272

[64] Carrell AA, Kolton M, Glass JB, Pelletier DA, Warren MJ. Experimental warming alters the community composition, diversity, and N_2 fixation activity of peat moss (*Sphagnum fallax*) microbiomes. Global Change Biology. 2019;**25**:2993-3004

[65] Li Y, Lv W, Jiang L, Zhang L, Wang S. Microbial community responses reduce soil carbon loss in Tibetan alpine grasslands under short-term warming. Global Change Biology. 2019;**25**:3438-3449

[66] Xue K, Xie J, Zhou A, Liu F, Li D. Warming alters expressions of microbial functional genes important to ecosystem functioning. Frontiers in Microbiology. 2016;7:668

[67] DeAngelis KM, Pold G, Topçuoglu BD, van Diepen LTA, Varney RM. Long-term Forest soil warming alters microbial communities in temperate forest soils. Frontiers in Microbiology. 2015;**6**:104

[68] Xiong J, He Z, Shi S, Kent A, Deng Y. Elevated CO_2 shifts the functional structure and metabolic potentials of soil microbial communities in a C4 agroecosystem. Scientific Reports. 2015;**5**:9316

[69] Chen YP, Liu Q, Liu YJ, Jia FA, He XH. Responses of soil microbial activity to cadmium pollution and elevated CO_2. Scientific Reports. 2015;**4**:4287

[70] Lladó S, López-Mondéjar R, Baldrian P. Forest soil bacteria: Diversity, involvement in ecosystem processes, and response to global change. Microbiology and Molecular Biology Reviews. 2017;**81**:e00063-e00016

[71] Bastida F, Torres IF, Hernández T, García C. The impacts of organic amendments: Do they confer stability against drought on the soil microbial community? Soil Biology and Biochemistry. 2017;**113**:173-183

[72] Veach AM, Zeglin LH. Historical drought affects microbial population dynamics and activity during soil drying and re-wet. Microbial Ecology. 2020;**79**:662-674

[73] Tiedje JM, Bruns MA, Casadevall A, Criddle CS, Eloe-Fadrosh E, Karl DM, et al. Microbes and climate change: A research prospectus for the future. Journal of Clinical Microbiology. 2022;**13**:e00800-e00822

[74] Tiwari P, Bose SK, Park KI, Dufossé L, Fouillaud M. Plant-microbe interactions under the extreme habitats and their potential applications. Microorganisms. 2024;**12**(3):448

[75] Lugtenberg B, Kamilova F. Plant-growth-promoting rhizobacteria. Annual Review of Microbiology. 2009;**63**:541-556

[76] Denton B. Advances in phytoremediation of heavy metals using plant growth promoting bacteria and fungi. MMG 445 Basic Biotechnology. 2007;**3**:1-5

[77] Rangel-Castro JI, Killham K, Ostle N, Nicol GW, Anderson IC, Scrimgeour CM, et al. Stable isotope probing analysis of the influence of liming on root exudate utilization by soil microorganisms. Environmental Microbiology. 2005;7:828-838

[78] Ma J, Wang S, Xue L, Liu D, Zhu X, Ma J, et al. Research of the impact of elevated CO2 on soil microbial diversity. Energy Procedia. 2017;**114**:3070-3076

[79] Okubo T, Liu D, Tsurumaru H, Ikeda S, Asakawa S, Tokida T, et al. Elevated atmospheric CO_2 levels affect community structure of rice root-associated bacteria. Frontiers in Microbiology. 2015;**6**:136

[80] Compant S, Clément C, Sessitsch A. Plant growth-promoting bacteria in the rhizo-and endosphere of plants: Their role, colonization, mechanisms involved and prospects for utilization. Soil Biology and Biochemistry. 2010;**42**:669-678

[81] Sharma N, Tapwal A. Mycorrhizal symbiosis in Taxus: A review. Mycorrhiza. 2024:1-8

[82] Newsham KK, Fitter AH, Watkinson AR. Multi-functionality and biodiversity in arbuscular mycorrhizas. Trends in Ecology and Evolution. 1995;**10**:407-411

[83] Staddon PL, Ramsey CB, Ostle N, Ineson P, Fitter AH. Rapid turnover of hyphae of mycorrhizal fungi determined by AMS microanalysis of 14C. Science. 2003;**300**:1138-1140

[84] Garcia MO, Ovasapyan T, Greas M, Treseder KK. Mycorrhizal dynamics under elevated CO_2 and nitrogen fertilization in a warm temperate forest. Plant and Soil. 2008;**303**:301-310

[85] Rogers A, Ainsworth EA, Leakey AD. Will elevated carbon dioxide concentration amplify the benefits of nitrogen fixation in legumes. Plant Physiology. 2009;**151**:1009-1016

[86] Kannadan S, Rudgers JA. Endophyte symbiosis benefits a rare grass under low water availability. Functional Ecology. 2008;**22**:706-713

[87] Lyons PC, Evans JJ, Bacon CW. Effects of the fungal endophyte Acremonium coenophialum on nitrogen accumulation and metabolism in tall fescue. Plant Physiology. 1990;**92**:726-732

[88] Rodriguez RJ, Henson J, Van Volkenburgh E, Hoy M, Wright L, Beckwith F, et al. Stress tolerance in plants via habitat-adapted symbiosis. The ISME Journal. 2008;**2**:404-416

[89] Brosi GB, McCulley RL, Bush LP, Nelson JA, Classen AT, Norby RJ. Effects of multiple climate change factors on the tall fescue–fungal endophyte symbiosis: Infection frequency and tissue chemistry. New Phytologist. 2011;**189**:797-805

[90] Godbold DL, Berntson GM. Elevated atmospheric CO_2 concentration changes ectomycorrhizal morphotype assemblages in *Betula papyrifera*. Tree Physiology. 1997;**17**:347-350

[91] Fransson PM, Taylor AF, Finlay RD. Mycelial production, spread and root colonisation by the ectomycorrhizal fungi *Hebelomacrustuliniforme* and *Paxillusinvolutus* under elevated

atmospheric CO_2. Mycorrhiza. 2005;**15**:25-31

[92] Okubo T, Tokida T, Ikeda S, Bao Z, Tago K, Hayatsu M, et al. Effects of elevated carbon dioxide, elevated temperature, and rice growth stage on the community structure of rice root–associated bacteria. Microbes and Environments. 2014;**29**:184-190

[93] Meena H, Ahmed MA, Prakash P. Amelioration of heat stress in wheat, Triticum aestivum by PGPR (*Pseudomonas aeruginosa* strain 2CpS1). Bioscience Biotechnology Research. 2015;**8**:171-174

[94] Mosier AC, Li Z, Thomas BC, Hettich RL, Pan C, Banfield JF. Elevated temperature alters proteomic responses of individual organisms within a biofilm community. The ISME Journal. 2015;**9**:180-194

[95] Alexandre A, Oliveira S. Most heat-tolerant rhizobia show high induction of major chaperone genes upon stress. FEMS Microbiology Ecology. 2011;**75**:28-36

[96] Estrada B, Barea JM, Aroca R, Ruiz-Lozano JM. A native *Glomus intraradices* strain from a Mediterranean saline area exhibits salt tolerance and enhanced symbiotic efficiency with maize plants under salt stress conditions. Plant and Soil. 2013;**366**:333-349

[97] Jatav SS, Parihar M, Patra A, Singh SK, Chitara MK, Mohapatra KK, et al. Soil microbes in plant growth promotion and for mitigation of abiotic stress of drought. In: Soil Microbiomes for Sustainable Agriculture: Functional Annotation. Springer; 2021. pp. 175-201

[98] Lenoir I, Fontaine J, Sahraoui ALH. Arbuscular mycorrhizal fungal responses to abiotic stresses: A review. Phytochemistry. 2016;**123**:4-15

[99] Hottiger T, Boller T, Wiemken A. Rapid changes of heat and desiccation tolerance correlated with changes of trehalose content in Saccharomyces cerevisiae cells subjected to temperature shifts. FEBS Letters. 1987;**220**:113-115

[100] Wu QS, Xia RX. Arbuscular mycorrhizal fungi influence growth, osmotic adjustment and photosynthesis of citrus under well-watered and water stress conditions. Journal of Plant Physiology. 2006;**163**:417-425

[101] Figueiredo MV, Burity HA, Martinez CR, Chanway CP. () alleviation of drought stress in the common bean (*Phaseolus vulgaris* L.) by co-inoculation with *Paenibacilluspolymyxa* and *rhizobium tropici*. Applied Soil Ecology. 2008;**40**:182-188

[102] Xue S, Yang X, Liu G, Gai L, Zhan C, Ritsema CJ, et al. Effects of elevated CO_2 and drought on the microbial biomass and enzymatic activities in the rhizospheres of two grass species in Chinese loess soil. Geoderma. 2017;**286**:25-34

[103] Sanaullah M, Blagodatskaya E, Chabbi A, Rumpel C, Kuzyakov Y. Drought effects on microbial biomass and enzyme activities in the rhizosphere of grasses depend on plant community composition. Applied Soil Ecology. 2011;**48**:38-44

[104] Or D, Smets BF, Wraith JM, Dechesne A, Friedman SP. Physical constraints affecting bacterial habitats and activity in unsaturated porous media–a review. Advances in Water Resources. 2007;**30**:1505-1527

[105] Zeeshan M, Prasad SM. Differential response of growth, photosynthesis,

antioxidant enzymes and lipid peroxidation to UV-B radiation in three cyanobacteria. South African Journal of Botany. 2009;**75**:466-474

[106] Bjerke JW, Gwynn-Jones D, Callaghan TV. Effects of enhanced UV-B radiation in the field on the concentration of phenolics and chlorophyll fluorescence in two boreal and arctic–alpine lichens. Environmental and Experimental Botany. 2005;**53**:139-149

[107] Zhang C, Xi N. Precipitation changes regulate plant and soil microbial biomass via plasticity in plant biomass allocation in grasslands: A meta-analysis. Frontiers in Plant Science. 2021;**12**:614968

[108] Allison SD, Martiny JBH. Resistance, resilience, and redundancy in microbial communities. National Academy of Sciences of the United States of America. 2008;**105**:11512-11519

[109] Amundson R, Berhe AA, Hopmans JW, Olson C, Sztein AE, Sparks DL. Soil and human security in the 21st century. Science. 2015;**348**:1261071

[110] Cavicchioli R, Ripple WJ, Timmis KN, Azam F, Bakken LR. Scientists' warning to humanity: Microorganisms and climate change. Nature Reviews Microbiology. 2019;**17**:569-586

[111] Lal R. Soil carbon sequestration impacts on global climate change and food security. Science. 2004;**304**:1623-1627

[112] Campbell EE, Paustian K. Current developments in soil organic matter modeling and the expansion of model applications: A review. Environmental Research Letters. 2015;**10**:123004

[113] Jansson JK, Hofmockel KS. Soil microbiomes and climate change. Nature Reviews Microbiology. 2020;**18**:35-46

[114] Sulman BN, Phillips RP, Oishi AC, Shevliakova E, Pacala SW. Microbe-driven turnover offsets mineral-mediated storage of soil carbon under elevated CO_2. Nature Climate Change. 2014;**4**:1099-1102

[115] Smith A, Johnson B, Anderson C. Climate change adaptation in agriculture: Role of soil microbiome in enhancing resilience. Environmental Microbiology Reports. 2022;**14**:215-230

[116] IPCC. Climate change (2014) impacts, adaptation, and vulnerability. In: Contribution of Working Group II to Fifth Assessment Report of the Intergovernmental Panel on Climate Change. Cambridge, United Kingdom; New York, NY, USA: Cambridge University Press; 2014

[117] Esmaeel Q, Miotto L, Rondeau M, Leclère V, Clément C, Jacquard C, et al. *Paraburkholderiaphytofirmans* PsJN-plants interaction: From perception to the induced mechanisms. Frontiers in Microbiology. 2018;**9**:2093

[118] Anderson JA, Buchanan DW, Stall RE, Hall CB. Frost injury of tender plants increased by *pseudomonas syringae van Hall1*. Journal of the American Society for Horticultural Science. 1982;**107**:123-125

[119] Tiryaki D, Aydın İ, Atıcı Ö. Psychrotolerant bacteria isolated from the leaf apoplast of cold-adapted wild plants improve the cold resistance of bean (*Phaseolus vulgaris* L.) under low temperature. Cryobiology. 2019;**86**:111-119

[120] Khan AL, Hussain J, Al-Harrasi A, Al-Rawahi A, Lee IJ. Endophytic fungi:

Resource for gibberellins and crop abiotic stress resistance. Critical Reviews in Biotechnology. 2015;**35**:62-74

[121] Omomowo IO, Fadiji AE, Omomowo OI. Assessment of bio-efficacy of *Glomus versiforme* and *Trichoderma harzianum* in inhibiting powdery mildew disease and enhancing the growth of cowpea. Annals of Agricultural Sciences. 2018;**63**:9-17

[122] Mayak S, Tirosh T, Glick BR. Plant growth-promoting bacteria that confer resistance to water stress in tomatoes and peppers. Plant Science. 2004;**166**:525-530

[123] Egamberdieva D. *Pseudomonas chlororaphis*: A salt-tolerant bacterial inoculant for plant growth stimulation under saline soil conditions. Acta Physiologiae Plantarum. 2012;**34**:751-756

[124] Kang SM, Radhakrishnan R, Khan AL, Kim MJ, Park JM, Kim BR, et al. Gibberellin secreting rhizobacterium, *pseudomonas putida* H-2-3 modulates the hormonal and stress physiology of soybean to improve the plant growth under saline and drought conditions. Plant Physiology and Biochemistry. 2014;**84**:115-124

[125] Li T, Hu YJ, Hao ZP, Li H, Chen BD. Aquaporin genes GintAQPF1 and GintAQPF2 from *Glomus intraradices* contribute to plant drought tolerance. Plant Signaling and Behavior. 2013;**8**:e24030

[126] Saravanakumar D, Samiyappan R. ACC deaminase from Pseudomonas fluorescens mediated saline resistance in groundnut (*Arachis hypogea*) plants. Journal of Applied Microbiology. 2007;**102**:1283-1292

[127] Li J, McConkey BJ, Cheng Z, Guo S, Glick BR. Identification of plant growth-promoting bacteria-responsive proteins in cucumber roots under hypoxic stress using a proteomic approach. Journal of Proteomics. 2013;**84**:119-131

[128] Shintu PV, Jayaram KM. Phosphate solubilising bacteria (*bacillus polymyxa*)-an effective approach to mitigate drought in tomato (*Lycopersicon esculentum* mill.). Tropical Plant Research. 2015;**2**:2349-9265

[129] Jahromi F, Aroca R, Porcel R, Ruiz-Lozano JM. Influence of salinity on the in vitro development of *Glomus intraradices* and on the in vivo physiological and molecular responses of mycorrhizal lettuce plants. Microbial Ecology. 2008;**55**:45-53

[130] Sangiorgio D, Cellini A, Donati I, Pastore C, Onofrietti C, Spinelli F. Facing climate change: Application of microbial biostimulants to mitigate stress in horticultural crops. Agronomy. 2020;**10**:794

[131] Jaemsaeng R, Jantasuriyarat C, Thamchaipenet A. Positive role of 1-aminocyclopropane-1-carboxylate deaminase-producing endophytic *Streptomyces sp*. GMKU 336 on flooding resistance of mung bean. Agriculture and Natural Resources. 2018;**52**:330-334

[132] Itakura M, Uchida Y, Akiyama H, Hoshino YT, Shimomura Y, Morimoto S, et al. Mitigation of nitrous oxide emissions from soils by *Bradyrhizobium japonicum* inoculation. Nature Climate Change. 2013;**3**:208-212

[133] Beed F, Benedetti A, Cardinali G, Chakraborty S, Dubois T, Garrett K, et al. Micro-organism genetic resources for food and agriculture and climate change. In: Coping with Climate Change—The Roles of Genetic Resources for Food and Agriculture. Rome: FAO; 2015. pp. 87-101

[134] Ibrahim AE, Khalid E. Climate change impact on vegetable crops and potential for adaptation: A review. Hortscience Journal of Suez Canal University. 2021;**10**(1):13-30

[135] Rai N, Rai SP, Sarma BK. Prospects for abiotic stress tolerance in crops utilizing phyto-and bio-stimulants. Frontiers in Sustainable Food Systems. 2021;**5**:754853

[136] Ma Y, Freitas H, Dias MC. Strategies and prospects for biostimulants to alleviate abiotic stress in plants. Frontiers in Plant Science. 2022;**13**:1024243

[137] Nephali L, Piater LA, Dubery IA, Patterson V, Huyser J, Burgess K, et al. Biostimulants for plant growth and mitigation of abiotic stresses: A metabolomics perspective. Metabolites. 2020;**10**:505

[138] Zaid A, Mohammad F, Fariduddin Q. Plant growth regulators improve growth, photosynthesis, mineral nutrient and antioxidant system under cadmium stress in menthol mint (*Mentha arvensis* L.). Physiology and Molecular Biology of Plants. 2020;**26**:25-39

[139] Duc NH, Csintalan Z, Posta K. Arbuscular mycorrhizal fungi mitigate negative effects of combined drought and heat stress on tomato plants. Plant Physiology and Biochemistry. 2018;**132**:297-307

[140] Fadiji AE, Babalola OO, Santoyo G, Perazzolli M. The potential role of microbial biostimulants in the amelioration of climate change-associated abiotic stresses on crops. Frontiers in Microbiology. 2022;**12**:829099

[141] Weng Z, Van Zwieten L, Singh BP, Tavakkoli E, Joseph S, Macdonald LM, et al. Biochar built soil carbon over a decade by stabilizing rhizodeposits. Nature Climate Change. 2017;**7**:371-376

Chapter 14

Utilizing Alternative Carbon Sources for Biofloc System for Growth and Survival of Pacific Whiteleg Shrimp (*Litopenaeus vannamei*)

Arien Jean M. Lopez, Marlyn B. Llameg, John Paul R. Pacyao and Godofredo P. Lubat Jr

Abstract

Selection of a suitable carbon source is crucial in the advancement of biofloc technology. This study aims to evaluate the usability and effect of leftover bread and surplus rice as carbon source for the biofloc system for growth and survival of Pacific whiteleg shrimp (Litopenaeus vannamei). Biofloc was developed 2 weeks before stocking using probiotics (0.03 g/L) with carbohydrate (0.1 g/L) source. The experimental treatments: T0 (clear water), T1 (leftover bread), T2 (surplus rice), and T3 (50% surplus rice+50% leftover bread) replicated three times. Stocked with 12 postlarvae (PL10) weighing 0.028–0.052 g each (6PL/L), parameters were regularly monitored. During a 30-day culture period, shrimp were fed twice daily with a diet containing 55% protein, and the carbon source was maintained at a C:N ratio of 15:1. One-way ANOVA of sampling data (collected every 15 days) indicated no significant differences ($p > 0.05$) in growth performance, survival rate, and feed conversion ratio (FCR) across treatments. Physicochemical parameters of the water are at the optimum; the leftover bread biofloc attained more ideal parameters (dissolved oxygen: 7.54, temperature: 27.30, salinity: 29.10, and pH: 7.6) due to simpler starch structure. Poor fermentation and slow degradation properties of leftover bread and surplus rice on biofloc formation were the reason why biofloc has no effect on shrimp growth and survival. Leftover bread shows promising results among the BFT treatments; for further study with greater area and period, additional parameters are encouraged.

Keywords: *Litopenaeus vannamei*, biofloc technology, indoor shrimp aquaculture, nutrient recycling, sustainable aquaculture

1. Introduction

The global consumption of fish has been drastically increasing at an average annual rate of 3.1% recorded from 1961 to 2017, which doubled the annual world population growth rate (1.6%) in the same period [1]. With this documented higher demand, aquaculture firms attempt to intensify their production. However, maximizing aquaculture produces higher aggregates of organic materials, resulting in toxic effects and long-run environmental risks [2]. Enhancing productivity is the anticipated solution and simplest method addressing this heightened demand, which is also one of the focal concerns in improving aquaculture and shrimp farming [3–6].

Sustainable aquaculture enhancement and responsible use of resources realizing the U.N. 2030 Agenda for Sustainable Development Goals principally: ending poverty (SDG 1), ending hunger, achieving food security and improved nutrition (SDG 2), and promoting sustained, inclusive, and sustainable economic growth (SDG 8) [7]. Furthermore, embracing the modernization of the aquaculture industry, particularly through the economic management of biofloc systems, is crucial for local development.

Shrimp farming utilizing biofloc technology exhibits gains compared to traditional fish farming, where there is zero water exchange and more negligible ecological effect – minimized contaminant on the production system. The nitrogen compounds generated from the unconsumed 20–40% incorporated feeds were recycled and converted into bacterial biomass, which can be used as supplemental feeds [3, 8].

Subsequently, the use of biofloc shrimp is a low cost, sustainable, and environment-friendly technology. The shrimp industry has been a promising venture as market potential and sustainability supporting food security [8]. Biofloc systems have been a potential aquaculture technology improving shrimp farming, and introducing this to the locale with the available raw materials can be a capable livelihood opportunity. Moreover, biofloc technology was productively utilized in nearly 20–25% of the shrimp aquaculture farms. Most Asia's highest shrimp production industries are adapting to biofloc – Singapore, Indonesia, Myanmar, and Malaysia [9, 10]. Shrimp farming shows a promising future through the nutrient recycling principles of biofloc technology. BFT has previously been initially tested in three farms in Luzon, two in the Visayas, and two in Mindanao, employing whiteleg shrimp (*L. vannamei*) as cultured species [11].

This study highlights the shrimp culture practices using biofloc technology through utilizing alternative carbon sources. The BFT has been adapted worldwide, and through this research, the locale is informed on the technology and a supplementary livelihood opportunity. The result of this study will serve as an alternative carbon source on maintaining the biofloc sources reducing production cost and at the same time reduction of wastage of leftover food. This study will serve as a baseline information to those who plan on adapting biofloc technology on culturing whiteleg shrimp. The data sought can be a guide in support to the continuous improvement of Bureau of Fisheries and Aquatic Resources and other non-government agencies on their implementation of BFT as program. The result of this research is useful to the academe in need of further studies in modifying biofloc systems as well as to the people in fisheries sector visioning for the development of shrimp industry.

2. Materials and methods

2.1 Description of the study site, experimental species, treatment, and setup

This study was conducted at Balasinon, Sulop, Davao del Sur, Philippines. The location has the convenience of available materials and equipment needed and laboratory for water parameter analysis from Davao del Sur School of Fisheries, taking into consideration the status of the COVID-19 Protocols of the locale. The experiment species used in the study were 150 specific pathogen-free (SPF) Pacific whiteleg shrimp postlarvae (*L. vannamei*) with an approximate length of 5.0 mm, procured from available accredited hatchery from Sta. Cruz, Davao del Sur. The 5 L containers were used with 2 L water with 12 postlarvae (PL10) [12]. **Figure 1** shows the experimental setup.

The experiment was done indoors using 12 cylinder containers 15 cm in diameter and 18 cm in height with 5 L capacity filled with 2 L of seawater arranged in a Complete Randomized Design (CRD). The four treatments were replicated three times: T0 (clear water system), T1 (leftover bread), T2 (surplus rice), and T3 (bread and rice) as shown in **Table 1**. The containers were stocked with 15 days postlarvae at a density of 6 PLs/liter and were constantly aerated throughout the study.

Figure 1.
A. Experimental setup for biofloc system for L. vannamei with constant aeration and lighting; B. Daily water parameter monitoring for pH, salinity, dissolved oxygen, temperature, and turbidity; C. Second sampling activity measuring the weight and length of the shrimp.

Treatment	Carbon content needed (g)	Carbon content g/100 g (NHS, 2016; FAO, 199)	Carbon incorporated to maintain C/N ratio (g)
T_0			
T_1 LB*	1.06	50	2.12
T_2 SR*	1.06	30	3.18
T_3 LB + SR	0.53 + 0.53	50+ 30	1.06 + 1.59

**LB = leftover bread, SR = surplus rice.*

Table 1.
The different treatments used in the study.

Treatment	Sea water (L)	Carbohydrates (0.1 g/L)	Probiotics (0.03 g/L)	Lime* (0.05 g/L)
T_0: clear water	7	—	—	—
T_1: bread	7	0.7	0.21	0.35
T_2: rice	7	0.7	0.21	0.35
T_3: bread + rice	7	0.35 × 2	0.21	0.35

**Addition done if pH decreases below.*

Table 2.
The formulation of different treatments used in the study.

2.2 Development of biofloc water, carbon addition, and feeding management

The biofloc was inoculated a week prior to the procurement of the experimental species. The process flow of developing biofloc water presented in, where there are four treatments including a control treatment and three biofloc preparation tanks: T1 (leftover bread), T2 (surplus rice), and T3 (bread and rice) that were filled with approximate 7 L seawater. The proportion of the solution was 0.1 g/L carbohydrate source and 0.03 g/L probiotics to the exact amount per treatment as shown in **Table 2**. The pH of the prepared water was monitored after 24 hours. Application of lime at 0.05 g/L was applied to treatment 3 with bread and rice as carbon sources as the pH level dropped below 5. Aeration was maintained to the optimum dissolved oxygen requirements 4.0 mgL^{-1} and at least 60% of saturation [4, 13]. Carbon addition was done a week before the procurement of the seeds, and star commercial feeds were utilized following the feeding rates: 3% biomass (0–20 days of culture) and 4% feeding rate (21–40 days of culture) [14, 15]. The culture species were fed twice a day (6 am and 6 pm) with 55% protein (5% lipid and 3% fiber) **Figure 2** [4, 10]. The formulation of the biofloc was computed as shown in **Table 2**.

2.3 Physicochemical monitoring

Reading of the temperature, pH, turbidity, and dissolved oxygen were done and carried out at 7:00–8:00 am and 4:00–6:00 pm. Measuring ammonia and observing the biofloc through microscope were done every week. The physico-chemical parameters: dissolved oxygen, pH, temperature, and salinity using a 5-in-1 multifunction water quality test probe, while turbidity was measured through identifying Secchi depth using Secchi disk (four-inch diameter). Ammonia was measured by a colorimetric ammonia kit [17].

2.4 Sampling and data gathering

Growth measurement was done every day after 15 days, and the increase in total length and weights were determined. The total length of shrimp was determined using ruler, while weight was determined using the electronic weighing scale. Sampling was conducted in the morning. Samples of five shrimps per container were taken. The shrimps were held in restraint container and then measured as soon as possible to prevent the stocks from being stressed. The total length was measured from the tip of the rostrum to the tip of the [18]. The shrimps were weighed individually. After the sampling, the shrimp will be immediately returned to the containers. The

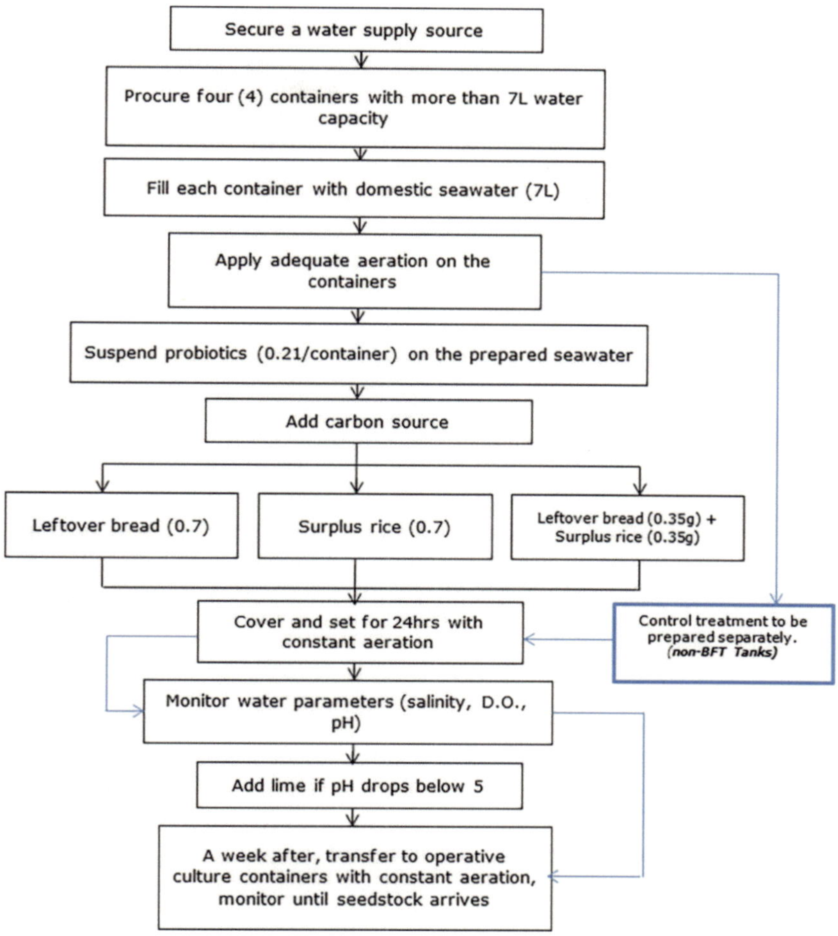

Figure 2.
Schematic procedure in preparing and developing biofloc water [16].

procedure was done simultaneously per individual container to minimize stress on the cultured organisms [17].

2.5 Statistical analysis

Data were statistically analyzed using a one-way ANOVA at 5% significance level to identify significant difference using Statistical Package for Social Sciences (SPSS) version 22.

3. Results

3.1 Growth performance

The use of different biofloc carbon sources in shrimp culture (*L. vannamei*) using biofloc systems obtained comparable results among all treatments ($p > 0.05$) in terms of growth parameters (weight and length increments and specific growth rate).

	T_0 – CW	T_1 – LB	T_2 – SR	T_3 – LB + SR
IBW	0.040 ± 0.018	0.037 ± 0.012	0.033 ± 0.016	0.027 ± 0.013
ITL	0.593 ± 0.234	0.620 ± 0.174	0.527 ± 0.205	0.440 + 0.118
FTW^{ns}	0.242 ± 0.135	0.180 ± 0.129	0.166 ± 0.082	0.144 ± 0.052
FTL^{ns}	3.250 ± 0.625	3.139 ± 0.668	2.389 ± 0.772	2.528 ± 0.475
SGR^{ns}	0.68	0.48	0.44	0.39
DGR^{ns}	0.023	0.016	0.015	0.013

Mean values (±SD) in the same row are not significantly different (ns); p > 0.05. Abbreviations used: IBW = initial weight (g); ITL = initial total length (cm); FTW = final weight (g); FTL = final total weight (cm); SGR = specific growth rate (%); DGR = daily growth rate (%); CW = clear water; LB = leftover bread; SR = surplus rice.

Table 3.
Growth performance of Pacific whiteleg shrimp postlarvae (Litopenaeus vannamei) grown on biofloc systems with different carbon sources for the 30 days culture period.

Among the BFT treatments, the leftover bread has a higher final mean weight and length output (0.180 g ± 0.129; 3.139 cm ± 0.668). The same trend was observed on specific growth rates where all treatments were comparable as shown in **Table 3**.

3.2 Survival rate

The effect of different treatments on the survival of *L. vannamei* reared in biofloc system is presented in **Table 4**. The highest survival rate (58.33%) was observed in treatment 1, while the lowest was attained in treatment 2 (47.22%) on the biofloc system with surplus rice as carbon source. But statistically, the survival rate of the different treatments was not significantly different ($p > 0.05$).

3.3 Feed conversion ratio

Table 5 shows the final mean FCR of Pacific whiteleg shrimp postlarvae *(Litopenaeus vannamei)* for 30 days of culture. Results showed that the lowest feed conversion ratio in treatment 1 obtained 2.208 with leftover bread as carbon source among the treatments. Highest feed conversion ratio in treatment 2 was obtained in those Pacific whiteleg shrimp postlarvae that were grown on the biofloc system with surplus rice as carbon source, but statistically there were no significant difference, and all treatments were comparable ($p > 0.05$).

Treatment	Number of stock	$Survival^{ns}$	Percentage $(\%)^{Ns}$
0CW	36	19	52.78
ILB	36	21	58.33
IISR	36	17	47.22
III SR + LB	36	18	50

Values in the same column are not significantly different (ns) p > 0.05. Abbreviations used: CW = clear water; LB = leftover bread; SR = surplus rice.

Table 4.
Survival rate of pacific whiteleg shrimp postlarvae (Litopenaeus vannamei) grown on biofloc systems with different carbon sources.

		Sampling period		
Treatment		**Initial sampling to 1st sampling^ns^**	**1st sampling to 2nd sampling^ns^**	**Average^ns^**
0	CW	0.23	0.40	0.32
I	LB	0.31	0.32	0.32
II	SR	0.28	0.44	0.36
III	SR + LB	0.26	0.44	0.35

Values in the same column are not significantly different (ns); p > 0.05. Abbreviations used: CW = clear water; LB = leftover bread; SR = surplus rice.

Table 5.
Feed conversion ratio of Pacific whiteleg shrimp postlarvae (Litopenaeus vannamei) grown on biofloc systems with different carbon sources.

3.4 Physicochemical parameters

Results showed that there was a statistically significant difference in dissolved oxygen levels on biofloc systems for shrimp using alternative carbon sources ($p < 0.005$). The level of dissolved oxygen in treatment 1 is significantly higher than in treatment 2 and treatment 3, and the difference is statistically significant. **Figure 3** shows a declining trend of dissolved oxygen level; however, all the treatments were within the normal range for biofloc system.

The water temperature range in tanks was observed at 26.51–27.62°C for the entire duration of the experiment. Water temperature was highest during the fourth week of the experiment, with treatment 1 where carbon source with leftover bread as the highest (27.72 ± 0.72°C) and carbon source with treatment 2 where combined surplus rice and leftover bread (27.41 ± 0.65°C) as the lowest. The recorded value shows significant difference ($p < 0.05$), yet all treatments attained the normal range for temperature for biofloc (**Figure 4**). However, the consequent week shows a drop of temperature in the biofloc systems with the means (T0: 27.08 ± 0.73°C; T1: 27.62 ± 0.92°C; T2: 27.36 ± 0.78°C; T3: 27.16 ± 0.72°C).

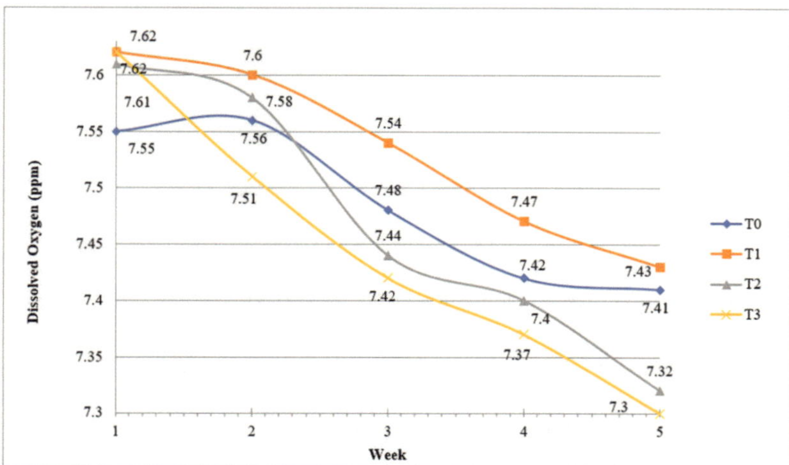

Figure 3.
Mean dissolved oxygen (mgL^{-1}) levels of biofloc systems with different carbon sources for culturing Pacific whiteleg shrimp postlarvae (Litopenaeus vannamei).

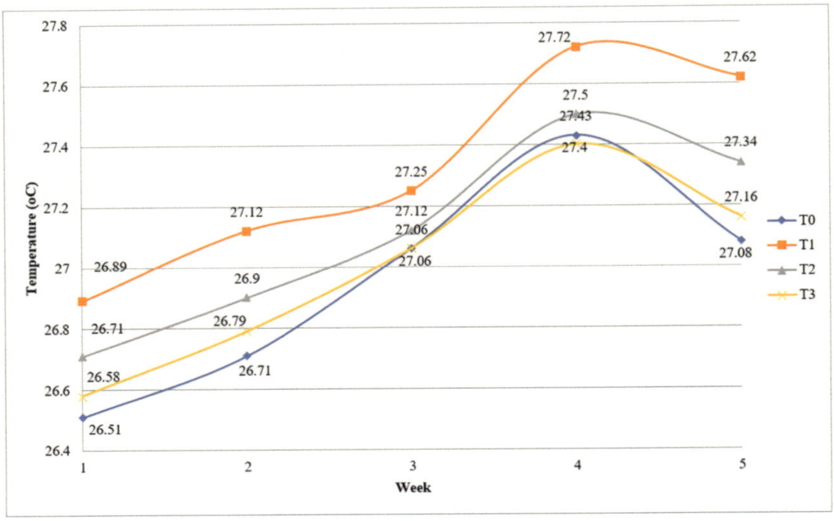

Figure 4.
Mean temperature (°C) of biofloc systems with different carbon sources for culturing pacific whiteleg shrimp postlarvae (Litopenaeus vannamei).

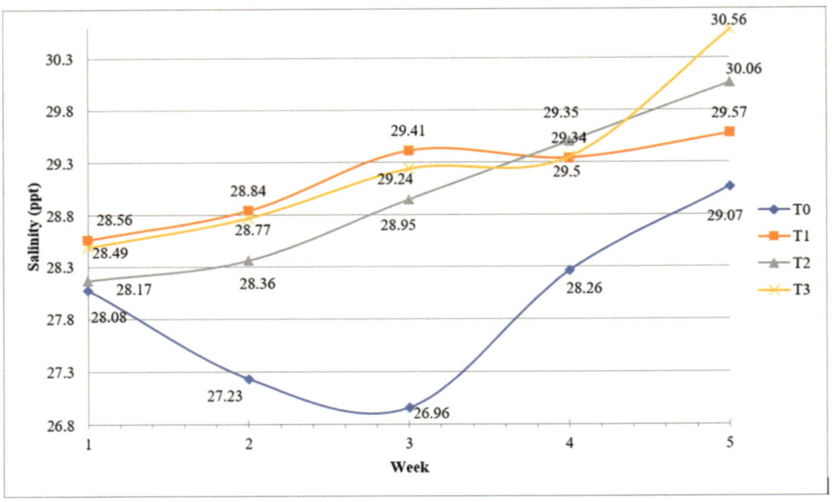

Figure 5.
Mean salinity (ppt) of biofloc systems with different carbon sources for culturing Pacific whiteleg shrimp postlarvae (Litopenaeus vannamei).

The water salinity shows an inclining trend during the 30 days of experiment. However, in treatment 0, clear water system recorded fluctuating ppt levels, where in the first week, it started to decline and then rise on the fourth week as shown in **Figure 5**. The lowest mean salinity was 26.90 ± 0.68 ppt during the third week on treatment 0. The salinity measurements in the first week of the experiment ranged at 28.08 ± 0.59–28.56 ± 0.78 ppt and rose at the levels 29.07 ± 0.49–30.56 ± 0.48 ppt.

The level of pH on the control treatment is significantly higher among the biofloc treatments; however, the pH of leftover bread biofloc is significantly higher among

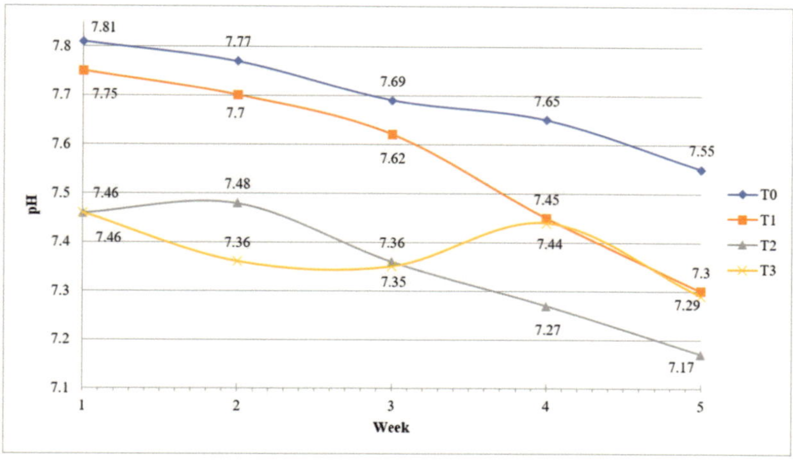

Figure 6.
Mean pH of biofloc systems with different carbon sources for culturing Pacific whiteleg shrimp postlarvae (Litopenaeus vannamei).

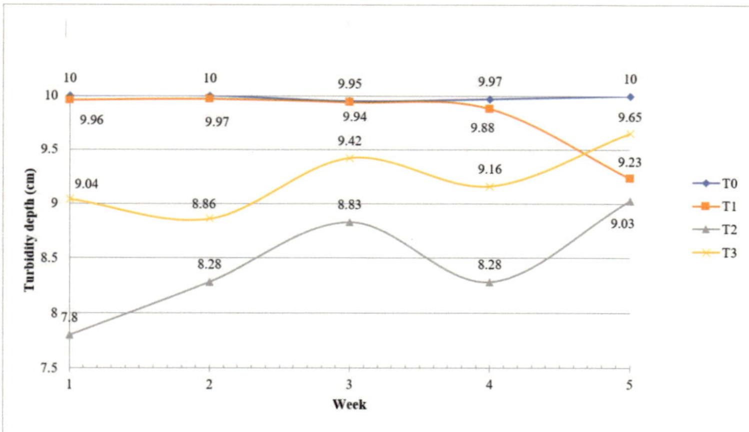

Figure 7.
Mean turbidity (cm) of biofloc systems with different carbon sources for culturing Pacific whiteleg shrimp postlarvae (Litopenaeus vannamei).

other treatments of carbon sources. The level of pH level in T0 is significantly higher than in T1, T2, and T3, while T1 is significantly higher than T2 and T3. Shown in **Figure 6**, the recorded pH levels indicate they are within the ideal range for shrimp biofloc systems. The pH levels show a gradual acid transition with lowering of pH measurements. In first sampling, treatment 3 recorded as the most acidic with mean 7.44 ± 0.08 during day and 7.39 ± 0.42 during night, while on the second sampling, treatment 2 shows more acidity than the other treatments with mean of 7.26 ± 0.20 and 7.26 ± 0.15 at night and day, respectively.

Figure 7 shows the trend of turbidity on the experimental biofloc systems. There is significant difference ($p < 0.05$) in the turbidity of *L. vannamei* biofloc system with different alternative carbon sources applied. The level of turbidity in T0 significantly higher than in T2 and T3, while T1 is significantly higher than T2 and T3, and T3 is

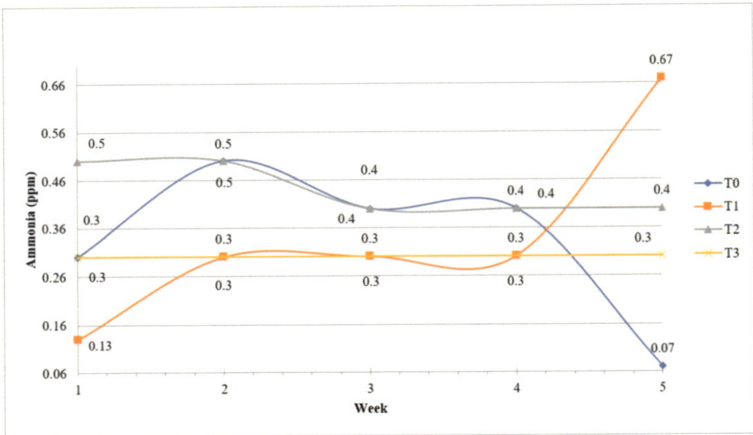

Figure 8.
Mean ammonia (ppm) of biofloc systems with different carbon sources for culturing Pacific whiteleg shrimp postlarvae (Litopenaeus vannamei).

significantly higher than T2. Obviously, the control treatment is significantly less turbid than the biofloc system, while the most turbid biofloc treatment is the surplus rice and the least turbid biofloc system is the treatment with leftover bread as carbon source followed by the 50% surplus rice +50% leftover bread, and the differences are statistically significant.

The maximum ammonia measurement observed was 0.51 ± 0.0 mgL^{-1}, while the minimum observed ammonia was 0.18 ± 0.2 mgL^{-1}. The flow of pattern in ammonia level did not follow of any regular pattern as demonstrated in **Figure 7**. The biofloc systems as culture medium were within the safe threshold (mgL^{-1}) (**Figure 8**) [19].

4. Discussion

The complex carbohydrates needed more digestion duration and less fermentation time required [20]. Leftover bread was easier to breakdown, hence easier for bacterial aggregates and manageable parameters to develop as to compare the structure of surplus rice and the combination of CR + LB [19, 21]. The inadequate fermentation of biofloc with leftover bread and surplus rice has no significant difference as carbon source as these are classified as complex starches with longer duration to be fully fermented [19, 22]. Hapsari [23] also observed that weight gain was better with fermented biofloc compared to a clear water system and unfermented biofloc, with no significant difference between the latter two.

The BFT treatments using the experimental carbon sources manifested sticky and gel-like texture due to its soluble proteins. The occurrence of floc attaching on the part of the container wall on the BFT treatments caused the decomposition process to not maximally compare to clear water systems making it unable to achieve its optimal yield [24, 25]. Monosaccharides with less protein and lipids attained the best increase in biomass in BFT farming, and this can be the reason for low value output from the more complex starches [21, 26]. Combining carbohydrates affects its digestibility base on its glycemic index and can reduce its conversion to simpler sugar, and hence, it can explain the observance of low mean weight gain on the LB + SR carbon source biofloc system [21, 27].

Fermentation was done for a short period of time, aggregates were not fully matured as mentioned by Ogello et al. [28], small quantities of bacteria cannot form substantial flocs in the culture system, and this can further explain the high yield on leftover bread as simpler starch; however, the combination of these carbohydrates can drastically affect its digestibility and reducing it, making it difficult to convert into simpler sugars [27, 29] relating this to the low growth of biofloc systems with 50% leftover bread + 50% surplus rice.

High mortality rate was evident in the experiment. This could be due to cannibalism, which is induced by the absence of hides. This is a problem for *L. vannamei* species in the nursery and early juvenile stages, which is understandable given their high metabolic rates and frequent molting [30]. The issue of low survival may also be linked to natural response to captivity, lack of shelter, ecdysis, increased density, lack of food, and poor water quality [31].

The parameters were maintained within the normal range; however, the significant difference in turbidity between treatments can be connected to the survival yield between the BFT systems and the clear water systems. Comparing the results, the turbidity of the BFT system using only leftover bread as a carbon source was 9.94 cm, and it had a higher survival rate (58.33%) compared to the other treatments. As explained by Kathyayani et al. [32], high water turbidity in farming of *Litopenaeus vannamei* is one of the environmental stressors that hinder survival of cultured species due to gills blockage as a compensatory reaction to the disruption of osmotic and ionic balance produced by high turbidity in the rearing medium.

Simpler starches, such as leftover bread, are easier to digest and convert into biomass compared to complex chains [26, 33]. In the context of this study, surplus rice facilitated experimental investigations into different carbon sources in biofloc systems, where simpler sugars resulted in slight improvements in survival rate, FCR, and water quality [34–37]. The choice of carbon sources is critical to the biofloc system's performance, as it can affect the availability, palatability, and digestibility for the cultured species [20].

It was also mentioned by Peiro-Alcantar et al. [38] that despite on the desirable water quality, survival, and growth yield of the biofloc, there were no perceived significant differences among treatments and the control. Another biofloc study by Arias-Moscoso et al. [39] revealed that while the microbial community's dependence on floc palatability is influenced by the carbon sources used, this palatability affects the metabolization of harmful nitrogen compounds. However, the study found that probiotics have no effect on crop production.

The observed dissolved oxygen and temperature of all the recorded value during the whole duration of the study were still within the tolerable range ideal for biofloc culture system [4]. This could be caused by the simpler starch structure of bread compared to rice which is simpler to digest and attain optimal parameters [24, 29, 40]. The optimum temperature for the growth of microorganisms is around 28–32°C that was manifested on biofloc with leftover bread carbon source. Thus, more microorganisms thriving on the culture medium are consequences of the greater metabolism [19].

Low dissolved oxygen on biofloc with combined rice and bread is possible because of the digestibility characteristics of the combined carbon source [27]. This can be due to the metabolic activity of the heterotrophic and autotrophic components of the biofloc formed and more microorganism thriving on the culture medium as consequences of the greater metabolism. The leftover bread is simpler starch; then, it can be digested easier and able to facilitate a more desirable BFT system to better yield as can be seen on its survival output [24, 26].

Betanzo-Torres et al. [41], Manan et al. [42], and Ray et al. [43] pointed out that the presence of planktons consuming the available dissolved oxygen in the medium the levels was differentiated due to the difference of carbon source, whereas the simpler starch yields high in terms of its consumability, yet the D.O levels were maintained with the constant aeration as requirement for biofloc sustenance [3, 10].

The complex starch structure of rice matched to bread which is easier to convert simpler sugar [24, 29, 40]. Becerril-Cortés et al. [19] observed that the optimal temperature for the growth of microorganisms is around 28–32°C. This was also confirmed in an experiment using a biofloc system with leftover bread as the carbon source, which resulted in favorable survival rates. This can be associated with Nugroho et al.'s study [44] which states that temperature differs for colonies of various compositions, different algae species, and different groups at their ideal temperature. The biomass of community phytoplankton exposed to various temperature settings is influenced by abiotic factors. The limited water exchange might be the reason of increasing salinity levels, and this can be associated with evaporation loss though. This is controlled by adding water to conform to the setup stocking density [38, 43, 45].

Different salinity can be derived from the level of complexity of polysaccharide properties of bread and rice due to the saturation of the water evaporation which slows down, and hence the salinity increases [24, 40, 46]. The disparities which observed salinity levels of culture systems sought no significance was found in terms of survival and growth performance. The use of more complex starches caused slow degradation; hence, metabolic systems affect the parameters of the culture medium. The presence of organic matter from unfermented floc from complex starches gained more H^+ ions making it slightly acidic than others, as to be observed on the leftover bread and compared to the control, rice, and its combination (rice and bread) [21, 24, 47]. However, recorded levels are within the normal range.

Leftover bread as carbon source gained a lesser turbid biofloc among the biofloc treatments,and this can be connected to the degradation properties of the used carbon source sought by El-husseiny et al. [47] stating that complex starch tends to have slow degradation, and hence, high total suspended solids and salinity follows. The turbidity of the biofloc treatment with leftover bread was less turbid among the other treatments depicted in **Figure 9**. As discerned by Kathyayani et al. [32], raised water turbidity in the intensive farming of *Litopenaeus vannamei* is one of the environmental stressors due to gills blockage as a compensatory reaction to the disruption of osmotic and ionic balance produced by high turbidity in the rearing medium. Biofloc with moderate turbidity and floc aggregate density are more reliable in *L. vannamei* growing systems – favorable element in the maintenance of efficient and stable production [40].

Minimal variation in ammonia observed in all treatments during the whole duration of the study. The values were all within the ideal range (1 mgL^{-1}) [4]. The current study shows that the supplementation of different carbon source in biofloc system has no effect on the ammonia levels on culturing whiteleg shrimp. The short duration of fermentation and the complex polysaccharide structure of the carbon source can be the reasons [24, 29, 47]. The reason of lesser area used can also affect

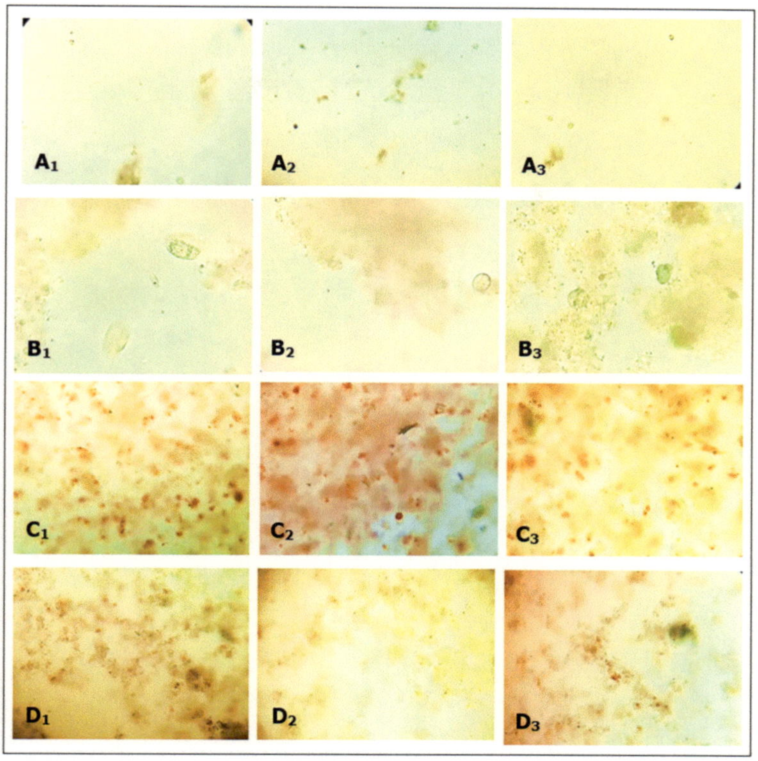

Figure 9.
Images of biofloc from different carbon sources on their fourth week of DOC, seen 20× magnified under the compound microscope. (A1–2. T0 – clear water system, B1–3. T1 – leftover bread, C1–3. T2 – surplus rice, D1–3. T3–50% surplus rice +50% leftover bread).

the development of biofloc and the supposed progress of nitrification of ammonia for microbial aggregates [28].

5. Conclusion

The current study recommends that the use of leftover bread and surplus rice is recommended as alternative carbon sources in biofloc systems. Further investigation for leftover bread as probable carbon sources is recommended. Conducting a similar study using a higher stocking density for a longer duration of culture period and other physico-chemical parameters is recommended to validate the result of this study.

Acknowledgements

The author acknowledges the support of Institute of Fisheries and Marine Sciences of Southern Philippines Agribusiness and Marine and Aquatic School of Technology, Bureau of Fisheries and Aquatic Resources – XI, and Davao del Sur School of Fisheries.

Author details

Arien Jean M. Lopez*, Marlyn B. Llameg, John Paul R. Pacyao
and Godofredo P. Lubat Jr
Institute of Fisheries and Marine Sciences, Southern Philippines Agribusiness and
Marine and Aquatic School of Technology, Philippines

*Address all correspondence to: ajmlopez@spamast.edu.ph

References

[1] FAO. Summary of the impacts of the Covid-19 pandemic on the fisheries and aquaculture sector - addendum to the state of world fisheries and aquaculture 2020. In: Addendum to the State of World Fisheries and Aquaculture 2020. FAO; 2020. pp. 1-4

[2] Ahmad I, Babitha Rani AM, Verma AK, Maqsood M. Biofloc technology: An emerging avenue in aquatic animal healthcare and nutrition. Aquaculture International. 2017;**25**(3):1215-1226. DOI: 10.1007/s10499-016-0108-8

[3] Avnimelech Y. Biofloc Technology: A Practical Guide Book. New Orleans: World Aquaculture Society; 2015

[4] Emerenciano MGC, Martínez-Córdova LR, Martínez-Porchas M, Miranda-Baeza A. Biofloc technology (BFT): A tool for water quality Management in Aquaculture. In: Water Quality. London, UK: InTech; 2017. DOI: 10.5772/66416

[5] FAO. The State of World Fisheries and Aquaculture 2020. FAO; 2020. Available from: https://www.fao.org/3/ca9229en/ca9229en.pdf

[6] Piedrahita RH. Reducing the potential environmental impact of tank aquaculture effluents through intensification and recirculation. Aquaculture. 2003;**226**(1-4):35-44. DOI: 10.1016/S0044-8486(03)00465-4

[7] FAO. Indicators: Measuring up to the 2030 agenda for sustainable development. In: Sustainable Development Goals. FAO; 2017. Available from: http://www.fao.org/3/a-i6919e.pdf%5C

[8] Bossier P, Ekasari J. Biofloc technology application in aquaculture to support sustainable development goals. Microbial Biotechnology. 2017;**10**(5):1012-1016. DOI: 10.1111/1751-7915.12836

[9] van Beijnen J, Yan G. Ten Easy Steps towards Biofloc Production of Shrimp or Tilapia. The Fish Site; 2019. pp. 1-16 Available from: https://thefishsite.com/articles/ten-easy-steps-towards-biofloc-production-of-shrimp-or-tilapia

[10] Taw N. Shrimp Farming in Biofloc System: Review and Recent Developments. 2014. Available from: https://www.aesweb.org/presentations/biofloc/adelaide2014/Shrimpfarminginbioflocsystem-reviewandrecentdevelopments.pdf

[11] Calpe AT. Biofloc Technology to Reduce Risk of Diseases and Enhance Nutrition in Shrimp Farms. Philippine Council for Agriculture, Aquatic and Natural Resources Research and Development; 2016. pp. 1-2

[12] FAO Fisheries and Aquaculture Department. 2. Sampling and Analysis 12.1. Estimating the Number of Shrimp in a Pond. FAO Fisheries and Aquaculture Department. Available from: http://www.fao.org/3/ac006e/AC006E12.htm

[13] CABI. Invasive Species Compendium Detailed Coverage of Invasive Species Threatening Livelihoods and the Environment Worldwide *Litopenaeus vannamei* (Whiteleg Shrimp) Toolbox Horizon Scanning Tool Invasives Open Data Country Pest Alerts. CABI; n.d. Available from: https://ckan.cabi.org/data/

[14] Antunes CRN, Da Silva Ledo CA, Pereira CM, Dos Santos J. Evaluation of feeding rates in the production of *Litopenaeus vannamei* shrimp

using artificial substrates. Ciencia Animal Brasileira. 2018;**19**:1-11. DOI: 10.1590/1809-6891v19e-50805

[15] Quagrainie K. Profitability of Indoor Production of Pacific White Shrimp (*Litopenaeus vannamei*): A Case Study of the Indiana Industry. Kwamena Quagrainie; 2015

[16] Chandan. Biofloc Fish Farming. Biofloc Fish Farming; 2020. Available from: www.bioflocfishfarming.com/water-preparation-for-biofloc/

[17] FAO Fisheries and Aquaculture Department. *Penaeus vannamei* (Boone, 1931). FAO Fisheries and Aquaculture Department; 1931. Available from: http://www.fao.org/fishery/culturedspecies/Penaeus_vannamei/en

[18] Bailey-Brock JH, Moss SM. Chapter 2 - Penaeid taxonomy, biology and zoogeography. In: Fast AW, Lester LJB-T-MSC, editors. Developments in Aquaculture and Fisheries Science. Elsevier; 1992. pp. 9-27. DOI: 10.1016/B978-0-444-88606-4.50008-4

[19] Becerril-Cortés D, del Carmen M-DM, Emerenciano MGC, Castro-Mejía G, Beatríz Sofia Bermúdez S, Correa GV. Effect on nutritional composition of produced bioflocs with different carbon sources (molasses, coffee waste and rice bran) in biofloc system. International Journal of Fisheries and Aquatic Studies. 2018;**6**(2):541-547

[20] de Schryver P, de Bossier P, Ekasari J, Azhar MH, Surawidjaja EH, Nuryati S. Shrimp Grown in Biofloc Systems with Different Carbon Sources. Global Aquacult Advocate; 2019

[21] Foster-Powell K, Holt SHA, Brand-Miller JC. International table of glycemic index and glycemic load values: 2002. The American Journal of Clinical Nutrition. 2002;**76**(1):5-56. DOI: 10.1093/AJCN/76.1.5

[22] Booth DA, Mather P, Fuller JAN. Starch content of ordinary foods associatively conditions human appetite and satiation, indexed by intake and eating pleasantness of starch-paired flavours. Appetite. 1982;**3**(2):163-184. DOI: 10.1016/S0195-6663(82)80009-3

[23] Hapsari F. The effect of fermented and non fermented biofloc inoculated with bacterium bacillus cereus for catfish (*Clarias gariepinus*) juveniles. AACL Bioflux. 2016;**9**(2):334-339. Available from: https://www.researchgate.net/publication/302445106_The_effect_of_fermented_and_non_fermented_biofloc_inoculated_with_bacterium_bacillus_cereus_for_catfish_Clarias_gariepinus_juveniles

[24] Adipu Y, Lumenta C, Mangindaan REP, Manoppo H. Growth performance of *Litopenaeus vannamei* grown in biofloc system produced from different carbohydrate sources. AACL Bioflux. 2019;**12**(2):472-479

[25] Kerr RM, Rayas P. Adding Value to OKLAHOMA Fiber: Bulk of Life. Food Technology Fact Sheet; 2005. pp. 1-8

[26] Moongngarm A. Chemical compositions and resistant starch content in starchy foods. American Journal of Agricultural and Biological Sciences. 2013;**8**(2):107-113. DOI: 10.3844/ajabssp.2013.107.113

[27] Sheng W, Jiu Y. Effect of food composition of mixed food on glycemic index. Wei Sheng Yan Jiu. 2015;**28**(6):356-358. Available from: https://pubmed.ncbi.nlm.nih.gov/12016989/

[28] Ogello EO, Outa NO, Obiero KO, Kyule DN, Munguti JM. The prospects of biofloc technology (BFT) for sustainable aquaculture development. Scientific African. 2021;**14**:e01053. DOI: 10.1016/j.sciaf.2021.e01053

[29] Harvard School of Public Health. Carbohydrates and Blood Sugar. Harvard T.H. Chan School of Public Health; 2020. Available from: https://www.hsph.harvard.edu/nutritionsource/carbohydrates/carbohydrates-and-blood-sugar/

[30] Romano N, Zeng C. Cannibalism of decapod crustaceans and implications for their aquaculture: A review of its prevalence, influencing factors, and mitigating methods. Reviews in Fisheries Science & Aquaculture. 2017;**25**(1):42-69. DOI: 10.1080/23308249.2016.1221379

[31] Reyes AW. Management of the Interaction and Cannibalism of Postlarvae and adults of the freshwater shrimp Cryphiops caementarius (Molina, 1782). In: Crustacea. London, UK: IntechOpen; 2020. p. 13. DOI: 10.5772/intechopen.87438

[32] Kathyayani SA, Muralidhar M, Kumar TS, Alavandi SV. Stress quantification in *Penaeus vannamei* exposed to varying levels of turbidity. Journal of Coastal Research. 2019;**86**(sp1):177-183. DOI: 10.2112/SI86-027.1

[33] Bello-Pereza LA, Flores-Silva PC, Agama-Acevedo E, Tovar J. Starch digestibility: Past, present, and future. Journal of Organizational Behavior. 2018;**28**(3):303-325. DOI: 10.1002/j

[34] Dauda AB, Romano N, Ebrahimi M. Different carbon sources affects biofloc volume, water quality and the survival and physiology of African catfish *Clarias gariepinus* fingerlings reared in an intensive biofloc technology system. Fisheries Science. 2017;**83**(6):1037-1048. DOI: 10.1007/s12562-017-1144-7

[35] Davis DA, Amaya E, Venero J, Zelaya O, Rouse DB. A case study on feed management to improving production and economic returns for the semi-intensive pond production of Litopenaeus vannamei. Avances En Nutrición Acuícola VIII. In: VIII Simposium Internacional de Nutrición Acuícola. Monterrey, Mexico: Universidad Autónoma de Nuevo León; 2006. pp. 282-303

[36] Effendi I, Lukistyowati I, Tang UM, Muchlisin ZA, Suharman I, Putra I, et al. Effect of different biofloc starters on ammonia, nitrate, and nitrite concentrations in the cultured tilapia *Oreochromis niloticus* system [version 3; peer review: 2 approved]. F1000Res. 2020;**9**:1-13

[37] Rajkumar M, Pandey PK, Aravind R. Effect of different biofloc system on water quality, biofloc composition and growth performance in *Litopenaeus vannamei* (Boone, 1931). Moriarty. 1997;**2015**:1-13. DOI: 10.1111/are.12792

[38] Peiro-Alcantar C, Rivas-Vega ME, Martínez-Porchas M, Lizárraga-Armenta JA, Miranda-Baeza A, Martínez-Córdova LR. Effect of adding vegetable substrates on *Penaeus vannamei* pre-grown in biofloc system on shrimp performance, water quality and biofloc composition. Latin American Journal of Aquatic Research. 2019;**47**(5):784-790. DOI: 10.3856/vol47-issue5-fulltext-7

[39] Arias-Moscoso JL, Espinoza-Barrón LG, Miranda-Baeza A, Rivas-Vega ME, Nieves-Soto M. Effect of commercial probiotics addition in a biofloc shrimp farm during the nursery phase in zero water exchange. Aquaculture Reports. 2018;**11**:47-52

[40] Gomes Vilani F, Schveitzer R, Da Fonseca AR, Do Nascimento Vieira F, do Manoel Espírito Santo C, Quadros Seiffert W. Strategies for water preparation in a biofloc system: Effects of carbon source and fertilization dose on water quality and shrimp performance. Aquacultural Engineering. 2016;**74**:70-75. DOI: 10.1016/j.aquaeng.2016.06.002

[41] Betanzo-Torres EA, de los Ángeles Piñar-Álvarez M, Sandoval-Herazo LC, Molina-Navarro A, Rodríguez-Montoro I, González-Moreno RH. Factors that limit the adoption of biofloc technology in aquaculture production in Mexico. Water. 2020;**12**(10):1-24. DOI: 10.3390/w12102775

[42] Manan H, Amin-Safwan A, Kasan NA, Ikhwanuddin M. Effects of biofloc application on survival rate, growth performance and specific growth rate of Pacific whiteleg shrimp, *Penaeus vannamei* culture in closed hatchery system. Pakistan Journal of Biological Sciences. 2020;**23**(12):1563-1571. DOI: 10.3923/pjbs.2020.1563.1571

[43] Ray AJ, Seaborn G, Vinatea L, Browdy CL, Leffler JW. Effects of biofloc reduction on microbial dynamics in minimal-exchange, superintensive shrimp, *Litopenaeus vannamei*, culture systems. Journal of the World Aquaculture Society. 2012;**43**(6):790-801. DOI: 10.1111/j.1749-7345.2012.00607.x

[44] Nugroho LA, Masithah ED, Satyanrini WH. Evaluation of biofloc technology: The risk of giving different commercial probiotics to C:N and N:P ratio and quality of seawater. AACL Bioflux. 2019;**12**(2):706-715

[45] de Lima ECR, de Souza RL, Girao PJM, Braga ÍFM, Correia ES. Culture of Nile tilapia in a biofloc system with different sources of carbon. Revista Ciencia Agronomica. 2018;**49**(3):458-466. DOI: 10.5935/1806-6690.20180052

[46] Silva UL, Falcon DR, Pessôa MNDC, Correia EDS. Carbon sources and C:N ratios on water quality for nile tilapia farming in biofloc system. Revista Caatinga. 2017;**30**(4):1017-1027. DOI: 10.1590/1983-21252017v30n423rc

[47] El-husseiny OM, Goda AMA, Mabroke RS. Complexity of carbon sources and the impact on biofloc integrity and quality in tilapia (*Oreochromis niloticus*) tanks. AACL Bioflux. 2018;**11**(3):846-855